The Urban Political Economy and Ecology of Automobility

Driving Cities, Driving Inequality, Driving Politics

Just how resilient are our urban societies to social, energy, environmental, and/or financial shocks, and how does this vary among cities and nations? Can our cities be made more sustainable, and can environmental, economic, and social collapse be staved off through changes in urban form and travel behaviour? How might rising indebtedness and the recent series of financial crises be related to automobile dependence and patterns of urban automobile use? To what extent does the system and economy of automobility factor in the production of urban socio-spatial inequalities, and how might these inequalities in mobility be understood and measured? What can we learn from the politics of mobility and social movements within cities? What is the role of automobility, and auto-dependence, in differentiating groups, both within cities and rural areas, and among transnational migrants moving across international borders? These are just some of the questions this book addresses.

This volume provides a holistic and reflexive account of the role played by automobility in producing, reproducing, and differentiating social, economic, and political life in the contemporary city, as well as the role played by the city in producing and reproducing auto-mobile inequalities. The first section, titled Driving Vulnerability, deals with issues of global importance related to economic, social, financial, and environmental sustainability and resilience, and socialization. The second section, Driving Inequality, is concerned with understanding the role played by automobility in producing urban socio-spatial inequalities, including those rooted in accessibility to work, migration status and ethnic concentration, and new measures of mobility-based inequality derived from the concept of effective speed. The third section, titled, Driving Politics, explores the politics of mobility in particular places, with an eye to demonstrating both the relevance of the politics of mobility for influencing and reinforcing actually existing neoliberalisms, and the kinds of politics that might allow for reform or restructuring of the auto-mobile city into one that is more socially, politically, and environmentally just. In the conclusion to the book Walks draws on the findings of the other chapters to comment on the relationship between automobility, neoliberalism, and citizenship, and to lay out strategies for dealing with the urban car system.

Alan Walks is an associate professor of urban geography and planning at the University of Toronto. His research explores issues and policies related to urban socio-spatial inequality, urban development processes, and political ideology. He is a co-editor of the book *The Political Ecology of the Metropolis*.

Routledge studies in urbanism and the city

This series offers a forum for original and innovative research that engages with key debates and concepts in the field. Titles within the series range from empirical investigations to theoretical engagements, offering international perspectives and multidisciplinary dialogues across the social sciences and humanities, from urban studies, planning, geography, geohumanities, sociology, politics, the arts, cultural studies, philosophy and literature.

Published:

The Urban Political Economy and Ecology of Automobility
Driving Cities, Driving Inequality, Driving Politics
Edited by Alan Walks

Forthcoming:

Cities and Inequalities
Edited by Faranak Miraftab, Ken Salo, and David Wilson

Mega-Urbanization in the Global South
Fast cities and new urban utopias of the postcolonial state
Edited by Abdul Shaban and Ayona Datta

The Urban Political Economy and Ecology of Automobility

Driving Cities, Driving Inequality, Driving Politics

Edited by Alan Walks

Routledge
Taylor & Francis Group

LONDON AND NEW YORK

First published 2015
by Routledge
2 Park Square, Milton Park, Abingdon, Oxon OX14 4RN

and by Routledge
711 Third Avenue, New York, NY 10017

Routledge is an imprint of the Taylor & Francis Group, an informa business

British Library Cataloguing in Publication Data
A catalogue record for this book is available from the British Library

Library of Congress Cataloging in Publication Data
Urban political economy and ecology of automobility: driving cities,
driving inequality, driving politics / edited by Alan Walks.
    pages cm. – (Routledge studies in urbanism and the city)
    1. Automobiles–Social aspects. 2. Cities and towns. 3. Urbanization.
    I. Walks, Alan II. Title: Urban political economy and ecology of
    automobility.
    HE5611.U73 2014
    306.4'819091732–dc23   2014003081

ISBN: 978-0-415-70615-5 (hbk)
ISBN: 978-1-315-76618-8 (ebk)

Typeset in Times New Roman
by Wearset Ltd, Boldon, Tyne and Wear

Printed and bound in the United States of America by Publishers Graphics,
LLC on sustainably sourced paper.

Contents

Figures

Tables

Contributors

Ron N. Buliung, Associate Professor, Department of Geography, University of Toronto Mississauga, 3359 Mississauga Road North, Mississauga, ON, Canada, L5L 1C6.

Guy Faulkner, Associate Professor, Faculty of Kinesiology and Physical Education, University of Toronto St. George, 55 Harbord Street, Toronto, ON, Canada, M5S 2W6.

Caroline Fusco, Associate Professor, Faculty of Kinesiology and Physical Education, University of Toronto St. George, 55 Harbord Street, Toronto, ON, Canada, M5S 2W6.

Jason Henderson, Associate Professor, Department of Geography and Environment, San Francisco State University, San Francisco, CA, USA.

Paul Hess, Associate Professor, Department of Geography and Program in Planning, Sidney Smith Hall, University of Toronto St. George, 100 St. George Street, Toronto, ON, Canada, M5S 3G3.

Helen Hao Wen Huang, Department of Geography and Program in Planning, Sidney Smith Hall, University of Toronto St. George, 100 St. George Street, Toronto, ON, Canada, M5S 3G3.

Jeffrey Kenworthy, Professor, Curtin Sustainability Policy Institute (CUSP), Curtin University, Box U1987, Perth, Western Australia 6845, Australia.

Kristian Larsen, Department of Geography, Sidney Smith Hall, University of Toronto St. George, 100 St. George Street, Toronto, ON, Canada, M5S 3G3.

George T. Martin, Professor, Montclair State University, 308 Dickson Hall, Montclair, NJ, USA 07043.

Pablo Mendez, Assistant Professor, Department of Geography, Carleton University, Ottawa, ON, Canada, K1S 5B6.

Raktim Mitra, Assistant Professor, School of Urban and Regional Planning, Ryerson University, Toronto, ON, Canada.

Markus Moos, Assistant Professor, School of Planning, University of Waterloo, Waterloo, ON, Canada, N2L 3G1.

Peter Newman, Professor and Director, Curtin Sustainability Policy Institute (CUSP), Box U1987, Perth, Western Australia 6845, Australia.

Rebecca Osolen, Department of Geography and Program in Planning, Sidney Smith Hall, University of Toronto St. George, 100 St. George Street, Toronto, ON, Canada, M5S 3G3.

Emily Reid-Musson, Department of Geography and Program in Planning, Sidney Smith Hall, University of Toronto St. George, 100 St. George Street, Toronto, ON, Canada, M5S 3G3.

Linda Rothman, Child Health Evaluative Sciences, The Hospital for Sick Children, Toronto, 55 University Avenue, Toronto, ON, M5G 1X8, Canada.

Annya C. Shimi, Department of Geography and Program in Environment, University of Toronto Mississauga, Mississauga, ON, Canada.

Matti Siemiatycki, Associate Professor, Department of Geography and Program in Planning, Sidney Smith Hall, University of Toronto St. George, 100 St. George Street, Toronto, ON, Canada, M5S 3G3.

Matt Smith, Director, Research & Engagement, Stratcom, 1179 King Street W., Suite 202, Toronto, ON, Canada, M6K 3C5.

Matt Talsma, Lecturer, Institute of Communication, Culture and Information Technology, University of Toronto Mississauga, Room 3014, CCT Building, 3359 Mississauga Road North, Mississauga, ON, Canada, L5L 1C6.

Paul Tranter, Associate Professor, School of Physical, Environmental and Mathematical Sciences, UNSW Canberra, ACT 2600, Australia.

Mirej Vasic, Department of Geography and Program in Planning, Sidney Smith Hall, University of Toronto St. George, 100 St. George Street, Toronto, ON, Canada, M5S 3G3.

Alan Walks, Associate Professor, Department of Geography, University of Toronto Mississauga, 3359 Mississauga Road North, Mississauga, ON, Canada, L5L 1C6.

Preface and acknowledgements

No person's life is untouched by automobility. I often accompanied my father, who dodged the Brazilian draft and immigrated to Canada before the 1964 military coup, at the gas station where he worked as a mechanic, and later when he picked up broken cars with his tow truck and delivered them to various locations around the Toronto region. He was able to support a large family, and we were able to live in our own house in the suburbs, through his hard work fixing and towing the automobiles of others (often working both day and night shifts). Although growing up my sisters and I felt no different from our neighbours (well, maybe a little bit), we learned that some of them would complain to the police about the unsightliness of the tow truck parked in our driveway, even though my father would use it to plough the snow from *their* driveways each winter. This is where I learned to drive. Unfortunately, my father could never return to his place of birth (his Canadian passport explicitly noted it was not recognized by Brazil). Eventually over-work took its toll in a broken family, and everyone ended up the poorer after the divorce. My father never gave up his belief in modernity, however. He was diagnosed with terminal cancer shortly before I began researching the suburban links to neoliberalism.

It seems to me this story is as good a metaphor as any for the trajectory of fossil-fuelled auto-mobile capitalism. Good (particularly for those raised in its bosom) while it lasted, but unsustainable, and leading directly to a new slate of problems that subsequently developed. Le Corbusier wrote of the then-emerging auto-mobile future: "We shall use up tires, wear out road surfaces and gears, consume oil and gasoline. All of which will necessitate a great deal of work … enough for all" (cited in Duany and Plater-Zyberk 2000), and he was correct. The growing auto-industrial complex consumed much labour and oil, extended lifespans and quality of life, and supported a rapidly growing population, the vast majority of whom came to live in the city and, in the most developed nations, to take automobility for granted. At the same time, as chapters in this volume remind us, the rise of automobility has led to approximately 1.2 million fatalities and upwards of 50 million injuries from traffic accidents every year. Such is the brave new world spawned by automobility, an unprecedented historical development producing novel forms of inequalities and vulnerabilities, yet one that has often been left unquestioned, naturalized, and unpoliticized. It is impossible to return to the place whence we came. We now must deal with the significant legacy begat by auto-mobile capitalism and our many varied and powerful adaptations to it, and that includes the auto-mobile city. My wife's father was killed while out bicycling by a driver who fell asleep at the wheel – another pertinent metaphor, this time for where our current predicament might lead. But while the diagnosis may be terminal, we still have time to figure out how to build socially just, resilient, environmentally sustainable post-automobile cities; cities that model new systems that allow for meaningful and fulfilling working lives, and where we might finally take real democratic responsibility for each other and the world we are leaving to our own children.

While international and comparative in scope, a significant portion of the research contained in this book was produced as part of the research project *Automobility and Urban Canada: Politics, Spatial Mismatches and Implications for Citizenship*, based at the University of Toronto and funded by a grant from the Social Sciences and Humanities Research Council of Canada (SSHRC), for which the editor served as the principal investigator. Chapter 6 in this volume by Pablo Mendez, Markus Moos, and Rebecca Osolen is also partly based on work the first two of these authors conducted as part of the Major Collaborative Research Initiative (MCRI) project *Global Suburbanisms: Governance, Land, and Infrastructure in the 21st Century*, also funded by SSHRC. All the Canadian researchers who contributed to this book, including the editor, would thus like to thank SSHRC for their generous support, without which this book would not have been possible. Thanks also to my co-investigators on the *Automobility and Urban Canada* project, Drs Ron Buliung, Paul Hess, and Matti Siemiatycki, who helped review the chapters in this volume, and whose work is represented in some of the chapters. In addition, Peter Newman and Jeffrey Kenworthy are very grateful for the tireless and professional assistance given to them by Mr Phil Webster in getting the presentation of their chapter up to a publishable standard. Pablo Mendez and Markus Moos would like to thank Geoff Chase, Tristan Wilkin, and Robert Walter-Joseph for research assistance related to Chapter 6. Ron Buliung, Annya Shimi, and Raktim Mitra wish to acknowledge contributions to Chapter 8 from Md. Musleh Uddin Hasan at Bangladesh University of Engineering and Technology, and Bayes Ahmed at University College London, and would like to thank Debra Efroymson for granting them permission to adapt and update her mapping of the timing and location of Dhaka's rickshaw street bans. Thanks to Dylan Simone for providing invaluable assistance in formatting and preparing the materials in this book for publication. I dedicate my own work in this book to my sister Jodi, one of the many indirect victims of the system and culture of automobility, who unfortunately passed away before she learned to drive.

Dr Alan Walks
University of Toronto

Introduction

1 Driving cities

Automobility, neoliberalism, and urban transformation

Alan Walks

The period since the late 1800s, what Arrighi (1994) has called the "Long Twentieth Century", has been witness to an unprecedented level of social, economic, cultural, and political change. Typically understood as the pre-eminent era of modernity and progress, such changes have been associated with the power of capitalism to remake the social and economic order in its own image (as they were also with socialism between the 1930s and the 1990s). In common discourse, such changes are often summarized simply as "development". At the core of such changes is the rise of a fossil-fuelled industrialized society, and with it, a particularly complex division of labour and set of social relations (see Huber 2008, 2013; Scott 2008). The social economies that such a transformation have wrought have been able to support a burgeoning population, more than quadrupling since 1901, when an estimated 1.6 billion people inhabited the planet, to more than seven billion by 2012 (UN 2013). The main industries at the heart of such changes are virtually all linked in some way to the automobile industry, including much of the mining, oil extraction, steel production, trucking and shipping, road building, advertising and marketing, fuel distribution, suburban land development, and accessories. Even computing and finance have been primarily put to work expanding and restructuring what Freund and Martin (1993) call the "auto-industrial complex". This "long" twentieth century might be labelled, among other things, the century of the automobile.

In his book *The Urban Revolution* (2003/1970), Henri Lefebvre presented a novel thesis: the world was (becoming) completely urbanized, and urbanization was replacing industrialization as the dominant mode of production. While this thesis continues to be debated, the extent and rapidity of global urbanization is not in question: from only 13 per cent of the total population living in cities in 1900 (just 220 million people) to 52 per cent in 2011 (3.6 billion people) (UN 2012). In 1900 there were fewer than 15 metropolitan areas with over one million inhabitants. By 2011 the number of such cities had grown to 457, which together housed approximately 45 per cent of the world's urban population (ibid.). Of the latter metros, 23 recorded populations of more than ten million in 2011, up from zero in 1900 (ibid.). Perhaps most tellingly, the urban population in 2011 was 43 per cent larger than was the *entire* global population just 60 years earlier (just over 2.5 billion in 1950).

Regardless of how it is measured, automobiles, and automobility, have been central to such an urban revolution, expanding the logics of urbanism into rural areas and new territories, and transforming pre-automobile societies into highly urbanized, automobilized ones. Current estimates provided in the chapter by Martin in this volume indicate that at least one billion registered automobiles currently grace the planet, virtually one per extended family. In 1950, total vehicular travel was estimated at 2.8 trillion passenger-km. In only 50 years (by the year 2000), total vehicular travel had risen to 32 trillion passenger-km, an increase of more than 1,042 per cent, with most such travel occurring via automobiles and trucks (Moriarty and Honnery 2008).

Since the publication of Newman and Kenworthy's important early work (1989, 1999) the automobile and its influence on changing urban forms, public transit viability, and urban sustainability has received increasing attention (Low and Gleeson 2003; Banister 2005; Ryan and Turton 2007; Mees 2001, 2009; Sperling and Gordon 2009; Black 2010; Filion 2010; Schiller et al. 2010). Although the automobile was largely invented in Europe, it was in the United States (US) that automobility was adopted most quickly and most thoroughly, inspiring a number of insightful studies of the importance of the automobile, and of automobility, in the US context (Flink 1975, 1988; Rajan 1996; Volti 1996; Norton 2008; Packer 2008; Seiler 2008; Wells 2012; Lewis 2013). Recently, the conceptual development around what is often called the "new mobilities paradigm" (Adey 2010; Elliott and Urry 2010; Kaufmann 2002; Kellerman 2006, 2012; Merriman 2012; Sheller and Urry 2000; Urry 2000, 2004, 2007) has facilitated an interrogation of the social, political, and cultural significance of global automobility. The progenitor and inspiration for much of this work is Freund and Martin's seminal book *The Ecology of the Automobile* (1993), which outlined an integrated, holistic view of the sociological, cultural, ecological, ideological, and phenomenological implications of automobility among the most auto-dependent nations. In addition to those studies mentioned above, recent works building on this shared legacy include important volumes by Miller (2001), Featherstone et al. (2005), Bohm et al. (2006), Best (2006), Larsen et al. (2006), Paterson (2007), Redshaw (2008), Conley and McLaren (2009), Ohnmacht et al. (2009), Lucas et al. (2011), Pelligrino (2011), and Grieco and Urry (2012).

While this profusion of recent work has significantly extended the range of theorizing concerning the automobile and automobility, the relationships between automobility and *urban transformations* remain underexplored. This chapter sets out a framework for approaching scholarly explorations of the urban social and political economies and ecologies of automobility. The first section of this chapter delineates the concept of automobility, and explores the parameters of the system of automobility that has emerged. The second section delves into questions concerning the politics of automobility, and explores the relationship between the cultures and practices of Fordist and post-Fordist automobilities, and the ideology of neoliberalism. In the third section, the specific relationship with the city is explored. A key thesis that infuses this book is that the contemporary city is not only a product of automobility, but the field upon which struggles originating under automobility are contested and transformed, producing new forms of vulnerability, inequality, and politics under contemporary capitalism. Any lasting reforms will thus have to involve changes not only to travel behaviour, but to the physical, social, economic, and political structure of the city as well. The chapter concludes with discussion of the organization of this book and its rationale.

Automobility and its system

The concept of automobility is far broader than simply the notion of mobility via the car. The term itself, as well as the core idea that automobility is not just a form of transportation but "an ideologically and symbolically loaded cultural phenomenon", goes at least back to writings from the early 1920s (Furness 2010: 6). The concept of automobility has subsequently been developed through the work of Freund and Martin (1993), Rajan (1996), Beckmann (2001), and Urry and his co-authors (Sheller and Urry 2000; Urry 2000, 2004, 2007), among others. Automobility is conceptualized as a complex, path-dependent non-linear system with its own evolving coherent logic of movement, production, and consumption (Sheller and Urry 2000). This encourages a melding of the functions and practices of autonomous humans with machines (cars, trucks, etc.), thus creating new social hybrid "car-drivers" that are co-constituted by the roads, signs, cultural practices, and daily activities that bind them (ibid.). Automobility constitutes a historical assemblage of social, economic, cultural, and technological-material practices, and the power relations that they have spawned, which

> not only facilitate and accelerate automobile travel but also help to reproduce and ultimately normalize the cultural conditions in which the automobile is seen, and made to be seen, as a technological savior, a powerful status symbol, and a producer of "modern" subjectivities.
>
> (Furness 2010: 6)

One important theoretical implication of automobility stems from the rise of automobile dependence (Newman and Kenworthy 1989), and the specific forms of flexibilized mobility it creates and subsequently compels. While the automobile has expanded greatly the mobility of those able to access one, and thus has provided significant benefits to certain sections of the population, its popularity has also meant a refashioning of the very space within which it travels. The speed, volume, and smoothness of flow required to enhance efficiency of automobile travel in the city requires changes in urban form that significantly reduce the efficiency and viability of other modes (Kramer 2013). The low-density spread-out urban form begat by automobility produces the ubiquitous and (by now) somewhat familiar "auto-space" of highways and multi-lane arterials, shopping mall, and office parks surrounded by parking lots, low-density suburbs, and retail strips (Freund and Martin 1993). It is this auto-space of the contemporary city, rather than the automobile itself, which "coerces people into an intense flexibility" (Urry 2004: 28). One has little real choice but to drive, and to use the car to weave together ever more disparate trips and activities, as well as to keep abreast of ever more distant family and friends (see Farber and Paez 2009).

The kinds of behaviour compelled by auto-dependence produces a key feature in the experience of modernity. Through its "power to remake time and space", auto-space

> forces people to orchestrate in complex and heterogeneous ways their mobilities and socialities across very significant distances.... It forces people to juggle fragments of time so as to deal with the temporal and spatial constraints that it itself generates ... constraining car "users" to live their lives in spatially stretched and time-compressed ways.
>
> (Urry 2004: 28)

It is through the workings of such a coercive flexibility that the auto-mobile city has made the automobile an object of "compulsory consumption", creating hybrid car-driver subjects that are "driven to drive" (Soron 2009). Urry (2004: 28) argues that the "car is the literal 'iron cage' of modernity, motorized, moving and domestic". Yet ironically, because of this, automobility "produces desires for flexibility that so far only the car is able to satisfy" (ibid.: 29). Such is the Janus-faced duality of automobility – it is constructed as both problem and solution to the issue of mobility in the contemporary city.

A related issue concerns the path-dependent institutional power created by and through the system of automobility. With the rising dominance of automobility, the political stakes are high for powerful interests invested heavily in the manufacturing technologies, labour-force practices, and socio-political institutions undergirding the "steel and petroleum" car industry. Such invested interests are compelled to work to "lock in" political support for the continuation of the "automobile-industrial complex" (Freund and Martin 1993), and to actively expel competing realities and dismiss alternative normative visions, and thus other forms of human movement (Urry 2004). The automobile-industrial complex is the key articulation through which development under fossil-based capitalism has been materialized (Huber 2008). As the core of the contemporary hydro-carbon mode of production, automobility "is the internal combustion engine of late modernity itself" (Beckmann 2001: 594). Sheller and Urry (2000) outline key features that produce and distinguish automobility's "character of domination" and give it such immense social, cultural, and political power. The automobile is the quintessential manufactured object producing by the leading industries, a major item of consumption (after housing) providing social status, and provides the predominant global form of "quasi-private" mobility (ibid.). Automobility defines the dominant culture and is at the heart of discourses of what represents the good life, and constitutes a powerful complex of social and technological linkages with other industries, which together make up the largest cause of the environmental resource use, pollution, and carbon emissions (ibid.). The automobile constitutes a "maximum commodity" (Dawson 2011). Maximum commodities hold the potential for sales opportunities "that are deep and wide enough to form the basis for a self-perpetuating capitalist industrial complex" and are pursued by firms for the years of profits guaranteed once they become established in the market (ibid.: 271). Maximum commodities combine high levels of complexity, fetishizability, and fragility (which protects monopoly profits of firms and forces constant new purchases and trade-ins as older models lose their status or break down) with high levels of labour intensity, saleability, secondary business implications (once automobility is established, numerous other industries become necessities), and most importantly, continued practical necessity (ibid.).

The political efforts of invested players, in combination with the structured ways that auto-space compels behaviour across diverse fields of activity – enforcing the "continued practical necessity" of the automobile – leads to what Freund and Martin (1993) term "auto hegemony". Not only does the latter structurally define questions challenging the dominant system as outside the realm of possibility, but as a system it reproduces the conditions necessary for its own self-expansion. "Auto" in this case channels the idea of auto-poeisis – the ability of automobility to self-organize and self-generate, due to the forces of propulsion automobility itself creates. It also channels the idea of automation. Once locked in, such a process, while originating in a historically contingent fashion, becomes automatic as the forces promoting automobility become ever more powerful and the institutions upholding it ever more ingrained. And as the automobile city compels hybrid car-drivers (or "carsons" – car-persons) to drive (Bohm *et al.* 2006: 12), the latter become the "automatons" of the system, the units whose mobility and fluidity creates the flows giving shape and strength to the system, and for which the system is increasingly geared to perpetuate. The automobile is thus not significant on its own: automobility is co-constructed socially, spatially, culturally, and politically through the ways that dominant practices and affordances of the system are materialized, psychologically internalized, and political supported.

However, the work performed by the concept of automobility goes further than just structuring and normalizing conditions for the continued expansion of automobile hegemony. It also involves particular citizen subjectivities and cultural phenomena that can be extended and applied beyond the car, to other automobilities (including those related to the truck, aeroplane, bicycle, etc.). The prefix ("auto") is also extended to refer to the autobiographical and autonomous aspects of the modern mobile self. According to Urry (2004: 29), automobility coerces people into temporal schedules that "assemble complex, fragile and contingent patterns of social life, patterns that constitute self-created narratives of the reflexive self". New, flexible "subjective temporalities" replace the modern clock-time and work schedule of the pre-automotive era. "This produces a reflexive monitoring of the self" in which contingent biographical narratives must be continually revised, partly because "personal times are de-synchronized from each other" (in comparison with the regimented schedules of train travel, etc.), and partly because high speeds and autonomous travel reduce the intimacy of face-to-face interactions experienced during the act of collective mobility (ibid.: 29–30). In the modern, auto-mobile world, "people become anonymized flows of faceless ghostly machines ... [while] those living on the street are bombarded by hustle and bustle and especially by the noise, fumes and relentless movement of the car that cannot be mastered or possessed" (ibid.). The mass of traffic in the city thus works to help construct an image of the modern individualized self as an atomized identity set against faceless others with which one is in competition, directing social interaction into more restricted and private spaces. At the same time, through the act of driving, the driver feels the vehicle as an extension of her/his self, producing new subjectivities that incorporate the machine in the construction of one's identity (as does the mobile phone).

Seiler (2008) and Packer (2008) extend this thinking further, arguing that automobility has set the context, as well as the rules, delineating and governing conceptions of citizenship. The rules of the road, safety standards, and driving etiquette form the basis of the moral education at the heart of (particularly, "American") citizenship. As one ages, one is taught to respect the power of the car, to keep away from the road, to always look both ways when crossing, making the driving licence and the skills required to attain it not only a valuable commodity but a social rite of passage. As discussed further in Chapter 14 by Talsma, this "driver education" involves formal and informal rules, the recognition and obeying of formal and casual signs, and both formal surveillance and self-discipline. Formal codes invoke threats of punishment and reprimand (for careless or dangerous driving, for instance), while informal codes compel drivers and thus citizens to be courteous, attentive, and responsible.

Seiler (2008) likewise documents how automobility shaped, and was shaped by, a raced, gendered, and classed set of ideas regarding the meanings of American individualism and freedom, informing and structuring the mutual obligations and boundaries of trust among citizen subjects. However, while automobility is tied up with ideological constructions of freedom that have become naturalized within (American) citizen subjectivities, these do not match with the reality of increasingly rigid and limiting systems of governance that are needed to maintain the system. As Ker and Tranter (2003) note, the independence often attributed to automobility is an illusion. Automobility appears to provide autonomy and independent travel, yet drivers are fundamentally dependent on a range of services, functions, and commodities produced by a host of others, and require that others always follow the rules. The Republic of Drivers is one characterized by the necessity of increasing institutionalization, regulation, and routinization of both the driver-citizen and of the larger system of supports on which its depends, yet one for which the solution is always discursively constructed in terms of fewer regulations and greater personal liberty (Seiler 2008; see also Beckmann 2001). Truly autonomous mobility (apart from walking) is conceptually impossible, due to the set of factors that must be in place, rendering any vehicles' movement "beyond the control of an individual subject given its systematic interdependencies. Traffic is itself a socially negotiated phenomenon where trajectories cross and intersect in a complex but never independent movement" (Bohm *et al*. 2006: 12).

Regimes of automobility, automentality, and neoliberalism

The discussion of auto-mobile citizen subjectivities and those interests vested in their perpetuation raises questions concerning the political contours of automobility. Bohm *et al*. (2006) argue that automobility should be considered a regime with its own historically contingent and specific relations of power, rather than simply a ubiquitous "system". "The notion of system tends to underplay collective human agency in the production of automobility and to avoid the political questions about the shaping of the automobile 'system'", and thus to naturalize and reify automobility "as a closed loop reproducing its logics relentlessly" (ibid.: 5–6). Bohm *et al*.'s stance calls into question the assumed universality of automobility, and provides a refined approach in dealing with alternatives to automobility other than the range of those that can be generated within the confines of the system itself. This point is important not only for understanding the relationship between automobility and neoliberalism, but also for contemplating alternative automobilities (the subject of the other chapters in this volume).

Positing automobility as a regime allows for the extension of governmentality perspectives into analyses of automobility, and with it more nuanced understandings of the local contingencies involved in "actually existing automobilities". This is undertaken to good effect in Seiler's (2008) genealogical account of the rise of automobility and its importance in the socio-political constructions of democracy, self-reliance, self-responsibility, and "free society" in "American" culture. Automobility is central to the making of autonomous liberal American subjects, through the effects of the act of motorized driving on the "open road", variously perceived and experienced as "liberating, individuating, revivifying, equalizing" (ibid.: 130). This gives the *impression* of autonomy and self-determination, that one is "free to live as one wills", yet such a life is circumscribed by the form and extent of the highway system, the risks and rules of the road, the reality of congestion and the costs of driving, and the necessity to be constantly self-policing and monitoring oneself in order to conform to the ideal of autonomy (Packer 2008; Rajan 1996, 2006; Seiler 2008). Seen in this light, automobility should be construed as a premier technology of contemporary liberalism, disciplining subjects and constructing a particular form of individualized democratic ethos, while steering political expression (and dissent) through narrow pre-designated institutional channels (Rajan 2006).

Shifting the focus from system to regime likewise allows for a fuller understanding of the political economy of automobility. There is by now a well-established literature theorizing regimes of accumulation and the restructuring of such regimes since the heyday of Fordism, which of course takes its name from the practices of the Ford Motor Company in the early twentieth century. Fordism involved a policy commitment to full employment and regional equalization, Keynesian fiscal and monetary policies (including high tariffs protecting, and even state subsidization or ownership of, key sectors), the promotion of mass consumption, unionization and collective bargaining, standardization and economic regulation, the application of Taylorist practices, and tolerance of oligopolistic industrial practices (foremost, among the big automobile manufacturers) (Martin 1988). Production and consumption of the automobile was absolutely essential to the success of Fordism, while one of its primary results – support for extension of the family-supportive wage – allowed many in the working class to afford a car and a house in the suburbs (Paterson 2007). For all intents and purposes, the early system of automobility and the Fordist regime of accumulation are essentially the same thing.

From auto-mobile Fordism to neoliberalism?

Scholars often juxtapose neoliberalism against Fordism, seeing the former as the "putative successor" to the latter, resulting from the rigidities and exhaustion of Fordism as if they represent two distinct and opposed sets of policies or regimes of accumulation (Leitner *et al.* 2007: 2). By this logic, neoliberalism might be expected to entail a negation or transcendence of the precepts of Fordism, and by extension should undercut the continued extension of (automobile-dependent forms of) automobility (Filion and Kramer 2011). Neoliberal shifts accompanying the rise of "flexible accumulation" or "post-Fordism" (so called because of uncertainty regarding whether a truly new regime of accumulation had yet arisen) include the jettisoning of Keynesianism for the triumvirate of monetarism, privatization, and deregulation, as well as financialization, trade liberalization, attacks on collective bargaining rights, and renewed tolerance for mass unemployment and rising inequality (see Harvey 2005). By extension, it might be expected that the advent of post-Fordist neoliberalism should entail a break with the dominance of the auto-industrial complex and auto-dependent urban forms and mobilities.

However, while neoliberalism clearly came to dominance as the primary response to the perceived failings of Keynesianism under Fordism, the construction of such an oppositional dualism misses key attributes of, and continuities between, Fordism, post-Fordism, and neoliberalism. Importantly, the restructuring of production and consumption under the post-Fordist regime of accumulation, as well as the neoliberal attack on the welfare state and labour rights, all arose in order to reinvigorate a waning automobile-industrial complex. The new production techniques that typify post-Fordism (just-in-time delivery, small-batch and custom design, lean production, organizational fragmentation, etc.), as well as changing labour arrangements (flexible workplaces, temporary and contract work, redeployment on the shop floor, for instance), were all meant to enhance global automobile production and re-establish the competitiveness and profitability of the auto sector (Kawahara 1997; Paterson 2007).

The kinds of policies pursued under neoliberalism, including tax cuts, trade liberalization, the extension and guarantee of private property rights, deregulation of finance, and the attack on organized labour, have all been demanded by those capitalists benefiting from expansion of the global system of automobility. Such policies facilitate the extension of resource extraction and production into new territories, and the opening up of new markets for the products and cultural artefacts of the auto-industrial complex. Neoliberal labour policies compel workers in one location to compete on wages with those located in other countries, equalizing and protecting the profitability of firms' investments in global production networks. The deregulation of finance stimulates the development of carry trades that fund the offshoring of production facilities, and has produced the credit necessary to bolster demand for automobiles and new houses in the face of the growing inequality and stagnant real incomes. To this end, financialized neoliberalism can be thought of as a form of "privatized Keynesianism" (Crouch 2009), in that it is no longer the state but private households who are doing the borrowing to stimulate demand in the face of economic stagnation. The restructuring of the Fordist political economy under a globalized post-Fordist neoliberalism represents an attempt at maintaining the prevailing regime of automobility at all costs, not its transcendence.

(Ir)rationalities of governance, neoliberalism, and automobility

There is a tight ideological symbiosis between the values promoted by automobility (individual freedom and autonomy) and the rationalities of neoliberalism as formulated by Hayek (1944, 1945). For Hayek, a fundamental opposition exists between a political economy of state planning and one that develops according to the preferences of, and competition among, agents that are "free". Under planning, the government is presuming to know not only what everyone values and desires, but also what policies will produce the greatest utility and welfare. This knowledge cannot but be biased, and involves the imposition of the will of a small elite on the real but unknown interests of the majority. Alternatively, a system of free competitive markets, according to Hayek, allows everyone's preferences to be mediated through the price mechanism; hence one can know what people really want from simple examination of their revealed preferences in the market. At work in Hayek's treatise are a number of false equivalencies, particularly that between political and economic freedom, implying that any attempt to plan on behalf of a majority, or otherwise intervene in the free market, must result in a reduction of both political freedom and general economic welfare, even when such decisions are made by a democratically elected state (see Peck 2010).

According to Hayek's logic, the just society is one in which individuals are completely "free" to pursue their desires without state interference. Furthermore they cannot be wrong about what they desire or value – individual self-knowledge is defined by Hayek as instinctive and situated prior to any formal learning or collective influence (Davies and McGoey 2012). The only proper position for the state to adopt is a "scientific-moral agnosticism" concerning what people value and the manner in which they express it in the marketplace, and the "most significant fact about this system is the economy of knowledge with which it operates or how little the individual participants need to know in order to be able to take the right action" (Hayek 1945, cited in Davies and McGoey 2012: 70). The Hayekian neoliberal world is thus one in which "ignorance-is-fairness" (ibid.: 79), and the "right" action is simply the one people desire most as demonstrated by their willingness to pay. It is this stance that undergirds Hayek's lumping of all forms of political collectivities and limitations on market freedom in the same category as authoritarianism and social engineering. The counterpart to ignorance is, of course, denial, and supporters of such a Hayekian worldview typically employ denial in order to accept, for instance, the false equivalencies noted above, and/or to dismiss the problem that the democratic election of left-leaning governments poses to the theory.

The Hayekian defence of such "rationalities of ignorance" (Davies and McGoey 2012: 79) fits well with the long lineage of pro-automobility political writings and movements. Germaine to the ideas of defenders of automobility, including Bruce-Briggs (1977), Lomasky (1997), Johnston (1997), and Dunn (1998), is the assumption that the automobile, and by association the infrastructures that arose to support its use, emerged as a result of the combined preferences and agency of autonomous individuals exercising their freedom of choice in an open and free market. Automobility is thus what people "want". It is the aggregate expression of individual desires, and as such, the autonomous mobility provided by the car should be considered a right, on par with any exercise of personal and political freedom. Within this logic, any restrictions on car use, and attempts by the state to tax the automobile or gasoline, or to modify commuting behaviour or modal shares, let alone impose speed limits, road tolls, traffic calming, or bicycle lanes, involve the imposition of state power against the expressed free will and preferences of individual subjects, and thus constitute a perverse form of social engineering (see Paterson 2007). Attempts to set pollution standards, or regulate fuel efficiency, are similarly viewed as state interference into the workings of the market, and because they might affect affordability they are painted as elitist and undue limitations on personal freedom of choice, even if demanded by low-income households who cannot afford a car (Rajan 1996; Huber 2009). The "war on the automobile", represented either by state regulation of automobility or alternatively by political opposition to its expansion on behalf of environmental and other groups, is thus equated with a war on personal freedom, the political will of the majority, and (neoliberal) conceptions of democracy (Bruce-Briggs 1977).

I term this disposition "automentality", in that like automobility, the knee-jerk impulse to such rationalities of ignorance and denial is somewhat automatic and auto-poeitic, structured by a necessity to make reality conform to ideology. It is this automentality that is central to both the neoliberal and pro-automobile rejection of state-supported public transit, on the grounds that if people wanted transit they would have provided sufficient demand for it in the marketplace (see Dunn 1998). Freund and Martin's (1993) discussion of the "ideology of automobility", and Paterson's work on the pro-automobile political backlash (2007), provide excellent explorations of how this automentality operates (discussed further in Chapter 11 in this volume).

Such a symbiosis between the dominant ideology of automobility and the philosophical tenets and citizen subjectivities of neoliberalism has been present from the start in places like the US, as documented by Seiler (2008). Neo-liberalism and automobility are co-travellers. Neoliberalism in its various guises, it might even be suggested, *is* the contemporary ideological face of automobility. The relationship between the ideologies of neoliberalism and automobility is rooted in their similar value structures and rationalities. It has not arisen as a result of the restructuring of the welfare state, although the more aggressive pro-nouncements and claims of the pro-automobility literature have been clearly invigorated by the political opportunities and challenges resulting from the crisis of Keynesianism. Neoliberalism cannot therefore be considered a solution to automobility, nor can there be a post-neoliberalism, or after-neoliberalism, as long as the regime of automobility remains dominant.

As Peck (2010) notes, under neoliberalism each new crisis of capitalism and failure of the market is seized upon as evidence of the negative unintended con-sequences of state intervention, justifying ever deeper neoliberalization and encouraging ever more brash claims in its favour. This characteristic of neo-liberalism is a key aspect of its structured rationalities of ignorance. Of course, the truth is that the entire system of automobility itself has been the product of an intense and persistent, yet often veiled, agenda of pro-automobile social engi-neering, planned *by the state* in conjunction with the leading industries of Fordism, while alternatives were systematically denied and driven out. Yet this fact is not only ignored by defenders of automobility such as Dunn (1998) and Lomasky (1997), it *must* be denied, lest it disrupt the logical (ir)rationalities seemingly cohering in the dominant neoliberal automentality. Instead, the active participation of the state at various scales in the planning of the auto-city is rationalized as the cumulative outcome of the will of individual "homevoters" (Fischel 2001), expressing their desires both through the ballot box and through their propensity to "vote with their feet" (or rather wheels) via the Tieboutian logic of local municipal political fragmentation (Beito *et al.* 2002; Peck 2011).

Of course, neoliberalism is not monolithic but contested, grounded in place, and built on/from pre-existing legislative and political legacies, leading to a number of diverse "actually existing neoliberalisms" in practice (Brenner and Theodore 2002). In Chapter 12 of this volume, Henderson distinguishes between two dominant variants of neoliberalism as constructed in contemporary US polit-ical discourses related to automobility. The first, linked to a "neo-conservative" ethos, has internalized and naturalized such an automentality, defining automo-bility as an inherent political right requiring the protection of the state, paid for through the roll-back of state spending elsewhere. The second dominant form involves a more "pragmatic" neoliberalism that seeks to deepen market relations, reduce state subsidies and taxes, and rely on price signals to inform policy making, and as a result is more ambivalent towards automobility. A similar dichotomy has arisen in Toronto, where duelling neoliberal approaches become articulated in mayoral electoral contests (Chapter 11 in this volume). Conflicts among different neoliberal factions then get inscribed in the socio-political space of the city, which itself comes to exert its own influence on policy and the reproduction of politics.

Automobility and the auto-city

While it may seem obvious, it bears highlighting that the form and structure of auto-space, and its effects in reproducing the contours of various automobilities, is directly related to how cities are planned and how the social geography of the city maps onto patterns of urban development. Thus far the city has not received sufficient attention in the literature on automobility. Identifiable by such salient forms as the shopping mall, parking garage, cloverleaf, motel, parking lot, gas station, and suburban cul-de-sac, such an auto-space has become the dominant feature of contemporary settlement, the "quintessential non-places of super-modernity" (Sheller and Urry 2000: 746, referencing Augé 1995). Although Sheller and Urry (2000: 746) define such places of "pure mobility" as less-than-urbane, "neither urban nor rural, local nor cosmopolitan", and evidence of the "victory of liquidity over the urban", the reality is that auto-space is a central, if not *the primary*, articulation of contemporary capitalist urbanism and of the victory of its logic over its alternatives. If ever there were needed evidence backing up Lefebvre's claims in *The Urban Revolution* (2003/1970) of the subversion of the globe to the logic of urbanization, it can be found in the virtual ubiquity of auto-space across all forms of settlement, whether urban or rural, modern or pre-modern, and the increasing dependence of developed-nation economies on the continued expansion and build-out of such an auto-space. Indeed, it could be argued that it is precisely through automobility that urbanization has become a primary productive force – a mode of production – in its own right, displacing industrial and agrarian modes under contemporary capitalism (ibid.).

This is not to judge such developments in normative terms. As noted above, the system of production built around automobility and fossil-fuelled capitalism has been able to support vastly larger populations, at much higher standards of living, than at any time in history. The secular decline in rates of inter-personal violence since the Enlightenment, and particularly since the end of the Second World War (Pinker 2011), for instance, is surely related to the fact that machines now do much of the heavy work, releasing many humans in the urban industrialized North from the need to rely on might, or to till the soil, and allowing for the nurturing of "cultural-cognitive" innovation and its expression in cities (Scott 2008). Yet, such developments have brought with them distinctly new ways of living, new inequalities, and new forms of politics.

At the "meso" level of the urban, the reproduction of the various components of auto-space is implicated in producing a new form of city, and with it new urban functions. In trying to grasp the profound transformation of the metropolis, urban theorists have come up with a litany of conceptual monikers to describe and theorize this new city, including the dispersed city (Bunting and

Filion 1999), technoburbia and technopolis (Fishman 1987), global sprawl (Keil 1994), post-modern urbanism (Dear and Flusty 1998; Ellin 1996), post-suburbia (Lucy and Phillips 1997), post-metropolis and exopolis (Soja 2000), and metroburbia (Knox 2008). However, none of these terms gets at the truly distinguishing feature of the contemporary city, and each contains significant drawbacks. Use of the prefix "post" (as in post-modern urbanism, post-suburbia, and post-metropolis) relies on a negative signifier, which does little to indicate the actual attributes of the new city, and can be incorrectly taken to imply that the contemporary urban structure represents a break from that of high modernism and auto-dependent forms. Imputing technological and/or suburban (or exurban) characteristics for the entire metropolitan structure, meanwhile, misses the significant diversity of urban forms and functions: not all contemporary cities rely on technological development, and there are a number of different kinds of possible suburbanisms, including those that do not involve automobility, or that imply alternative, multifunctional mobilities (Walks 2013a). Furthermore, a number of different suburbanisms (domesticity, marginality, mono-functionality, interiority, etc.) have *always* historically been co-present in the city, including the pre-automobile city (ibid.). The concepts of dispersed city and global sprawl come closest to identifying important dynamic processes at work, but they highlight the outcome (dispersal) rather than the logic or agents producing this form (automobility and the auto-industrial complex). Furthermore, they miss the point that the auto-mobile city need not involve low-density dispersion: witness the simultaneous concentration of the population *and* automobilization of cities in East Asia, for instance.

In the place of such concepts, the alternative "auto-city" is here proposed. Not only does this term get at the single most important (both universalizing and differentiating) feature of the contemporary city – automobility – it allows for the multiple dimensions and resonances related to the prefix "auto" to be theorized in terms of the dynamic social, economic, cultural, and political processes producing and reproducing its form and structure. Through its effect on the compulsion to drive, the auto-city inscribes and promotes particular ideas of individual autonomy, as well as constraining and compelling their expression via particular forms of consumption and the limited political participation constrained through (auto-mobile, and remote) forms of social interaction. As Lefebvre noted (2003/1970), the modern city produces particular biographical relations of the mobile self connected to the hard infrastructure of the extended transportation system, replacing the experience of being rooted in place gleaned from a life spent living locally, with the compulsion to remain connected in space to facilitate continued access to the resources of the city in the face of growing vulnerability and contingency.

The auto-city is auto-poeitic in a number of ways, self-generating not only because of its importance as the site of the powerful and vested automobile industry, and not only due to the physical difficulties of retrofitting the automobile-dependent city for alternative mobilities (Mees 2009), but also because of the political support it produces for maintaining systems and regimes of automobility. In the face of austerity, itself mainly resulting from the crisis and bankruptcy of auto-mobile capitalism (see Chapter 4), many residents desperately oppose attempts to direct resources to alternative mobilities, and cling to the ideology of auto-mental neoliberalism and its rationalities of ignorance. The more auto-dependent the city, the more automatic the interrelated functioning of such processes and the more such an automentality becomes accepted and naturalized as "common sense", as the alternatives (both alternative mobilities and alternative production systems) are experienced as transgressive of local sensitivities and seen, quite literally, as "out of place" (see Cresswell 1996, 2006).

The auto-city is also notable for producing specific forms of inequality and exclusion. On the one hand, these involve the general exclusions directly related to disparate mobilities and the often dispersed, mono-functional patterns of development encouraged by automobility. In the auto-city, most of those without access to a car are disenfranchised, including not only low-income households who cannot afford a car, but the disabled who cannot drive, marginalized racialized groups who may be concentrated far from work, children, and the elderly who are not legally or physically capable of driving, most of whom are then often compelled to use residual mobility systems (Hine 2011; Lucas 2004). Among car drivers themselves, those without the income for frequent fill-ups or who depend on older, less-reliable vehicles are disadvantaged in the auto-city, as this restricts their range for job seeking, social contacts, and access to amenities, among other things (Cahill 2010; Farber and Paez 2009). The compulsion for low-income households to purchase automobiles in order to attain minimum standards of mobility can itself lead to impoverishment, as there may not be enough disposable income to cover daily needs, yet such households may face an even worse situation without a car (Currie and Delbose 2011; Gleeson and Randolph 2002). It is thus partly due to the inequalities and disadvantages produced by the auto-city that even poor subjects "freely" choose to be auto-mobile.

On the other hand, there is an ecology to the auto-city that has its own independent effects in structuring social, economic, cultural, and political inequalities. Automobility facilitates new kinds of neighbourhoods, and new relationships between the central business district and the urban fringe. It became possible to plan new satellite suburban communities only accessible via automobile, effectively preventing anyone who could not afford a car, or the time or gasoline to search out such communities, from residing or even visiting there. Furthermore, this could be justified in neoliberal-Tieboutian terms that set the market against the more moderate and technocratic public aims of planners, under the claim that the market is merely building what people want (Peck 2011). Just as automobility can produce isolated atomized individuals, it facilitates the segregation of urban communities. While the idea of the gated community, perhaps the most salient neighbourhood form of the neoliberal city, may pre-date automobility (Webster 2002), its modern form – typically compelling entrance through garages and road gates – has been produced in accordance with it, dependent upon the fragmentations of auto-space. Even when communities are located close together, divided only by thoroughfares, levels of social interaction decline with every increase in traffic (Adams 1999).

Auto-mobile forms of planning became a major technology in their own right in producing class and race segregation in US cities, in turn constructing new suburban realities and images of white purity and urban modernity against racialized "chocolate" inner-city neighbourhoods (Avila 2004). Henderson (2006) theorizes such a socio-spatial-political practice as "secessionist automobility". Secessionist suburbanisms in the US have been further structured by the decentralization of firms and employment, producing *spatial mismatches* with the residential location of low-income and less-skilled labour, particularly of blacks, subsequently maintained through highly uneven public transit accessibility and housing affordability to poor inner-city neighbourhoods and less-accessible suburbs (Ihlanfeldt 1999). More recently, the rise of fuel costs, the suburbanization of poverty, and the gentrification of the inner city in a number of large metropolises has led to a shift in economic vulnerability to the automobiledependent suburbs (Gusdorf and Hallegatte 2007; Dodson and Sipe 2008; Kramer 2013, discussed in Chapter 4). Of course, actually existing automobilities and the spatial mismatches they produced are highly grounded in local realities and socio-political structures (see Chapter 6 in this volume).

The ecology of the auto-city structures new political cleavages, and foregrounds certain political struggles while obscuring others. Above all, it has produced a new politics of mobility, one that has been growing since the 1970s in which the battleground is increasingly over competing visions of mobility (including velo-mobility). Out of the counter-cultural resistance to the (automobile-) military-industrial complex has coalesced a new-left opposition to local and national regimes of automobility, constructed increasingly around environmental protection and green politics. Parties on the right, meanwhile, have adopted auto-mental neoliberal programmes attacking the welfare state, advocating for individual rights and privatization, and promoted populist anti-elite stances. Electoral support for these two poles is increasingly spatialized within the developed-world auto-city, with automobile-dependent suburbs increasingly leaning towards political parties of the new right, while urban neighbourhoods characterized by pre-war, pre-auto urban patterns become the bastions of the new left (Sellers *et al.* 2013). The dynamics and implications of such a process are explored in Chapter 11 in this volume.

Across the capitalist world, metropolitanization and suburbanization are associated with a general de-localization of politics, as the interiorizing of domestic life under automobility produces a technocratic and managerial approach to local issues, while heightening the ideological associations and stakes related to national issues and parties (Sellers *et al.* 2013). Sheller and Urry (2000) raise the question of an auto-mobile civil society in which the traditional agoras of public space are deprived of their socio-political centrality and civic functions within the larger public sphere, as detached auto-commuters bypass traditional public spaces in the central cities altogether and political debate moves into the virtual arenas of the television and Internet. The auto-city is thus one that undercuts the "stable associational life" of politics established in pre-automotive eras and traditionally performed in public spaces as envisaged by thinkers such as Habermas (Sheller and Urry 2000: 741). Such a shift in the dominant civitas is augmented by reforms to free speech legislation that limit "unwanted" or "inconvenient" social interaction and political communications, and with them the open and unsolicited exchange of ideas in public spaces, reproducing what Mitchell calls (in the US context) the "SUV model of citizenship" (Mitchell 2005). One response has been for political protests to shift to those public places in the auto-city where one is certain to find (and inconvenience) people: the highway (explored by Talsma in Chapter 14 of this volume).

The social and political ecologies of the auto-city combine in other ways that produce additional exclusions and inequalities. There is by now a well-established feminist critique of the post-war suburbs as being an instrument of patriarchal social relations, isolating and harmful to the social networks and life satisfaction of women and girls mainly due to unequal gendered access to automobiles (Fava 1980; England 1991). It remains unclear how the extension of automobile ownership and access, coupled with the gentrification of the city (not least of all by single women; see Kern 2010), might affect the spatial articulation of gender inequalities. The contemporary city is also now faced with rising flows of immigrants under globalization, some of whom have quite different experiences with mobility in their homelands, and who arrive in developed nations with varying levels of preparation in negotiating the auto-mobile city. With gentrification of the (pre-war and thus pre-automobile) inner city, which generally sports higher levels of walkability and accessibility by public transit, many new immigrants find themselves living in distant auto-dependent suburbs. How immigrants experience and adapt to such exclusions of automobility is taken up in Chapter 9 of this volume.

Despite the growing literature dealing with automobility, there remain many unanswered questions regarding the relationships between automobility, and urban processes. The system of automobility and fossil-fuelled capitalism have been the source of historically unprecedented population growth and wealth generation, and have massively transformed and extended the mobility of many people, producing benefits and improvements in quality of life that would have been unfathomable only 200 years ago. Yet, it is unclear whether and how this might continue, not least of all due to the problems of depending on a finite resource (oil) for fuel, but just as importantly due to the evolving contours and contradictions of the auto-city itself. As the city undergoes its own transformation, the automobile turns from luxury enabler to mundane necessity, becoming a dependent appendage of the self, the family, and the polity. As the urban revolution transforms the globe in the image of automobility, the relationships between forms of mobility, social welfare, and democratic politics are also transformed. There remain many unanswered questions regarding the sustainability and resilience of contemporary auto-mobile urbanism, the articulation of auto-mobilities with urban vulnerabilities and inequalities, and the potential of the city as a locus of political transformation. Such are the questions and relationships that are dealt with in this book.

Organization of the book

This book involves an exploration of the urban affects, impacts, functions, vulnerabilities, contingencies, and implications of contemporary forms of automobility. The volume is structured into three parts. The first part, *Driving Vulnerability*, is concerned with understanding the current state of global auto-mobility and the potential implications for social, economic, and political sustainability and resilience. Chapter 2 by Martin provides an automobile census, and outlines a number of the social, health, and environmental challenges posed by global automobility. Chapter 3 by Newman and Kenworthy analyses the sustainability and resilience of auto-mobile urban forms, and advocates significant reforms to enhance the resilience of contemporary cities. Chapter 4 by Walks examines the complex relationship between automobility, debt, and financial vulnerability. Chapter 5 by Buliung and colleagues explores the drive to school, and discusses the co-construction and reproduction of automobility among parents and children through school-based travel and social fears related to this.

The second part inquires into the relationships between the expression of urban automobility and the articulation and reproduction of social inequalities. Chapter 6 by Mendez, Moos, and Osolen examines how the restructuring of labour markets under post-Fordism, deindustrialization, and globalization has affected changing commuting patterns and mismatched mobilities. Chapter 7 by Walks and Tranter examines the issue of unequal mobilities in relation to effective speed, and shows how this concept can be used to measure effective mobility and motility. Chapter 8 by Buliung, Shimi, and Mitra moves the focus to the developing world, where non-motorized transport remains centrally important to circulation in the city, yet its use is challenged and users disadvantaged by the continued push towards auto-mobile forms of modernization. Chapter 9 by Hess, Huang, and Vasic brings the discussion back to the developed world, examining the experiences of different immigrant groups in negotiating and adapting to automobility in the contemporary auto-city. Chapter 10 by Reid-Musson examines how automobility produces its "others", in this case temporary, racialized, and often "illegal" migrant non-citizens, who bear the brunt of the irrationalities and dispossessions of the auto-mobile system.

The third part then explores the politics of the auto-city, and how automobility might be reformed or opposed. Chapter 11 by Walks examines the spatial bases of electoral support in the United Kingdom and Canada, and the role of a politics of mobility in shaping political campaigns and ideological contests. In Chapter 12 by Henderson, the political question of how to reform automobility is broached in relation to the removal of a freeway from central-city San Francisco. Chapter 13 by Walks, Siemiatycki, and Smith examines how alternatives to the auto-city, particularly velo-mobility, might be implemented given the political roadblocks put in place by regimes of automobility. Chapter 14 by Talsma examines the issue of political protest in the auto-city, revealing the rising importance of the highway as a contested public space of political importance and a site for political debate. Given the material set out in each part, the concluding chapter then turns its attention to the question of how the auto-city might be dealt with, politicized, and reformed to make it more sustainable, resilient, equitable, and just.

Part I
Driving Vulnerability

2 Global automobility and social ecological sustainability

George T. Martin

The social and material infrastructures of mobility are critical and ubiquitous features of human civilization. The first rudimentary settlements likely grew from places where footpaths crossed or ended at water's edge. Today, motorized transport is a routine activity as people regularly move by a variety of conveyances. Due to its scale and the technologies it uses this mobility has significant impacts on the lives of individuals, on the fabrics of their communities, and on their relations with nature.

The conveyances with the greatest social ecological impacts are motor vehicles. The automobile is the principal type of vehicle and it is an icon of modern civilization. In 2010, autos constituted 70 percent of global vehicles. The percentage is actually higher than this because the United States (US), home to about one-fourth of the world's vehicles, counts sports-utility (SUV) vehicles and minivans as trucks. Thus, "motor vehicle" is the inclusive category used by many national censuses for all roadway conveyances, including autos, buses, minivans, motorcycles, SUVs, and trucks.

The auto is a marvelous machine that affords the conveniences of individualized and flexible mobility. One can travel when one wishes, one can make linked journeys, and one can carry passengers and goods. Additionally, the auto can be used to signal social status. Its global adoption is understandable. In 1920, there were about 13 million motor vehicles in a world populated by 1.9 billion people – or 7 vehicles per 1,000 people. In 2010, in a world of 6.9 billion people there were just over one billion vehicles (topping the billion mark for the first time), for a density of 148 per 1,000 persons. In 90 years, then, humans increased their own numbers by 3.6-fold while increasing the number of their motorized vehicles by 77-fold.

The scale of auto use has become more intensive (more trips and miles driven) as it has become more extensive. Because of this scale, its principal technology, the petroleum-burning internal-combustion engine, consumes vast amounts of non-renewable energy and releases high levels of hazardous emissions. Additionally, the technology is risky to operate, leading to accidents that result in a high casualty toll for drivers, passengers, pedestrians, and cyclists.

Like other technological innovations, the drawbacks of the auto were only recognized over time as its use multiplied. One drawback emerged at its very introduction – the death and injury toll of roadway accidents. To this toll the auto's environmental effects were added as a drawback. The principal of these, air pollution, became a major public concern in the 1960s as it was featured in the emergent environmental movement. Since the 1990s, a specific pollutant, CO_2, has led the auto's critical queue – because it is the majority component of the greenhouse gases (GHG) complicit in global climate change. Motor vehicles contribute a substantial and growing share of CO_2. In 2010, road transport accounted for about one-sixth of emissions and fuel demand is projected to grow nearly 40 percent by 2035 (IEA 2012a: 9–10; 2012b: 9).

Science and policy have been mobilized to promote the mitigation of CO_2 emissions, which is now a major component in making our way of life sustainable into the future. Sustainability was put on the global agenda by the United Nations (UN) Conference on Environment and Development in Rio de Janeiro in 1992. A useful definition of transport sustainability is to "use substantially less rather than more carbon-based energy sources to provide the mobility needed for society to function efficiently" (Banister 2012: 1).

In addition to its road casualties, fuel use, and tailpipe emissions, the auto is associated with socio-economic inequalities. Its costs contribute to mobility disparities between social classes. The skill that its use requires rules out driving for large numbers of citizens, including the young and the infirm. In its extreme manifestation, the auto contributes to the segregation of sub-groups in society.

In this chapter's first section, "A Motor Vehicle Census," the global distributions of production and consumption are presented and analyzed. This is followed by a look at declining automobility in the more developed nations. The social ecological problems associated with automobility are the subject of the third section. The fourth section, "The Two Elephants in the Road," examines sustainability issues for the two dominant auto nations, China and the US. In the last section, the compact city is advanced as a model for transport and urban sustainability.

A motor vehicle census

Motor vehicles are mass-produced and mass-consumed commodities, and the spheres of production and consumption provide a useful basis for their historical enumeration. The over-arching trend of this history has been a sizeable absolute and per capita growth in the number of vehicles in the world.

Production

Automobiles were first mass produced in the early twentieth century as Henry Ford's adaption of assembly-line technology made autos cheap enough for their mass consumption. The history of motor vehicle production (see Table 2.1) between 1920 and 2010 can be divided into three periods.

As the data indicate, a major theme of production has been globalization. Its first period, from 1920 to the 1960s, featured five countries in North America

Table 2.1 World motor vehicle production (in millions) and leading producers, 1920–2010

Year	World production	Leading producer	Percentage	2nd and 3rd producers	Percentage of top 3	Nations with 2% or more
1920	2.4	US	94	Canada France	99	3
1930	4.1	US	81	France UK	99	4
1940	4.9	US	91	UK Germany	98	5
1950	10.6	US	76	UK Canada	87	6
1960	16.5	US	48	Germany[a] UK	72	8
1970	29.4	US	28	Japan Germany[a]	59	9
1980	38.6	Japan	27	US Germany[a]	58	10
1990	48.6	Japan	28	US Germany	58	13
2000	58.4	US	22	Japan Germany	48	14
2010	77.7	China	24	Japan US	46	14
2012	84.1	China	23	US Japan	47	14

Sources: American Vehicle Manufacturers Association, *Motor Vehicle Facts & Figures*, 1990, 1992, 1997; *World Motor Vehicle Data*, Detroit, 1995; International Organization of Motor Vehicle Manufacturers, *World Motor Vehicle Production*, Paris, 2011; US Census Bureau, International Database, Washington, 2013.

Note
a West Germany only.

and Western Europe: Canada, France, Germany, the United Kingdom (UK), and the US. In the next period, from the 1970s to the 2000s, an Asian manufacturer, Japan, became a top producer; it first led world production in 1980. The third and current period began in 2009 when China became the leading producer.

At the ten decennial comparison points the leading nation's share of world production declined steadily, from a peak of 94 percent for the US in 1920 down to 24 percent for China in 2010. The same is true for the three leading producers. In 1920, they commanded 99 percent of production; by 2010, their share was down to 46 percent. In addition to less dominance at the very top the record shows a broadening of the base of production. The number of countries with at least 2 percent of total production rose from only three in 1920, all in North

America and Western Europe, to 14 in 2010, one-half of which were in Asia and South America. Africa, by contrast, still does not have a major producer, while its leader, South Africa, accounted for only 0.6 percent of production in 2010.

The biggest shifts in this globalization occurred between 1950 and 1970. In those two decades, the share of production in the dominant nation declined from 76 to 28 percent, while the share of the top three producers declined from 87 to 59 percent. The globalization was first dispersed within the more developed nations and then into developing nations. It has not reached the less-developed nations, as the economic data in Table 2.2 indicate.

In 2010, 39 nations accounted for over 99 percent of global production. Only two, India and Uzbekistan, were in the lower range (less developed) of Gross Domestic Product (GDP) per capita. There were ten in the middle range (developing) and 27 in the higher range (more developed).

For developing nations vehicle production is a major asset that enhances their economic growth. The contemporary exemplars of this are the five BRICS – Brazil, Russia, India, China, and South Africa. They accounted for 35 percent of 2010's production, up from 10 percent in 2000.

Consumption

Motor vehicle consumption is measured in terms of the registrations required by governments. In 2010 there were 12 countries with at least 20 million registrations. The data in Table 2.3 show the changes in registrations among these 12 since 1921.

The shifts in the locus of vehicle consumption have been similar to those for production. For 1921 through the 1960s, registrations were concentrated in North America and Western Europe. In the 1970s, Japan became a leader, and in the 2000s, China. The diffusion of consumption is somewhat broader than that for production. For registrations, the share for countries outside the top 12 rose from 5 percent in 1921, to 18 percent in 1960, and to 32 percent in 2010. The

Table 2.2 Percentage distribution of motor vehicle producing and non-producing countries, by GDP per capita (in current US$), 2010

Countries	GDP per capita range[a]				
	Lower	Middle	Higher	Total	(N)
Auto producers	5	26	69	100	(39)
Non-producers	41	35	24	100	(150)
	(63)	(63)	(63)		(189)

Sources: World Bank, *World Development Indicators*, Washington, 2013; International Organization of Motor Vehicle Manufacturers, *World Motor Vehicle Production*, Paris, 2011.

Notes
a Ranges: Lower = $199–$2,400; Middle = $2,532–$8,781; Higher = $9,070–$103,574.

Table 2.3 1921–2010 motor vehicle registrations (in millions) for nations with at least 20 million registrations in 2010

	1921	*%*	*1960*	*%*	*2000*	*%*	*2010*	*%*
World	13	100	127	100	752	100	1,015	100
US	11	83	74	58	213	28	240	24
China	[a]		[a]		13	2	78	8
Japan	[a]		1	1	73	10	74	7
Germany	[b]	2	6	4	47	6	45	4
Russian Federation	[a]		[c]		25	3	41	4
Italy	[a]		2	2	36	5	41	4
France	[b]	2	7	6	34	5	38	4
UK	1	4	7	6	31	4	36	4
Brazil	[a]		1	1	19	3	32	3
Spain	[a]		[c]		21	3	28	3
Canada	1	4	5	4	18	2	21	2
India	[a]		[a]		8	1	21	2
Others	1	5	22	18	214	28	321	32

Sources: American Vehicle Manufacturers Association, *Motor Vehicle Facts & Figures*, Detroit, 1990, 1992, 1997; *World Motor Vehicle Data*, Detroit, 1995, 1996; Ward's Communications, *Ward's World Motor Vehicle Data*, Southfield, 1992, 2003, 2011; J. Dargay, D. Gately, and M. Sommer (2007), "Vehicle Ownership and Income Growth, Worldwide: 1960–2030," *Energy Journal* 28: 163–190; L. Gordon (2001), *Brazil's Second Chance*, Washington, DC: Brookings Institution Press; "World Motor Car Census," *New York Times*, February 26, 1922; European Motor Vehicle Parc, "Vehicles in Use, 2003–2008," 2008, Paris; UN, *World Population Prospects 2011*, 2012.

Notes
Columns may not total 100 because of rounding.
a None or negligible amount
b Less than 250,000
c Unavailable.

comparable figures for production were near zero in 1920, 2 percent in 1960, and 19 percent in 2010.

With regard to level of development, consumption follows the pattern of production in that it is concentrated in developing and more developed countries. In 2010, the ten nations with the lowest GDP per capita averaged just 34 vehicles per 1,000 persons while the ten nations with the highest GDP per capita averaged 604 (World Bank 2012). While there is an association between vehicle registrations and per capita income, it is not a linear one over time:

> Vehicle ownership grows relatively slowly at the lowest levels of per-capita income, then about twice as fast as income at middle-income levels, and finally, about as fast as income at higher income levels, before reaching saturation at the highest levels of income.
>
> (Dargay *et al.* 2007: 4)

It is in the middle-income (developing nations) level that consumption spurts upward, as it is doing today in the BRICS nations. These five countries increased their share of registrations from 10 percent in 2000 to 18 percent in 2010. Meanwhile, the numerical growth of motor vehicles may be coming to an end in more developed nations.

"At the end of the road?"

As Newman and Kenworthy note in their chapter in this volume, the steady growth of motor vehicle use over the course of the last century may be coming to an end in its birthplace (see also Goodwin and van Dender 2013; Puentes and Tomer 2009). Annual vehicle miles traveled (VMT) per licensed driver have been decreasing since 2005 in the US. In 2010 total VMT were about 10 percent below the long-term trend (FHA 2010, 2012). Similar changes have been noted in other more developed countries (Litman 2013: 8). Data show that "in most industrialized countries it can be seen that urban mobility and car traffic have stagnated since the early 2000s" (Madre *et al.* 2012: 5). In the UK, vehicle travel began to decrease after 2007 following steady increases since 1949 (Le Vine and Jones 2012: 2).

In a summary analysis of the data related to the issue, Goodwin (2012: 29) posited three hypotheses: (1) interrupted growth that will be resumed at a slowing rate; (2) saturation that produces a stable automobility; and (3) peak automobility that is an early sign of a long-term decline. His conclusion was that, "subject to research, it seems unlikely that any of the three hypotheses can be firmly ruled out with confidence and consensus over the next two or three years" (Goodwin 2012: 30).

There are several cultural and demographic trends implicated in declining automobility in more developed countries (Puentes 2012). They include the following:

1   There has been a decline in driving among youth that is perhaps related to the growth of social media; for example, the proportion of US 19-year-olds with a driver's license declined from 87 percent in 1983 to 70 percent in 2010 (Sivak and Schoettle 2012; Stokes and Lucas 2011).
2   The gradual entrance of women into the paid workforce over the last 50 years is ending; for example, in the US in 2005, the number of female drivers surpassed that of male drivers for the first time (FHA 2008: 24).
3   Populations are aging and driving peaks in middle age when work and family responsibilities are greatest (Le Vine and Jones 2012: 22).

There are likely to be national differences in the causes of reduced automobility. For example, a study in the UK found that most of the decline could be accounted for by a sharp fall in company car use, which is linked to large increases in taxation on fuel. The study concluded:

> if company car mileage is discounted, then there has been a pattern of continuing strong growth in private car use for those aged 30 and over, outside

London, up to the start of the economic downturn. This group represents approximately 70% of the population of driving age in Great Britain.

(Le Vine and Jones 2012: ii)

If automobility has reached the limits of its growth in more developed nations, it could be a watershed moment for the advancement of their sustainable transport. Still, however, the present levels of automobility around the world are the basis for a number of social ecological problems at all levels of economic development.

A motor vehicle social ecology

The social ecological impacts of motor vehicle transport are rooted in its cost, its polluting technology, and the riskiness of its operation. The impacts are not distributed evenly or fairly across populations. The adverse consequences of vehicular cost and emissions are borne disproportionately by lower-income persons and minority groups. The auto's risky operation precludes driving for people who are not sufficiently skilled, or not mature enough, or not physically able to drive. Additionally, road casualties disproportionately impact a certain age group.

Cost

Vehicle travel increases with income. For example, in the UK in 2005–2007, adults who were in the £50,000 and over band of personal income averaged 10,000 miles of driving per year, while those in the £0–9,999 band averaged only 2,000 miles (Le Vine and Jones 2012: 28). One study of auto inequality in the UK concluded that "patterns of travel behaviour relate strongly to incomes, with people in lower income households tending to be much less likely to have a car available, likely to make fewer trips per week, and to travel shorter distances per week" (Stokes and Lucas 2011: 3–4). Additionally, the researchers found that the connection between income and automobility may be reflexive. Lack of employment may rule out the cost of an auto – and lack of an auto may be an impediment to employment. In the US this problem is referred to as the spatial mismatch between inner-city residents and suburban jobs. The geographic disconnect is supported by transport systems that favor the auto, while only a minority of inner-city residents have access to one (see Wilson 2009).

In part because of the cost of owning and operating cars, urban expansion that is based on auto transport fosters "greater segregation of residential development according to income" (EEA 2006: 35). The result is to enhance inequality: "This differential distribution tends to overlay and intensify long-standing social class and racial-ethnic divisions within societies, as motorization develops its own social ecology of risk across metropolitan areas" (Martin 2009: 223). Thus, automobility can foster a convergence of social cleavage and geographic separation. As Calthorpe (1991: 51) has noted, the car "allows the ultimate segregations in our culture – old from young, home from job and store, rich from poor and owner from renter."

Risky operation

Driving a motor vehicle is a technical skill that requires training and licensing. It is commonly prohibited for several groups of citizens, including the young, the very old, and the impaired. There are others who may be licensed to drive but who should not be driving under some conditions, including inebriation, medication, and infirmity.

With safety improvements in vehicles, infrastructures, and regulations, the rates of roadway fatalities decline over time so that increasing numbers of vehicles and miles traveled do not necessarily result in more deaths, at least in more developed nations. For example, in the US, the greatest number of fatalities, 54,589, was registered in 1972. In 2011, deaths had declined to 32,367. The highest rate of fatalities was in 1923–1927 when the average was 18.8 deaths per 100 million VMT. The rate declined to just 1.1 by 2011 (DOT 2012). However, while deaths and death rates have declined, the toll of road accidents remains high. They are the leading cause of all accidental deaths in the US, and they take considerably more lives than do homicides (TSA 2012). They are a leading cause of death for 15–34-year-olds around the world. A World Health Organization study (Racioppi *et al.* 2004: 5) of Europe found that roadway deaths and injuries promise to increase among "young people in the Region and are predicted to increase in countries with low or medium income as they become more highly motorized."

Globally, road fatalities are a major public health problem, on a par with malaria. In 2013, there were about 1.2 million deaths. The toll is not distributed evenly among nations (see Table 2.4). Comparing deaths to each dollar of per capita income shows a stark divergence between, for example, India at 183 and a number of more developed nations at just 0.1. In low-income or less-developed countries, there were 18 deaths per 100,000 people; in middle-income or developing countries, 20; and in high-income or more developed countries, 9 (WHO 2013). Higher-income nations have the resources to invest in road infrastructure improvements and road safety regulations, while lower-income nations do not. Middle-income countries show the highest rate of road fatalities because they are motorizing rapidly while infrastructure and regulation upgrades are not keeping pace.

Emissions

Motor vehicles are the leading contributors to outdoor air pollution. Their emissions contain toxic substances that can contribute to morbidity and mortality, shortening lifespans (depending on the length and intensity of exposure) because they can cause or exacerbate asthma, pulmonary disease, and lung cancer (Paul and Miranda 2011). Despite improvement in air quality in more developed nations due to technical improvements in vehicle engines and fuels prompted by increasing state regulation, there is ongoing concern for public health, especially with regard to particulate matter in the air. The UK

Table 2.4 Vehicles per capita and road traffic deaths, 2010, for nations with at least 20 million vehicles

	Vehicles per 1,000 people	Estimated road traffic deaths	Deaths per 100,000 people	Deaths per GNI per capita*
World	148	1,240,000	18.0	123.8
US	797	35,490	11.4	0.8
Italy	679	4,371	7.2	0.1
Canada	624	2,296	6.8	0.1
Spain	593	2,478	5.4	0.1
Japan	591	6,625	5.2	0.2
France	580	3,992	6.4	0.1
Germany	572	3,830	4.7	0.1
UK	519	2,278	3.7	0.1
Russian Federation	289	26,567	18.6	2.7
Brazil	164	43,869	22.5	4.6
China	58	275,983	20.6	64.8
India	18	231,027	18.9	183.4
Others	102	601,194	18.6	60.0

Sources: Ward's Communications, *Ward's World Motor Vehicle Data*, Southfield, 2011; United Nations, *World Population Prospects*, New York, 2012; World Bank, *GNI per Capita*, Washington, DC, 2012; World Health Organization, *Global Status Report on Road Safety*, Geneva, 2013.

Note

* Deaths divided by per capita Gross National Income (US$).

Parliamentary Environmental Audit Committee has cited 2008 research that estimated that particulates from road traffic cause the equivalent of about 29,000 deaths per year (PEAC 2011).

Transport sector particulate emissions are highest in the large cities of developing nations: in Chinese cities, they average over 350 micrograms per meter; in Indian cities, about 275 – while just 50 in the cities of developed nations (Gorham 2002: 15). In developing countries, outdoor air quality continues to decline due to the growth of industry and vehicle use, coupled with insufficient environmental regulation. An example is Mexico City, where residents experience rising exposure to pollution from the large amount of industry and some three million autos. Concentrations of particulate matter in some areas far exceed both Mexican and international standards (Escamilla-Nunez *et al.* 2008).

In the US, the distribution of unhealthy emissions is uneven geographically and socially, so that "poor people and people of color are disproportionately impacted by air pollution" (Frumkin 2002: 209). Research indicates that particulates have more hazardous ingredients in non-white and low-income communities than in affluent white ones (Katz 2012). These sub-groups are more likely to live near hot spots – local areas where emissions are high (Freund and Martin 2007: 43). In addition to heavy roadway traffic, the hot spots are caused by proximity to petroleum refineries and storage units, and to vehicle refueling, parking, and maintenance facilities (Bae 2004).

Vehicular pollution was a major issue stimulating the environmental movement in the 1960s and it took a leading role in the emergence of the environmental justice movement in the US in the 1980s. Campaigners were successful in getting environmental justice put on the national agenda in the 1990s. In 1994, President Clinton signed an Executive Order for federal departments and agencies entitled "Federal Actions to Address Environmental Justice in Minority Populations and Low-Income Populations" (EPA 2013).

"The two elephants in the road"

Two countries at quite different levels of development, China and the US, dominate global automobility and its social ecological problems.

In 2010, China and the US were first or second in motor vehicle production and consumption, accounting for about one-third of both, while having about one-fourth the world's population. The two countries are also the principal producers of CO_2 emissions from all fossil fuels. China is now the largest emitter, contributing 24 percent of the world's total, while the US contributes 18 percent (IEA 2012a: 9). However, the US has been extensively emitting for far longer and it is the country most responsible for today's high levels. One estimate is that the US contributed 29 percent of emissions between 1850 and 2002, while China contributed 8 percent (Baumert *et al.* 2005: 32).

China will likely experience considerably more growth in vehicle consumption. The US is estimated to be close to its saturation level of 852 vehicles per 1,000 people (Dargay *et al.* 2007: 19). In 2010, this ratio stood at 797. China's estimated saturation level is 807 and it is projected to reach 269 by 2030, which is about five times its current level.

How China and the US deal with the challenges they face on their roads to sustainable transport will have an enormous influence in the rest of the world. For China, a large part of the challenge is to avoid the US pattern of auto-based low-density urban sprawl. CO_2 emissions from Chinese urban transport are currently projected to *quadruple* by 2030. A simulation modeling study of one Chinese city concluded that a dense, mixed-use urban expansion pattern coupled with the promotion of public transit would eliminate about one-half the emissions produced by a business-as-usual scenario (He *et al.* 2011).

A dense, compact urban pattern supports public transit that in turn reduces energy use as well as CO_2 emissions. In an analysis of data from 84 cities around the world, Newman and Kenworthy (2011a: 13) found that the auto's use of energy per passenger kilometer was 3.6 times that for the average of bus, metro, and tram.

While its high urban densities bode well for the development of sustainable transport, it is not a good omen that SUVs are gaining popularity in China. Between 2012 and 2013, SUV sales increased by 49 percent, while overall auto sales increased by 13 percent (Bradsher 2013). SUVs are responsible for about 18 percent more emissions than are sedans (DOT 2010: 15).

CO_2 emissions and energy consumption highlight the social ecological sustainability problems associated with motor vehicles in China and the US.

These problems are illustrated in the US development of urban sprawl and hyperautomobility in the latter half of the twentieth century. In China, current large-scale urbanization poses a threat to agricultural production, especially if urban expansion favors auto transport.

US urban sprawl, transport systems, and hyperautomobility

The present level of vehicle ownership and use in the US is, in sustainable transport terms, over-saturation. It has been described as hyperautomobility, in the sense of being unbalanced transport (Martin 1999; Freund and Martin 2007, 2009). While rail and bus (and cycling and walking) in older cities such as New York command a respectable share of the daily transportation modal split, the auto has come to dominate transport in newer, post-World War II urban development. The bulk of this development has taken place in the South and West of the US in urban agglomerations such as Atlanta, Dallas, and Los Angeles. This urban sprawl "was fuelled by the rapid growth of private car ownership" (EEA 2006: 5). Auto domination of transport modal splits is the basis of hyperautomobility.

Hyperautomobility is the excessive use of autos supported by transport systems built on auto use, and it can be defined by the share (or split) of urban journeys made by car. Following is a comparative schematic for levels of automobility in large cities of the world (data from ACS 2011; LTA 2011):

Hyperautomobility: More than three-quarters of journeys made by auto. Examples: Dallas (89 percent), Houston (88 percent), Los Angeles (78 percent).
High automobility: Between one-half and three-quarters of journeys made by auto: Sydney (69 percent), Toronto (67 percent), Rome (59 percent).
Moderate automobility: Between one-quarter and one-half of journeys made by auto: London (40 percent), New York (33 percent), Seoul (26 percent).
Low automobility: Less than one-quarter of journeys made by auto: Delhi (19 percent), Tokyo (12 percent), Hong Kong (11 percent).

The development of hyperautomobility is a major reason why the US leads the world in per capita CO_2 emissions; however, it was not an inevitable outcome of auto use. It is produced by transport systems that are based on the auto mode of travel. Hyperautomobility developed in the post-World War II boom in low-density suburban housing in the US. The resultant dispersion meant that auto use was necessary to travel from home to work and to other activity sites. Many cities in more developed nations have high auto use *but within* multi-modal transport systems.

Automobility and hyperautomobility are distinctively fostered by auto-centered transport systems comprising social and material infrastructures inhospitable to walkers, cyclists, and train and bus riders – all of whom travel in a more sustainable fashion than do motorists. It is at its zenith in contemporary US exurban developments where sidewalks, cycle lanes, and bus and train services often are not provided. This leaves one choice of daily transport for residents – the automobile.

Ewing *et al.* (2002) developed a sprawl index for US cities based on 22 specific variables centered on low-density development and dispersion and separation of homes, workplaces, and shops. Urban sprawl and automobility are closely linked:

> As sprawl increases, so do the number of miles driven each day; the number of vehicles owned per household; the annual traffic fatality rate; and concentrations of ground-level ozone, a component of smog. At the same time, the number of commuters walking, biking or taking transit to work decreases to a significant extent.
>
> (Ewing *et al.* 2002: 17)

The high levels of road traffic featured in hyperautomobility are associated with problems other than high energy use and emission levels, including roadway congestion and road rage. Road rage is defined as an event in which an angry driver tries to harm another driver after a traffic dispute. Urban road traffic congestion is a prime factor in eliciting disputes (Frumkin 2002: 207). The iconic auto-based US metropolis is Los Angeles, and it leads the nation in congestion delays; drivers there spent the equivalent of nearly four days stuck in traffic in 2003 (TTI 2005). As well as the environmental costs of more emissions and more fuel use, congestion delays have large and growing monetary costs for drivers and their employers, as well as costs in terms of time stolen from other necessary productive activities.

A growing public health problem in more developed nations that is associated with urban sprawl arises from the sedentary lifestyle that it promotes. These are the risks posed by obesity and its association with diabetes and cardio-vascular ailments (Freund and Martin 2007, 2008). A national study (McCann and Ewing 2003) in the US compared the sprawl index of urban areas with the health characteristics of their populations. The likelihood of obesity was higher in areas of greater sprawl, regardless of gender, age, education level, and smoking and eating habits.

China's land use squeeze

While both countries are similarly sized continental-scale countries, the US has considerably more arable land than does China. Urban growth generally is a consumer of farmland and this is a special threat to China (Martin 2007). In China (and other developing nations) urban expansion is taking prime agricultural land, while in more developed nations it is a threat to protected natural areas (Seto *et al.* 2011). In 2003 alone, China lost more than 2 percent of its farmland to urban expansion (Yardley 2004).

The loss of farmland to urban expansion will continue in China because of the ongoing urbanization of its population, but if the expansion develops in an auto-based direction the loss will be greater. Moreover, climate change presents the coincident potential of rising sea levels, which also pose a threat. China's population and farmland are concentrated in low-lying river valleys and deltas, and along its seaboard. Shanghai and its surrounding rich farmland epitomize this, as they are located in the Yangtze River delta on the East China Sea.

If China's urban expansion is tied to auto-based low-density development, its farmland loss will be heightened because automobility requires large dollops of land. Autos use multiple, dedicated parking spaces at living, working, shopping, and recreation sites. In addition, multiple-lane roadways and servicing facilities require large amounts of land. As a result, the auto requires more land on a per capita basis by multiples over other transport modes (Litman 2012).

China and the US are not the only countries in which the land take of urbanization poses potential problems. A meta-analysis of over 300 studies using remotely sensed images to map urban land conversion indicates that the containment of urban sprawl has become a global problem. Over the last three decades urban land expansion rates have been higher than or equal to urban population growth rates (Seto *et al.* 2011). Between 1970 and 2000 urban conversion took the land equivalent of Denmark. Between 2000 and 2030 it is projected to take 36 times as much, or the land equivalent of Mongolia.

Conclusion

Global automobilization is coincident with global urbanization. The world is now majority urban and increasingly so. The more developed nations have been majority urban since the Industrial Revolution, and they continue to urbanize at an incremental rate. The less developed and developing world is the locus of the urbanizing process today, especially rapidly developing nations such as China where the proportion of urban residents grew from 36 percent of the population in 2000 to 50 percent in 2010 (NBS 2011). This urbanization is taking place at the same time as the world's population continues to grow. The UN's (2011) medium variant projection is that world population will stabilize at nearly 11 billion in 2100. The growing and increasingly urbanized population is driving the emergence of scores of giant metropolitan areas, or *megacities*, of ten million or more people (UN 2009). The forms that urban expansion and its transport take in these megacities, especially in China, is a critical factor bearing on the fate of humanity's efforts to live in a sustainable fashion.

The sustainability of cities and their transport systems depends on having population densities at sufficient levels to enable multi-modal travel. Sufficient density, coupled with mixed land-use patterns, makes walking and cycling practical modes of local travel. It is also a necessary basis for frequent bus and rail transit. Even in the highly automobilized US, higher density is associated with lower auto use. New York City's high density (over 26,000 residents per square mile) supports a habitat in which only 44 percent of households own autos, while in low-density Los Angeles (under 8,000 residents per square mile) nearly twice the proportion of households own autos (TSA 2012).

Density is of value to sustainability in several ways. Socially, it provides a base for face-to-face interaction and for the development of social capital. Ecologically, it provides a base for more efficient use of energy and land. For example, a comprehensive study in the US found considerable efficiency savings of more compact city development as compared with market-driven sprawl – in land resources, in the construction of local roads, and in the provision of water and sewage facilities (Burchell *et al.* 2002).

Environmentally, density can provide for lower GHG emissions. A Toronto study comparing center-city apartments with suburban detached housing found significantly higher per capita CO_2 emissions from both housing stock and transport in the suburbs. The per capita emissions of suburban housing stock and transport modes were more than twice those of city housing stock and transport modes (Norman *et al.* 2006). This assessment concluded that "the most targeted measures to reduce GHG emissions in an urban development context should be aimed at transportation emissions, while the most targeted measures to reduce energy usage should focus on building operations" (Norman *et al.* 2006: 10).

The development of a denser, more compact urbanization has been on the planning agenda since the 1990s in the US – in the form of such schemes as Smart Growth and Transit Oriented Development (Calthorpe and Fulton 2001; Handy 2005; TCRP 2002). There is growing evidence that sprawl is being contained. Smart Growth is successful in a number of US urban areas (Blair and Wellman 2011). Compact, mixed-use infill development was found to be effective in reducing vehicle miles traveled in four US cities (Zhang *et al.* 2012). Other data indicate that sprawl may have reached a plateau nationally. For the first time in the post-World War II period, US core cities grew at a slightly faster rate than their suburbs in 2010–2011 (Frey 2012).

The compact city is a model that is applicable around the world. In Europe, coordinating land use and transport in order to reduce energy use and emissions from transport has been integral in "Oslo's farewell to urban sprawl" (Naess *et al.* 2009: 135). Urban brownfield redevelopment has proven successful in Germany and the UK, using different approaches (Baing 2010). In China, a study of commuting and urban form in Beijing concluded that compact urban expansion policies contribute to the goal of containing the growth of vehicle miles (Zhao 2011: 522–523).

In a predominantly urban world, the best available practice for a socially and ecologically sustainable daily transport is to reduce the use of automobiles (or the rate of increase in their use) through the provision of built environments that maximize density. Density, at its minimum level, should be sufficient to support transport systems with inter-connected multiple modes of travel. In cities with such systems, the auto remains an option for travel along with cycling, public transit, and walking.

Automobility has a central role to play in a world at a crossroads as it faces uncertain climate changes while its growing population is increasingly urbanized. Global automobility is at an ambiguous and fractured juncture. While auto emissions make a significant contribution to global warming, the auto's convenience and status make it a popular consumer good. While auto-centered transport systems support a sprawled version of urban expansion that makes for comparatively cheap land at urban peripheries, the dispersion consumes ecologically useful farmland and green space at the same time that it promotes socially inequitable travel. While the more developed nations may be reaching auto saturation, automobility is growing rapidly in the developing nations. While auto production plays a major role in industrialization for developing nations, it also stimulates domestic consumption faster than auto infrastructure and traffic management can be developed, resulting in high levels of traffic deaths and toxic vehicle emissions.

China and the US are the outsized protagonists in this unfolding drama and their actions weigh heavily in shaping automobility's roles in the world's confrontation with climate change, population growth, and increasing urbanization. However, every nation faces critical policy decisions in these matters – individually and in blocs like the European Union. The pool of all nations provides a basis for the emergence and testing of innovations, both technological and social, that are needed in order for global automobility to be developed in a sustainable direction.

3   Automobility and resilience

A global perspective

Peter Newman and Jeffrey Kenworthy

Automobility is highly related to automobile dependence, the idea we coined in 1989 to describe the way we have created cities and regions where automobiles have become essential to daily life (Newman and Kenworthy 1989). The automobile is a very effective tool for enabling movement and connection, but when we are so dependent on it for daily life we have no other real choices. The use of automobiles should not be so fundamentally built into the structure of our cultures, our economies, our cities, that we are unable to free ourselves from them. It is our contention that cultures, economies, and cities that are fundamentally caught up in automobility and automobile dependence will not be resilient; that is, they will not have the ability to respond to the big issues of our age (Zhao *et al.* 2013).

Resilience is related to sustainability, but while the latter seeks to reduce resource needs and ecological footprints, the former concerns the capacity to respond and adapt to various stresses and shocks, both those of an environmental/ecological nature (climate change, natural hazards, water scarcity, higher summer temperatures) and those with implications for dynamic social conditions (liveability, rising fuel and food prices, industrial and employment change, disruptions in global trade, etc.) (Beatley and Newman 2013). This chapter examines the history of how cities in the developed North came to be so automobile dependent, the problems with automobility in terms of resilience, what is required of cities to become resilient in relation to transport policy, how cities compare in terms of resilience, what the new trends in cities might be as they face up to issues surrounding resilience, and what can be done to help facilitate more resilience in our cities.

Automobility and the history of urban form

Cities are shaped by many historical and geographical features, but at any stage in a city's history, patterns of land use can be changed by altering transportation priorities. Italian physicist Cesare Marchetti (1994) has argued that there is a universal travel time budget of around one hour on average per person per day. This *Marchetti constant* has been found to apply in every city in our Global Cities Data Base (Kenworthy and Laube 2001) as well as in data on UK cities for the last 600 years (Standing Advisory Committee for Trunk Road Assessment 1994). The biological or psychological basis of this Marchetti constant seems to be a need for a more reflective or restorative period between home and work, but it cannot go on for too long before people become frustrated due to the need to be more occupied rather than just "wasting" time between activities. Many functions are carried out in cars, as well as transit, biking, and walking, during travel time that are not considered to be wasted (e.g. family talk, phone contact, active exercise) but such activities are less oriented to primary functions of work and thus are often less valued.

The Marchetti constant helps us understand how cities are shaped and how activity patterns evolve within them (Newman and Kenworthy 1999, 2006). Importantly, cities grow to being "one hour wide" based on the speed with which people can move in them. Understanding this fundamental principle will enable us to see how automobility has developed. Over time, three distinct city types have emerged with their own urban fabrics, forms, and logics of mobility: walking cities, transit cities, and automobile cities. Of course, most cities reveal a mixture of all three urban fabrics on the ground. It is to these city types that we now turn.

Walking cities

Walking cities have existed for the past 8,000 years, because walking was the only form of transport available to enable people to get across their cities at speeds of around 5–8 km/hour. In turn, walking cities were, and remain, dense (usually over 100 people per ha), mixed-use areas with narrow streets, and were typically no more than 5–8 kilometres across. Substantial parts of cities such as Barcelona, Ho Chi Minh City, Mumbai, and Hong Kong retain the character of walking cities. In squatter settlements the urban fabric is usually a walking city. In wealthy cities like New York, London, Vancouver, and Sydney, the central areas are predominantly walking cities in character, though they struggle to retain the walking city urban fabric due to the competing interests of car users. Many cities worldwide are trying to reclaim the fine-grained street patterns associated with walkability in their city centres, finding that they cannot do this unless they adopt the urban fabric and form of ancient walking cities (Gehl 2011).

Transit cities

Emerging from 1850 to 1950, transit cities were based on trams and trains that could travel faster than walking – trams at around 10–20 km/hour and trains at around 20–30 km/hour. This meant cities could now spread out, up to 20 kilometres using trams, and up to 30 kilometres with trains. The resulting urban fabric and form was different with either linear tram-based development (trams being slower, and with closer spacing of stops, led to strips of walking urban fabric) or nodal dense centres along corridors (following faster heavy rail lines with walking city urban fabric at stations, like pearls along a string). Densities could be lower (around 50 people per ha) as activities and housing could be spread out further. Most European and wealthy Asian cities retain this urban fabric, as do the old inner cores in US, Australian, and Canadian cities. Many developing cities in Asia, Africa, and Latin America have this dense corridor form of a transit city, but they do not always have the transit systems to support them, so they become car-saturated. Singapore, Hong Kong, and Tokyo have high densities in centres based on mass public transit linkages. Cities like Shenzhen, Jakarta, and Dhaka have grown so quickly with dense, mixed-use urban fabric, based only on buses and para-transit, that the resulting congestion shows that their activity intensity demands higher-intensity mass public transit. Most of these emerging cities are now building the transit systems to suit their urban form, with, for instance, Shenzhen opening a metro system in 2004. China is building 82 metro rail systems and India is building 16 metros (Newman, Kenworthy, and Glazebrook 2013). Cities without reasonable densities around train stations are finding that they need to build up the numbers of people and jobs near stations, otherwise not enough activity is there to support such sustainable transport.

Automobile cities

As the most recent form, existing from the 1950s onward, automobile cities could spread beyond the 20-kilometre radius to as far as 50–80 kilometres, in all directions, and at low density because cars could average 50–80 km/hour when traffic levels are low. These cities disperse in every direction due to the flexibility of cars and with zoning that separated activities. These cities typically provide limited public transit, mostly bus services to support their sprawling suburbs, and within a generation such areas became not only the loci but the generators of automobility. Cities in the new world tended to leverage their growth to build automobile-dependent suburbs as their main urban fabric. Many European cities and Asian cities are now building such suburbs around their old public transit city fabric. Canadian, Australian, US, and New Zealand cities that were developed in this way are now reaching the limits of the Marchetti constant of a half-hour car commute as they sprawl outwards. The freeways that service such areas are full at peak times and commuters are unable to keep within a reasonable travel time budget (Newman 1995). This is now a serious political issue, as outer suburban residents are demanding fast rail links that can beat the traffic (McIntosh *et al.* 2013). Many have been leaving these areas and moving to locations with walking city and transit city urban fabric where they can live within their travel time budget. The first evidence (discussed below) is now appearing that these automobile cities are densifying, developing significant subcentres to provide more opportunities close by and reducing their car use – they are (re)discovering a more resilient city.

Cities of all kinds, whether growing megacities or rapidly sprawling ones, are constantly facing the need to adapt their land use or infrastructure to the travel time budget. According to the logic outlined above, if the Marchetti constant is exceeded, then markets and politics over time bring about changes, enabling people to adapt by moving closer to their work, finding better transportation options, or accessing more needs locally. The search for better options – in transport and land use – can form the basis of social movements that seek to provide more sustainable and resilient cities. Automobility is now being challenged in a much more fundamental way than even the most cogent critiques, dating back to as early as Lewis Mumford's *The City in History* (1961), where he rhetorically asks whether cities are for people or for cars.

Towards the resilient city

The problems of automobile dependence have been previously outlined in terms of the economic, environmental, and social issues that began emerging from the 1970s as the excessive dependence on cars grew (see Schiller *et al.* 2010). We have found that ideas of more sustainable cities need to merge with the increasing concerns for more resilient cities (Newman *et al.* 2009), where resilience is about the capacity for adapting to changes that the future will inevitably bring, especially to more automobile-dependent cities with their greater vulnerability to changes in oil price and supply and climate change.

The inter-linkages between social, economic, and environmental issues have now become very cogent. Highly automobile-dependent suburbs on the urban fringe are expensive to build and live in (Trubka *et al.* 2010) and are now highly vulnerable to oil and its issues with climate change and fluctuations in global oil prices (Newman *et al.* 2009). Some have argued that the global financial crisis that began in 2008 was caused by a housing bubble that burst because excessive housing was built in highly car-dependent locations: when the oil price suddenly tripled in 2008 the ability of people to pay mortgages was stretched and broke (see Chapter 4, this volume, for a full account). Whether true or not, the vulnerability of those living in automobile-dependent locations to oil price fluctuations and their impacts on mortgage repayments was predicted by Dodson and Sipe (2008), and their VAMPIRE maps outlined fringe suburbs in Australian cities most at risk from this phenomenon. In Australia, such suburbs are generally those that are not serviced by a suburban train line. There is a noticeable drop in vulnerability in inner suburbs and suburbs nearest to train lines, largely due to lower necessity for automobile ownership. Predictions by global agencies such as the Energy Information Administration suggest the era of cheap oil is over (EIA 2013).

Climate change is also adding to the issue of vulnerability, given the need to substantially phase out fossil fuel use by 2050 (World Bank 2013). The world has paid insufficient attention to the changes needed to enhance resiliency through reductions in the use of oil in the transport system. As outlined by Grove and Burgelman (2008: 1) in *The McKinsey Quarterly*:

> Our aim should not be total independence from foreign sources of petroleum. That is neither practical nor necessary in a world of interdependent

economies. Instead, the objective should be developing a sufficient degree of resilience against disruptions in imports. Think of resilience as the ability to absorb a significant disruption, bigger than what could be managed by drawing down the strategic oil reserve...

Our resilience can be strengthened by increasing diversity in the sources of our energy. Commercial, industrial, and home users of oil can already use other sources of energy. By contrast, transportation is totally dependent on petroleum. This is the root cause of our vulnerability.... Our goal should be to increase the diversity of energy sources in transportation. The best alternative to oil? Electricity. The means? Convert petroleum-driven miles to electric ones.

We would add that not only do we need more electric transport, we need less car use in general, since only a few of the problems of automobile dependence can be solved by simply switching propulsion systems. Unless the issue of automobile dependence is addressed along with oil vulnerability then the structural problems that cause a lack of resilience in our cities will remain.

The resilient city: compact, green, public transit-based polycentric centres

Overcoming automobility is not easily achieved in any city, especially given a predominantly automobile city urban fabric. The urban planning profession can now see the value in the walking city and the transit city as urban fabrics that enable a much greater range of options for people in their daily transport needs. But what can be done with the automobile city urban fabric? Urban planning is now suggesting there are ways to focus density in automobile cities in a range of centres throughout the suburbs linked by quality public transit systems. This model of a more resilient city is often called the polycentric city, a visualization of which appears in Newman and Kenworthy (1999, fig. 4.3). The rationale for this polycentric city model is to reduce automobile dependence by creating small, walkable cities in the suburbs that can be easily reached for local trips by walking, cycling, local buses, perhaps light rail transit and short car rides, but for more regional destinations the centres are linked by fast, quality rail transit.

Such centres need to be the focus of urban redevelopment rather than continuing with car-dependent suburbs on the fringe. With the availability of new green technology for power, water, and waste infrastructure ideally suited to precincts or centres (Rauland and Newman 2011; Newman *et al.* 2009), this new polycentric urban fabric can be the basis of green economy innovation (Glaeser 2010; EIA 2013). This model has multiple sustainability advantages in terms of reductions in car use and oil use and building energy use, better equity and health outcomes and stronger economies through agglomeration benefits (Schiller *et al.* 2010; Owen 2004; Newman and Kenworthy 1999; Newman *et al.* 2009).

How do cities compare in their resilience and what are the trends?

This section presents perspectives from an array of data on key variables from our global cities database. For 1995, data is available from the Millennium Cities Database for Sustainable Transport (Kenworthy and Laube 2001) for 84 cities worldwide. Certain key variables such as urban density and passenger transport energy use per capita are explored, with the range of cities available providing a truly global perspective. These data are being updated on an ongoing basis, with key data also presented for 2005, a full decade later, to see how certain factors that are important to our discussion of resilience have changed among cities in the United States (US), Canada, Australia, Europe, and two wealthier Asian cities (Singapore and Hong Kong).

Figure 3.1 from the Global Cities Database shows the huge range in per capita energy use for private passenger transport that characterizes cities across the world. They all have a combination of these three city types – walking, transit, and automobile cities. Energy use in a city is an important surrogate indicator of automobile dependence, and, in this case, it is easy to see how US, Canadian, and Australian cities are winning in that race.

The city highlighted near the right side of Figure 3.1 – Barcelona – uses just 8 GJ per person per year compared to 103 GJ in Atlanta, a difference of 13 times. And yet the GDP per capita in Atlanta was only 1.7 times more than Barcelona in 1995. The difference is that Barcelona is substantially a walking city with some elements of a public transit city and almost no trace of an automobile city, whereas Atlanta is almost completely an automobile city with just a little of the transit and walking city urban fabric.

The broader picture is expressed in Figure 3.2 where travel patterns (as reflected by either annual per capita car use or private passenger transport energy use) are exponentially related to urban population density. Atlanta is six persons per ha and Barcelona is 200 per ha.

The same patterns can be seen across cities where often the centres constitute walking cities, the inner to middle suburbs are public transit cities, and the outer suburbs are automobile cities. When data for Melbourne and Sydney are combined, covering transport greenhouse gases per person by suburb versus the density of residents and jobs per ha in each suburb, a very similar curve to Figure 3.2 is obtained with a very strong statistical fit (Trubka *et al.* 2010). Questions of wealth are not driving this phenomenon, as there is an inverse relationship between urban intensity and household income in Australian cities – outer suburbs are poorer and yet households in these areas can drive from three to ten times as much as households in the city centre. As the data on Melbourne indicates (see Table 3.1), poorer households are driving more, using public transit less, and walking less, largely because of where they live (Kenworthy and Newman 2001).

There are obviously complex interactions that influence the intensity of activity and how it impacts on transport patterns. Ewing and Cervero (2010) have outlined how each of the urban fabric elements, such as density, the mixture or

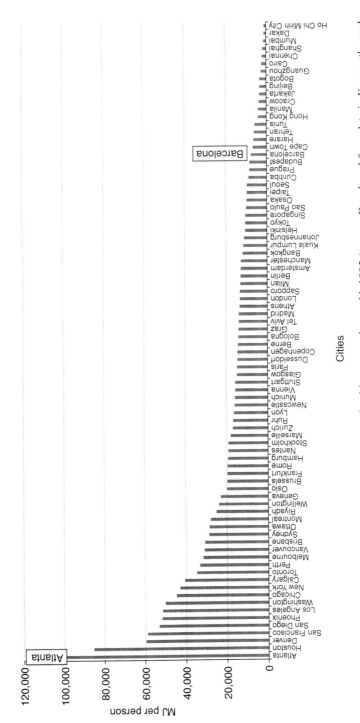

Figure 3.1 Private passenger transport energy use per person in cities across the world, 1995 (source: Developed from data in Kenworthy and Laube (2001)).

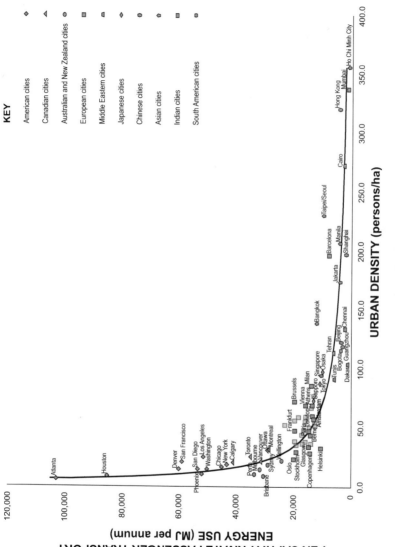

Figure 3.2 Private passenger transport energy use per person and urban density, 1995 (persons per ha) (source: Updated from Kenworthy and Laube (2001) (graph created by Mr Phil Webster)).

Table 3.1 Differences in wealth and travel patterns from the urban core to the urban fringe in Melbourne

	Core area	Inner area	Middle suburbs	Fringe suburbs
Percentage of households earning >$70,000/year	12	11	10	6
Car use (trips/day/person)	2.12	2.52	2.86	3.92
Public transit (trips/day/person)	0.66	0.46	0.29	0.21
Walk/bike (trips/day/person)	2.62	1.61	1.08	0.81

Source: Kenworthy and Newman (2001).

diversity of uses in a place, and design, influence travel. Other factors include the network of public transit services provided, income, fuel prices, cultural factors, etc., but all of these can also be linked back to the intensity of activity in various ways. Thus, although many discussions have tried to explain transport patterns in non-land-use terms (Brindle 1994; Mindali *et al.* 2004), the data suggest that the physical layout of a city has a fundamental impact on movement patterns: transport infrastructure priorities shape urban fabric and this in turn shapes transport. This chapter will now try to take the next step and explain how the relationship between transport and activity intensity works.

Table 3.2 provides a snapshot for 1995 and 2005 to show what has been happening in a sample of 33 cities in North America, Australia, Europe, and Asia, concerning some key factors related significantly to transport resilience (census data for Australia, Canada, and Hong Kong are actually from 1996 and 2006). Results for the US cities suggest they may suffer from the least levels of resilience in the great majority of factors, followed most often by the Australian cities. The Canadian cities are similar, but on most factors differentiate themselves in a positive way from the more extreme auto-dependent environments in the US and Australia (e.g. they have higher activity densities, their car use is slightly lower, and their public transit use higher). Canadian cities are a kind of bridge between the more extreme auto dependence of the US and Australia and the much lower automobile dependence of Europe. The Asian cities of Singapore and Hong Kong are a large step removed from all the other cities and are on most factors very much higher in their transport resilience.

While these tables can be examined by the reader for specific changes in each region on each factor between 1995 and 2005, for the purposes of this discussion it is more useful to present an overall view. Table 3.2 calculates the average value for each factor for 1995 and 2005 for all 33 cities in the sample. There are 16 factors examined and on average for these cities, 12 of the factors went in a positive direction from 1995 to 2005. Activity densities have turned around, the cities have reduced their parking availability in the CBD, and reserved public transit route per capita has increased, meaning that there is more public transit infrastructure protected from general traffic. Public transit systems have become faster compared to general road traffic, which is due to urban rail systems (see below for more detail). Public transit use is up regardless of the way it is

Table 3.2 Transport resilience factors in world cities, 1995 and 2005

Factor	Units	USA 1995	USA 2005	AUS 1995	AUS 2005	CAN 1995	CAN 2005
Activity density	jobs + people/ha	22.4	23.6	18.8	20.3	38.8	39.9
Proportion of jobs in the CBD	%	9.2	8.2	13.3	12.7	15.7	15
Length of freeway per person	m/person	0.156	0.156	0.086	0.083	0.122	0.157
Parking spaces per 1,000 CBD jobs	spaces/1,000 jobs	555	487	367	298	390	319
Length of reserved transit route per person	m/1,000 persons	49	72	170	160	56	67
Passenger cars per 1,000 people	units/1,000 people	587	640	591	647	530	522
Public transit seat kilometres of service per person	seat km/person	1,560	1,874	3,997	4,077	2,290	2,368
Public transit boardings per person	boardings/person	60	67	90	96	140	151
Public transit passenger kilometres per person	p.km/person	492	571	966	1,075	917	1,031
Passenger car passenger kilometres per person	p.km/person	18,155	18,703	12,114	12,447	8,645	8,495
Percentage of daily trips by non-motorized modes	%	8.1	9.5	14.9	14.2	10.4	11.6
Private transport passenger energy use per person	MJ/person	60,034	53,441	31,044	35,972	32,519	30,084
Proportion of car plus public transit passenger-km on transit	%	2.9	3.2	7.6	8.1	9.9	11.3

continued

Table 3.2 Transport resilience factors in world cities, 1995 and 2005

Factor	Units	USA 1995	USA 2005	AUS 1995	AUS 2005	CAN 1995	CAN 2005
Ratio of public transit versus car speeds	ratio	0.57	0.55	0.75	0.78	0.57	0.57
Ratio of public transit reserved route versus freeways	ratio	0.41	0.56	2.18	1.98	0.55	0.56
Transit-related deaths per 100,000 persons	deaths/100,000 persons	12.7	9.5	9.1	6.2	6.5	6.3

Factor	Units	EUR 1995	EUR 2005	HIA 1995	HIA 2005	All cities 1995	All 2005	% change	Resilience direction
Activity density	jobs + people/ha	69.3	69.3	317.7	330.6	59.4	60.9	2.4	Positive
Proportion of jobs in the CBD	%	19.6	17.5	11.4	9.2	14.6	13.2	−9.4	Negative
Length of freeway per person	m/person	0.086	0.099	0.025	0.027	0.109	0.118	8.6	Negative
Parking spaces per 1,000 CBD jobs	spaces/1,000 jobs	224	241	136	121	362	327	−9.6	Positive
Length of reserved transit route per person	m/1,000 persons	240	300	18	34	132	162	22.7	Positive
Passenger cars per 1,000 persons	units/1,000 persons	394	451	73	78	477	520	9.0	Negative
Public transit seat kilometres of service per person	seat km/person	5,599	6,893	6,882	7,267	3,757	4,377	16.3	Positive
Public transit boardings per person	boardings/person	304	330	477	450	190	203	6.4	Positive

Indicator	Units								
Public transit passenger kilometres per person	p.km/person	1,742	2,021	3,169	3,786	1,231	1,426	15.7	Positive
Passenger car passenger kilometres per person	p.km/person	6,737	7,232	1,978	1,975	10,849	11,213	3.4	Negative
Percentage of daily trips by non-motorized modes	%	30.5	35.8	25.2	26.1	18.5	21.0	13.5	Positive
Private transport passenger energy use per person	MJ/person	15,860	16,530	6,447	6,077	33,040	31,601	−4.4	Positive
Proportion of car plus public transit passenger-km on transit	%	20.6	21.9	63.3	65.3	14.6	15.6	6.6	Positive
Ratio of public transit versus car speeds	ratio	0.86	0.95	0.77	0.86	0.71	0.74	4.5	Positive
Ratio of public transit reserved route versus freeways	ratio	4.73	5.44	0.93	1.42	2.25	2.56	13.9	Positive
Transport-related deaths per 100,000 persons	deaths/100,000 persons	5.2	3.3	5.3	3.8	8.1	6.0	−26.0	Positive

Source: Kenworthy and Laube (2001) and authors' own update data for 2005.

Notes
Cities used for comparison:
USA: Atlanta, Chicago, Denver, Houston, Los Angeles, New York, Phoenix, San Diego, San Francisco, Washington, DC
Australia: Brisbane, Melbourne, Perth, Sydney
Canada: Calgary, Montreal, Ottawa, Toronto, Vancouver
Europe: Berlin, Copenhagen, Frankfurt, Hamburg, Helsinki, London, Manchester, Munich, Oslo, Stockholm, Stuttgart, Zurich

measured (boarding and passenger kilometres per capita, and the proportion of annual car plus transit travel undertaken by public transit). The proportion of daily trips undertaken by non-motorized modes of travel is also up and the per capita use of private passenger transport energy is, on average, down. The loss of life within the transport sector is also significantly and consistently down.

On the negative side, jobs in metropolitan areas have become less centralized in the CBDs of cities, though if they are clustering in sub-centres (see next section) this does not have to be a negative thing. If they are scattering it is a negative trend for resilience. Car ownership is up but car use measured by annual passenger kilometres per capita is only up by 3.4 per cent, which relates to the peak car use phenomenon discussed in detail below. Although freeway availability is also up overall, the ratio between reserved public transit routes and freeways has improved, meaning that quality public transit infrastructure is being added faster than new freeways in many places.

Summarizing briefly the distinctions between regions, trends in the US cities indicate that 11 out of 16 factors were positive and one neutral. In Australia only nine of the 16 factors measured were positive for resilience. In Canadian cities 13 of the factors were positive for resilience and one was neutral. In the European cities nine factors were positive and one neutral. For the two Asian cities 12 of the 16 factors experienced a positive change towards resilience. By this analysis it can be said that overall the Canadian cities shifted most strongly towards greater transport resilience and the Australian cities the least. A discussion of the comparative strength of changes for each variable in each region and their implications is not possible in the space available here (16 variables by five city groups). However, Table 3.2 shows the comparative values for each variable in each city grouping for 1995 and 2005 so readers can use the tables to determine for themselves the magnitude of trends for particular variables in regions of interest. What we can say is that in terms of the key factor in car dependence – actual car use – the Canadian cities and the two Asian cities were the only groups to experience a (very small) decline in car use per capita (1.7 per cent and 0.2 per cent drop in passenger kilometres per capita, respectively). Of the cities that increased in car use, the European cities increased the most with a 7.4 per cent increase (but from a comparatively low level compared to North American or Australian cities), while the US and Australian cities experienced virtually identical increases of 3 per cent.

Sub-centres and resilience

From the density relationships discussed previously and from data in areas where viable transit and walking occurs, there seems to be a critical minimum density at around 35 people and jobs per ha required for public transit, and around 100 people and jobs per ha for walking. How can these numbers be understood in terms of guidelines for development to ensure public transit and walking are viable options for more people? The detailed results of work on these questions are in Newman and Kenworthy (2006) and are summarized below in terms of how to create local centres and town centres with greater transportation resilience.

A pedestrian catchment area or 'ped shed', based on a ten-minute walk, creates an area of approximately 220 to 550 hectares for walking speeds of 5–8 km/hour. Thus for an area of around 300 hectares (a little over a square mile or about a radius of one kilometre or 0.6 miles) developed at 35 people and jobs per hectare, there is a threshold requirement of approximately 10,000 residents and jobs within this ten-minute walking area. The range would be from about 8,000 to 19,000 based on the 5–8 km/hour speeds. Some centres will have more jobs than others, but the important physical planning guideline is to have a combined minimum activity intensity of residents and jobs necessary for a reasonable local centre and a public transit service to support it. Other authors support these kinds of numbers for viable local centres and public transit services (Pushkarev and Zupan 1997; Ewing 1996; Frank and Pivo 1994; Cervero *et al.* 2004). The number of residents or jobs can be increased to the full 10,000, or any combination of these, as residents and jobs are similar in terms of transport demand. Either way, the number suggests a threshold below which public transit services become non-competitive without relying primarily on car access to extend the catchment area. The latter would mean "park and ride", which destroys the possibility of building density around public transit stops and ruins the quality of the public realm.

Many new car-dependent suburbs have densities more like 12 per hectare or less and hence have only a maximum of about one-third of the population and jobs required for a viable centre. When a centre is built for such suburbs it tends to have just shops with job densities little higher than the surrounding population densities. Hence the 'ped shed' never reaches the kind of intensity that enables a walkable environment to be created, one that can ensure viable public transit. The public realm of such centres is often dominated by roads and parking. Many new urbanist developments are primarily emphasizing changes to improve the legibility and permeability of street networks, with less attention to the density of activity (Falconer *et al.* 2010). As important as such changes are to the physical layout of streets, we should not be surprised when the resulting centres are not able to attract viable commercial arrangements and have only weak public transit. However, centres can be built in stages with much lower numbers to begin with, provided the goal is to reach a density of at least 35 per ha through enabling infill at higher intensities.

If a walking city centre is required, then a density of at least 100 per ha is needed. This gives an idea of the kind of activity that a town centre would need: approximately 30,000 residents and jobs within this ten-minute walking area (of 300 ha). The range again is from around 22,000 to 55,000 people and jobs. This number could provide for a viable town centre based on standard servicing levels for a range of activities. Fewer numbers than this means services in a town centre are non-viable, and it becomes necessary to increase the centre's catchment through widespread dependence on driving from much farther afield. This also means that the human design qualities of the centre are compromised because of the need for excessive amounts of parking and road infrastructure.

Of course, many driving trips within a walking ped shed still occur, but overall they constitute a very much lower percentage of total trips. However, if sufficient amenities and services are provided then only short car trips are needed, which is still part of making the centre less car dependent.

"Footloose jobs", including some of those related to the global economy, can theoretically go anywhere in a city and can make the difference between a viable centre or not. However, there is considerable evidence that such jobs are locating in dense centres of activity due to the need for networking and quick "face-to-face" meetings between professionals. High-amenity, walking-scale environments are better able to attract such jobs because they offer the kind of environmental quality, liveability, and diversity that these professionals are seeking. As Florida (2012) says:

> Economic growth and development, according to several key measures, is higher in metros that are not just dense, but where density is more concentrated. This is true for productivity, measured as economic output per person, as well as both income and wages.

Other cultural factors now seem to be associating with higher-density locations, providing for rising social integration, innovation, and talent levels. This holds for both the share of college graduates and the share of knowledge, professional, and creative workers (Glaeser 2010). Most citizens who experience car dependence, and have long commutes stuck in traffic, can understand the need for more sustainable options, since they directly feel and bear the economic, social, and environmental consequences of car dependence. Many want other options provided for them. As cities continue to evolve, the politics of sustainable transport will demand both more liveable and less car-dependent options for the future.

The key to this move towards sustainability is better provision of access to public transit that is faster than cars along corridors, and better provision for walking and cycling in local areas, associated with a supportive land-use structure of intensive centres with minimum land-use activity intensity of 35 people and jobs per ha. This is due to a fundamental need to ensure that the more sustainable transport modes have a competitive speed advantage for long trips (public transit) and for short trips (bike/walk) within centres. Such moves towards more sustainable transport will help to secure the resilience of sub-centres and the city as a whole.

Where are cities going?

In 2009 the Brookings Institution was the first to recognize a new phenomenon in the world's developed cities – declines in car use (Puentes and Tomer 2009). Peak car use suggests that we are witnessing the end of building cities around cars as the primary goal of planning – at least in the developed world – and probably the rediscovering of the compact city. Perhaps we are witnessing the demise of further automobile city building and the beginning of the end of automobility.

Peak car use

Puentes and Tomer (2009) picked up the trend in per capita car use starting in 2004 in US cities. They were able to show that this trend was occurring in most US cities and by 2010 was evident in absolute declines in car use. Stanley and Barrett (2010) found a similar trend in Australian cities and that the peak had come at a similar time – 2004 – and car use per capita at least seemed to be trending down ever since. We have since mapped this in all Australian cities, including small ones where congestion is no issue and relevant graphs and data can be found in Newman and Kenworthy (2011b). Millard-Ball and Schipper (2010) examined the trends in eight industrialized countries that demonstrate what they call "peak travel". They conclude that:

> Despite the substantial cross national differences, one striking commonality emerges: travel activity has reached a plateau in all eight countries in this analysis. The plateau is even more pronounced when considering only private vehicle use, which has declined in recent years in most of the eight countries.... Most aggregate energy forecasts and many regional travel demand models are based on the core assumption that travel demand will continue to rise in line with income. As we have shown in the paper, this assumption is one that planners and policy makers should treat with extreme caution.
>
> (pp. 16–17)

The Global Cities Database (Kenworthy and Laube 2001; Kenworthy *et al.* 1999) has been expanding its global reach since the first data were collected in the 1970s. While the 2005/2010 data are yet to be completed, the first signs of a decline in car use can be gleaned and were first recognized by us when it was seen that cities in the developed world grew in car use per capita in the 1960s by 42 per cent, in the 1970s by 26 per cent, and the 1980s by 23 per cent. The new data now show that the period 1995–2005 had a growth in car use per capita of just 5.1 per cent, which is consistent with the above data on peak car use (Newman and Kenworthy 2011b). The reductions have started since this decade and appear to be continuing (Gargett 2012).

Among the 26 cities that comprise the 1995–2005 percentage increase in car VKT per capita, some places actually witnessed declines in this period. Some European cities show this pattern: car use in London has declined 1.2 per cent, in Stockholm 3.7 per cent, Vienna 7.6 per cent, and Zurich 4.7 per cent. Declining car use is even more stark in the US: Atlanta went down 10.1 per cent, Houston 15.2 per cent (both from extraordinarily high levels of car use in 1995), Los Angeles declined 2 per cent and San Francisco 4.8 per cent. Accelerating decline is evident since that time (see also Goodwin and Van Dender 2013).

Peak car use appears to be happening. It is a major historical discontinuity that was largely unpredicted by most urban professionals and academics. What might be causing this to occur?

The possible causes of "peak car use"

The logic behind peak car use is spelled out by Metz (2010), who suggests that although the number of destinations that can be accessed tends to increase roughly in line with the square of distance, the additional utility provided by reaching each additional destination at the margin declines with distance. Given stable time travel budgets, this produces resistance to additional travel and the saturation of travel distances, which are hastened if trip speeds also decline. In this section, five potential factors leading to peak car use are outlined and examined. They all point to the possibility of a less automobile-dependent city emerging.

Hitting the Marchetti wall

As outlined above, the travel time budget matters. Freeways designed to get people quickly around cities have become car parks at peak hours. Travel times have grown to the point where cities based around cars are becoming dysfunctional. As cities have filled with cars the limit to the spread of the city has become more and more apparent with the politics of road rage becoming a bigger part of everyday life and many people just choosing to live closer in cities.

The trends in relative speeds are shown in Figure 3.3. The ratio of overall public transit system speed compared to general road traffic has increased from 0.55 to 0.70 between 1960 and 2005; the ratio of rail system speed to general road traffic has gone from rail being slower than cars in 1960 (0.88) to a situation in 2005 where rail was on average faster (1.13). Thus rail has become

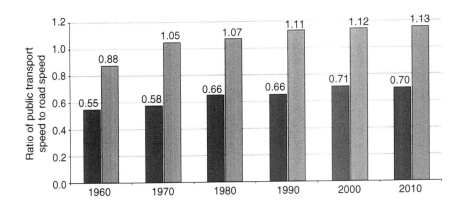

Figure 3.3 Relative speed of transit to car traffic and rail to car traffic in global cities, 1960 to 2005 (source: Newman *et al.* (2013)).

increasingly more viable as an option in the world's cities. And with it will be a greater emphasis on rail-induced compact land-use patterns and a growing move away from freeway-induced land-use scatter. The remaining data presented in this chapter supports this. The automobile city seems to have hit the Marchetti wall.

The growth of public transit

The extraordinary revival of public transit globally and especially in car-dependent American and Australian cities is demonstrated in Figures 3.4 and 3.5. Transport planners traditionally saw the growth in public transit as a small part of the transport task, and assumed that car use would continue unabated. However, there is an exponential relationship between car use and public transit use that indicates how significant the impact of public transit can be. By increasing public transit use per capita, the per capita use of cars is predicted to decline exponentially. This is the so-called "transit leverage" effect (Neff 1996; Newman *et al.* 2008). Even small increases in public transit can begin to put a large dent in car-use growth and eventually cause it to peak and decline.

The reversal of urban sprawl

The turning back in of cities leads to increases in density rather than the continuing declines that have characterized the growth phase of automobile cities in the past 50 years. Data on density suggest that the peak in decline has occurred and cities are now densifying faster than they are spreading out. Table 3.3 contains

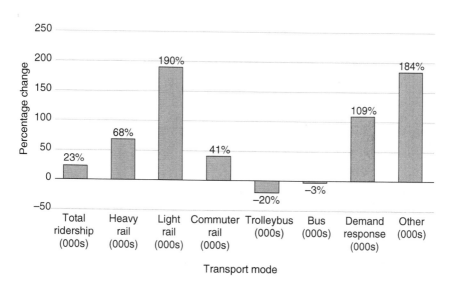

Figure 3.4 Growth in US transit use (source: American Public Transportation Association (2013)).

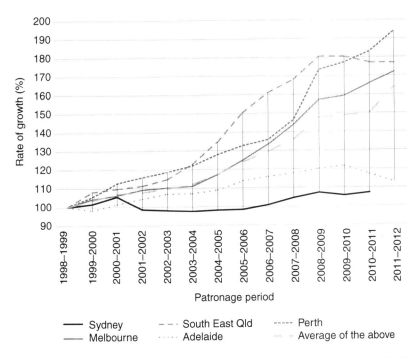

Figure 3.5 Growth in transit use in Australian cities since 1999 (source: Compiled by the authors).

data on a sample of cities in Australia, the USA, Canada and Europe showing urban densities from 1960 to 2005 that clearly demonstrate this turning point in the more highly automobile-dependent cities. In the small sample of European cities included in the table, densities are still declining due to *shrinkage* or absolute reductions in population, but the data clearly show the rate of decline in urban density slowing down and almost stabilizing as re-urbanization occurs. The relationship between density and car use is also exponential, as shown previously. If a city begins to slowly increase its density, then the impact on car use can be more extensive than expected. The compact city is being rediscovered.

The growth of a culture of urbanism

Commentators are increasingly picking up a renewed interest in living a more urban and less suburban lifestyle (Leinberger 2007; Newman and Newman 2006; Rosen and Walks 2013). Puentes and Tomer (2009) suggest this is not a fashion but a structural change based on the opportunities that are provided by denser urbanism. The cultural change associated with this shift is reflected in many aspects of popular culture, especially the use of mobile digital devices that enable freedom and connection without a car (Urry 2007; Florida 2010).

Table 3.3 Trends in urban density (persons per ha) in a sample of US, Canadian, Australian, and European cities, 1960–2005

	1960	1970	1980	1990	1995	2005
Brisbane	21.0	11.3	10.2	9.8	9.6	9.7
Melbourne	20.3	18.1	16.4	14.9	13.7	15.6
Perth	15.6	12.2	10.8	10.6	10.9	11.3
Sydney	21.3	19.2	17.6	16.8	18.9	19.5
Chicago	24.0	20.3	17.5	16.6	16.8	16.9
Denver	18.6	13.8	11.9	12.8	15.1	14.7
Houston	10.2	12.0	8.9	9.5	8.8	9.6
Los Angeles	22.3	25.0	24.4	23.9	24.1	27.6
New York	22.5	22.6	19.8	19.2	18.0	19.2
Phoenix	8.6	8.6	8.5	10.5	10.4	10.9
San Diego	11.7	12.1	10.8	13.1	14.5	14.6
San Francisco	16.5	16.9	15.5	16.0	20.5	19.8
Vancouver	24.9	21.6	18.4	20.8	21.6	25.2
Frankfurt	87.2	74.6	54.0	47.6	47.6	45.9
Hamburg	68.3	57.5	41.7	39.8	38.4	38.0
Munich	56.6	68.2	56.9	53.6	55.7	55.0
Zürich	60.0	58.3	53.7	47.1	44.3	43.0

Source: Newman and Kenworthy (2011).

The rise in fuel prices

The vulnerability of outer suburbs to increasing fuel prices was noted in the first fuel crisis in 1973–1974 and in all subsequent fuel crisis periods when fuel price volatility was clearly reflected in real estate values (Fels and Munson 1974; Romanos 1978). The return to "normal" after each crisis led many commentators to believe that the link between fuel and urban form may not be as dramatic as first presented by researchers like us (Newman and Kenworthy 1989, 1999). However, many commentators now believe that rising oil prices, and/or the urban forms that compel demand for oil, might themselves be a source of financial crises at various scales (see Chapter 4 by Walks for a full discussion).

Despite the global recession, the twenty-first century has been faced by a consolidation of fuel prices at the upper end of those experienced in the last 50 years of automobile city growth. Most oil commentators, including oil companies, now admit to the end of the era of cheap oil, even if not fully accepting the peak oil phenomenon (Newman *et al.* 2009). The compact city is being driven by transport factors outlined above. But fuel price volatility and uncertainty will certainly have added to the value in living closer to urban activity.

Mechanisms for facilitating resilient cities

Facilitating the more resilient polycentric city needs a range of new and old planning tools. Old tools like strategic plans linked to infrastructure are essential. Building fast rail out into the automobile city's suburbs has been shown to work very successfully when the speed of public transit is better than the clogged free-ways (McIntosh *et al.* 2013). New tools, such as financing public transit through value capture, builds the integration of dense centres into the building of public transit (McIntosh *et al.* 2013). New digital planning tools for assisting redevel-opment, especially in the revitalization of middle suburbs, can ensure that auto-mobile city urban fabric is upgraded to provide more resilient outcomes (Glackin *et al.* 2013; Newton *et al.* 2012). New forms of governance will be needed that can enable greater regional autonomy and more deliberative, participative pro-cesses (Briand and Hartz-Karp 2013; Bunning *et al.* 2013). In addition, the new techniques of biophilic urbanism with green walls and green roofs, which are appearing in the many compact cities of Asia, are also needed as more compact urban fabric demands new ways of bringing nature into the city (Newman and Matan 2013; Beatley and Newman 2013; see Chapter 15, this volume). Finally, for the car use that does remain, new vehicle technologies such as electric vehi-cles or plug-in hybrid electrics will provide additional environmental and livea-bility benefits. Much of this use may be through city car-sharing schemes, including those increasingly being pursued by auto manufacturers as they diver-sify their business models.

Conclusions

A new kind of city is being discovered, representing a combination of old tech-niques in creating walkable and transit-oriented urban fabric, along with new green technologies. The first signs of movement away from automobility are now appearing as part of this. Only time will tell if this is a truly structural change or a small shift in a longer-term continuation of automobility over the past century or so. The evidence presented above suggests it may be a structural change, and that a more sustainable and resilient city may finally be appearing to reduce the impact of automobile dependence.

4 Driven into debt?

Automobility and financial vulnerability

Alan Walks

The eruption of the global financial crisis in the late 2000s and its continuing aftermath have foregrounded questions surrounding the rise of household debt. A number of the explanations for high and increasing levels of indebtedness implicate automobility and the form taken by the auto-mobile city as important factors driving up levels of financial vulnerability and triggering financial crisis. However, there is much debate about the causes of both household indebtedness and financial crisis, and the precise relationships between the system of automobility, auto-dependent urban forms wrought by automobility, and contemporary financial vulnerabilities, remain unclear.

This chapter sheds light on these relationships, using comparative data at multiple scales of analysis. In doing so, it provides an explanation of the role played by automobility in the co-production of contemporary landscapes of debt at the global, metropolitan, and intra-urban levels. The chapter begins by examining and comparing trends among different nations, and assessing three key salient hypotheses linking automobility and financial vulnerability that have arisen in the literature, which I term: "peak oil", "peak car", and "peak suburbs". The limitations of these three hypotheses are critically analysed, and a synthetic theoretical explanation is put forward that takes into account the multi-scalar political-spatial-economic forces at work. In the final section, the chapter shifts gear to examine the relationships, and socio-spatial effects, of rising household indebtedness at the metropolitan scale, and among neighbourhoods within metropolitan areas. The chapter concludes by discussing the implications of this work for the evolving geography of the auto-city.

Automobility and debt among nations

The global financial crisis exposed a creeping fissure in the national economies and balance sheets of households and governments in a number of developed nations. Dismissed as unproblematic by neoclassical economists and such prominent figures as former chairman of the United States Federal Reserve Alan Greenspan, rapidly rising household indebtedness over the 2000s linked to credit cards, auto loans, and particularly mortgages ended up literally breaking many banks. In order to prevent the insolvency of financial institutions, created by aggressive and risky lending behaviour, from bringing on a wave of deflation and economic depression, central banks, national governments, and international organizations around the world were compelled to enact massive stimulus programmes coupled with bailouts of banks and large non-financial corporations, particularly large automobile manufacturers (Albo *et al.* 2010: ch. 5; Stanford 2010).

At first glance, there would appear much merit to the proposition linking automobility and its related factors to high levels of household debt. It was in the United States (US), the most auto-mobile large nation on the globe, that the financial crisis was triggered by defaulting mortgages and other loans, and much of the early media focus on bailouts and stimulus spending concerned actions of the US governmental functions. While data with which to examine such relationships is only available for the OECD and BRIC nations, among this group of countries there is a general trend in which higher levels of motor vehicle ownership accompany higher levels of household debt and, to a lesser extent, national government debt (Table 4.1). For this set of countries, the overall correlation between motor vehicles per 1,000 population and household debt as a percentage of GDP is fairly strong and positive ($r=0.59$), although lower for national government debt ($r=0.25$). Similarly, comparing changes in household debt over the decade of the 2000s with the level of motor vehicles in 2011 results in a moderate but positive correlation ($r=0.26$). Some of the most auto-mobile nations in the world, including New Zealand, Australia, and Canada, have high levels of household debt.

However, the strong relationship between levels of automobility and debt persists only when developed and developing nations are included together in the analysis. When the analysis is restricted only to Western developed nations, this relationship breaks down. Indeed, the straight correlation with household debt turns negative when Eastern Europe and the Asian and BRIC nations are excluded ($r=-0.29$), with a similar correlation for *change* in household debt ($r=-0.25$), and a strong negative relationship for government debt ($r=-0.55$). Furthermore, even when all countries are analysed together, when change in *both* household debt and motor vehicles is compared, the correlation is again negative, and weak ($r=-0.13$), and virtually non-existent for government debt ($r=-0.08$). Such findings beg the question: is there any real relationship between automobility and debt, and if so, what might it be?

The tri-peaks of auto-indebtedness?

Three partially related but nonetheless distinct hypotheses linking various aspects of automobility to rising indebtedness have been put forward in the literature. The first and most common emerges from the literature regarding resource depletion and peak oil, and is concerned with the effects of oil price rises on financial vulnerability and economic growth. In the second, it is automobile ownership and usage that are peaking, with ripple effects on local, national, and global economies. In the third hypothesis, financial crisis and indebtedness are rooted in urban form, particularly the inefficiencies, inflexibilities, and high costs of auto-dependent suburbs. These three hypotheses are critically analysed, their precise relationships to the city and processes of urbanization are discussed, and their limits highlighted.

Peak oil

The first hypothesis sees rising indebtedness and ultimately financial crisis and recession as resulting from resource depletion. The peaking of oil production, and the increasing cost of extracting oil from known recoverable reserves after the peak, signifies the "end of cheap oil" (Campbell and Laherrère 1998), the increasing foray into more expensive and lower-quality non-conventional sources of oil (McCabe 1998), and ultimately the "end of growth" (Rubin 2012). This not only has supply repercussions, but also costs implications across the spectrum of economic activities, including manufacturing processes, agricultural production, transportation, heating, mining, forestry, and other resource development, including fossil fuel extraction itself. Hubbert in 1956 correctly predicted that oil production would peak in the US around 1970, the point at which roughly half the total recoverable oil had been extracted, based on the ongoing rate of extraction at the time (Deffeyes 2001). Since then, the Association for the Study of Peak Oil (ASPO) and others sought to use various methods to predict the global peak, but with global complexity has come much more uncertainty and variability in the range of estimates. The best estimate by ASPO of the global peak year is 2010, whereas for the United States Geological Survey (USGS) it is 2037 (Bridge 2010).

Regardless of the actual date of the global peak, or the relative importance of each of the factors determining the rate of extraction (including "above-ground" geo-political factors; see Bridge 2010), the fact that oil is a non-renewable resource has direct implications for how economies, governments, and households deal with ongoing extraction and inevitable depletion. As oil becomes more costly to extract, this should filter into the price of many everyday commodities, making them more expensive. The difference has to be made up by either a reduction in the standard of living, as a greater proportion of everyone's income goes towards energy, or else articulated in growing indebtedness, or both. Rising interest rates, imposed to counter inflation, also lead to either higher debt burdens or reduced discretionary expenditures (Tverberg 2012). A direct link between rising costs of oil and increasing debt is often made by ecologically minded economists such as Rubin (2012: 51), who articulates the basic argument:

> Global oil consumption in 2000 was roughly 76 million barrels a day, with Brent crude averaging $28.50 a barrel; the world's annual oil bill was $792 billion. Skip ahead to 2010. World consumption was up to 87 million barrels a day, with Brent averaging $79.50 a barrel. The combination of higher prices and more demand had quadrupled the annual fuel bill to $2.5 trillion.

Table 4.1 Prevalence of motor vehicles and levels of indebtedness, OECD/BRIC nations

	Vehicles per 1,000 population				Household debt as a % of GDP			Government debt as % of GDP		
	1960	2002	2010	% diff. 2002–2010	2000	2011	Diff. 2000–2011i	2000	2011	Diff. 2000–2011
ANGLO SUBURBAN:	**310**	**656**	**702**	**7**	**68**	**95**	**27**	**40**	**50**	**11**
United States	411	799	797	0	72	87	15	45	80	35
New Zealand	271	612	711	16	67	98	31	32	31	−1
Australia	266	632	694	10	70	105	35	11	21	10
Canada	292	581	607	4	64	91	27	71	69	−2
ANGLO EUROPE:	**108**	**494**	**521**	**6**	**58**	**111**	**53**	**39**	**83**	**45**
Ireland	78	472	523	11	48	124	76	35	85	50
United Kingdom	137	515	519	1	68	98	30	42	81	39
NORTH-WESTERN EUROPE:	**107**	**533**	**551**	**3**	**64**	**86**	**21**	**55**	**58**	**3**
Norway	95	521	584	12	56	84	28	19	26	7
France	158	576	580	1	30	48	18	59	90	31
Austria	69	629	577	−8	47	55	8	61	66	5
Germany	73	586	572	−2	71	60	−11	59	83	24
Switzerland	106	559	566	1	109	120	11	46	37	−9
Belgium	102	520	559	8	39	54	15	99	97	−2
Netherlands	59	477	526	10	87	128	41	44	52	8
Sweden	175	500	519	4	50	82	32	57	34	−23
Denmark	126	430	480	12	91	140	49	55	40	−15
SOUTH-WESTERN EUROPE:	**24**	**537**	**601**	**12**	**35**	**71**	**36**	**83**	**105**	**22**
Italy	49	656	679	4	22	45	23	107	111	4
Greece	10	422	624	48	13	62	49	109	148	39
Spain	14	564	592	5	45	82	37	63	71	8
Portugal	N/A	507	509	0	59	94	35	52	88	36

EASTERN EUROPE:	**35**	**379**	**462**	**22**	**8**	**34**	**26**	**26**	**40**	**14**
Slovenia	N/A	492	567	15	15	30	15	26	36	10
Poland	8	370	537	45	7	35	28	36	50	14
Czech Republic	82	390	485	24	8	31	23	13	37	24
Estonia	N/A	427	476	11	9	46	37	3	3	0
Slovak Republic	N/A	288	364	26	4	27	23	24	39	15
Hungary	15	306	345	13	6	37	31	54	74	20
OECD ASIAN AND BRIC:	**10**	**206**	**248**	**58**	**23**	**33**	**10**	**56**	**68**	**13**
Japan	19	599	590	–2	74	67	–7	131	226	95
South Korea	2	293	363	24	48	81	33	13	33	20
Israel	25	303	321	6	41	49	8	83	75	–8
Russia	N/A	202	271	34	1	10	9	19	5	–14
Brazil	20	121	209	73	6	13	7	68	66	–2
Turkey	4	96	154	60	4	18	14	38	43	5
China	0	16	58	263	4	12	8	23	32	9
India	1	17	18	6	2	10	8	70	66	–4
ALL NATIONS	**140***	**449**	**472**	**5**	**51**	**73**	**22**	**42**	**56**	**14**

Source: OECD global statistics extract database (http://stats.oecd.org); World Bank Urban Indicators Database (http://data.worldbank.org/indicator/IS.VEH.NVEH.P3); McKinsey Global Institute (2012), Dargay *et al.* (2007); UN Habitat Global Urban Observatory, Urban Indicators Database (www.unhabitat.org/downloads/docs/global_urban_indicators).

Notes

Nations in each group are listed in order of vehicles per 1,000 population. Only nations with data are included.

OECD = Organization for Economic Cooperation and Development; BRIC = Brazil, Russia, India, China.

* Weighted average calculated only with those nations for which data was available.

Only a year later, Brent crude was averaging more than $100 a barrel. The price increase alone added more than $500 billion to what the world spends each year to keep the wheels turning. The extra money didn't fall from the sky.

According to this line of reasoning, it is rising oil costs, rather than predatory mortgage finance or other structural factors, that led to financial crisis and recession in 2007–2008 (Cortright 2008; Hamilton 2009; Rubin 2012). For Rubin (2012: 11), the

> US housing market wasn't responsible for blowing up the global economy. It was a symptom, not the cause. Federal reserve chairman Alan Greenspan was spurred to hike interest rates by soaring oil prices, which were stirring inflation. Higher interest rates pricked the housing bubble, and the rest of the world was dragged down when the bubble burst.

Similarly, according to Hamilton, the reason that rising oil prices in the first half of the 2000s did not lead to recession in the US was because households could afford to ignore them, but by "2007Q4 they no longer could" because by this time rising oil prices surpassed whatever remaining expenditure room remained on household balance sheets (2009: 258). To the extent that the urban phenomenon factors in such explanations, it is via rising demand for oil from developing nations such as China spurred on by rapid urbanization. In this story, it is the expansion of auto-mobile urban development to new areas, coupled with the limits of a finite resource, that has spurred rising oil prices and hence rising indebtedness.

However, there are problems with this peak oil-debt hypothesis. First of all, it assumes a direct and simple relationship between rising demand, rising oil costs, and rising debt. Yet as Labban (2010) has shown, prices for physical Brent crude became dislodged from actual demand with the development of sophisticated derivative and futures contracts after the first Gulf War in the early 1990s. Since this time, it has been the prices attained for the various oil-based swaps and forward contracts, driven by flows and patterns of speculation rather than actual use or demand, that have subsequently determined the prices for crude in physical markets, not the other way around. Oil prices have continued rising even during times of flattening or dropping demand, as "the spot price is driven by speculation on conditions in spot markets mediated by the reaction of financial markets to such speculation" rather than to actual demand (ibid.: 548). Second, according to an explanation rooted in peak oil, rising oil prices should be expected to lead to commodity-price inflation, which should theoretically increase interest rates, tighten credit availability, and squeeze the ability of households to consume and acquire assets. However, the 2000s were instead marked by the exact opposite: declining interest rates and easing credit conditions led to an increasing ability to consume and to rising asset-value inflation (particularly for assets such as financial derivatives and real estate), coupled with commodity-price *disinflation*, rapidly veering into deflation once the US mortgage market began to implode. It was *declining* interest rates that stimulated the take-up of credit and rising household indebtedness, and it was the threat of deflation, not inflation, that scared the central banks into unprecedented stimulus and monetary/credit easing policies as the financial crisis emerged (Foster and Magdoff 2009). The assumed direct relationship between rising demand, oil prices, and debt simply does not hold.

Peak car

Newman and Kenworthy (2011b) have shown that average household vehicle miles travelled peaked in the mid-2000s, and have since either levelled off, or even declined, among a number of automobile-dependent developed nations (see also Puentes and Tomer 2009; Metz 2010; Chapter 3, this volume). The condition of peak car is directly related to the way cities have been planned and built. As large cities have expanded outward in low-density fashion to accommodate mobility and accessibility via the automobile, they have hit the "Marchetti wall". That is, they have reached the limits of commuters' seemingly universal time travel budgets, which Marchetti (1994) argued averaged out at a maximum of roughly one hour of travel per day. Once the city becomes more than "one hour wide" by car, therefore, the factors spurring automobile dependence begin to break down and the population begins seeking other solutions, including living closer to work, relinquishing the automobile for other forms of mobility (high-speed transit and active transportation), and accepting higher densities that even have the potential to reverse urban sprawl (see Chapter 3 in this volume). Rising oil prices may also be a contributing factor helping to spur the peak in car use, but as Newman and Kenworthy note, "peak car use set in well before the 2008 peak of $140 a barrel" of oil (2011b: 38).

While Newman and Kenworthy do not discuss the potential implications of peak car for financial macro-economic stability, it is not difficult to ascertain the significant impact this might have on the profitability of the entire system of automobility. In many developed nations, automobile production and its related industries (marketing, sales, maintenance, repair, fuel distribution, road construction, financial services related to automobile purchase and leasing, etc.) are the source of a large number of jobs, as well as advertising revenues for media corporations and tax revenues for governments. A flattening or reduction in demand for automobiles has the potential to inflict serious economic damage on many national economies. Indeed, the hypothesis of peak car dovetails well with the explanation of the causes of the US crisis in 2008–2009 provided by Alessandria *et al.* (2010). In their account, the financial crisis and the collapse in global trade that occurred in the US is best explained as the result of an "inventory adjustment", in particular for automobile manufacturing and sales – the industries showing the largest declines in production and trade. Such changes in the inventory-to-sales ratios for automobiles in both the US and Japan seem to bear out the timing linking peak car use to the subsequent financial crisis (ibid.).

This hypothesis is intuitive and novel. As detection of this peak is still very recent, it remains unclear whether it is a temporary event, or a new trend that will persist over future decades. Similarly, it is unclear how slowing demand for automobiles may be related to changes in financial vulnerability over time, and whether their effects are only temporary. As Alessandria *et al.* (2010) note, once the great recession receded, automobile production and trade in the US recovered and surpassed previous levels, even as vehicle miles travelled apparently remained flat. As can be seen in Table 4.1, most nations, even if they may have experienced peak car use somewhere in the mid-2000s, still ended the decade with higher levels of motor vehicle ownership than they began. Household debt would appear to have risen both in countries experiencing rapid increases in vehicle ownership (Greece, South Korea, Poland), as well as in those with little or very minor increases in automobility (Canada, Sweden, Portugal). And the countries with the most rapid increase in automobiles (China, Brazil) reveal disproportionately small increments in household debt.

Thus, the role of household debt in this story remains unclear. Even among auto-mobile developed nations, did the peaking and then subsiding of car use lead to rising household debt, or did the squeeze of rising household debt lead households to reduce automobile consumption and travel? The econometric model developed by Alessandria *et al.* (2010: 292) "requires a global negative shock" to set off their inventory-induced feedback loops leading to financial crisis, recession, and collapsing global trade. In the global financial crisis, such a shock derived from the popping of the US housing bubble and the subsequent failures of large investment banks like Bear Stearns and Lehman Brothers. It is as yet unclear what the effect of shifting inventories and demand for automobiles might have been in the absence of such a shock. Furthermore, rising household indebtedness began well before the mid-2000s; hence peak car use would not appear to be able to explain the rise of household debt, even if it might help in explaining the contours of the financial crisis that followed.

Peak suburbs

Others place the city and urbanization more directly at the centre of explanations for rising indebtedness. For Dodson and Sipe (2009: 200) the financial crisis that emerged in the late 2000s is at its core a suburban crisis, rooted in "the synovial joints linking money, land, fuel, and transport technology", which together produce wasteful and outdated patterns of urban development destined to devalue (see also Harvey 2012: ch. 2). Of course there are many different kinds of suburbs (and suburbanisms), and not all of them are automobile dependent (Walks 2013a). However, it is the post-war auto-mobile suburbs, most prevalent in the Anglo nations of the new world, that are most singled out for critique. Peak oil factors in this story, since "petroleum, the great energizer of post-WWII suburbia, is no longer a reliable fuel for credit-based urbanization" (Dodson and Sipe 2009: 200). Likewise, according to Cortright (2008), the crisis is the result of a devaluing of the suburbs, "driven to the brink" by rising fuel costs, with the housing bubble popping once consumers got wise and refused to pay for over-valued auto-dependent locations. Urry (2009) accepts this explanation as the primary cause of the crisis, incorporating it into a larger theorization of the decline of the US Empire and of neoliberal capitalism. It might be noted that such a perspective ignores the role played by predatory finance in producing both the US housing bubble and its collapse, in which holders of the various adjustable rate, NINJA, balloon, jumbo, and other predatory mortgage forms were forced to default once they found out their interest rates were resetting to usurious levels, setting off a deflation that led to job losses and thus compounding ripple effects (see Engel and McCoy 2011; Immergluck 2009; Wyly *et al.* 2009).

Kunstler (2005) lays out perhaps the most dramatic hypothesis, arguing that predominant forms of automobile-dependent suburbs in the US represent "a pro-digious, unparalleled misallocation of resources", and "an armature for daily living that simply won't work without liberal supplies of cheap oil" and thus destined for abandonment (ibid.: 17). Rising indebtedness is not directly the result of peak oil here, but instead the exhaustion of a "high-entropy" fossil-fuelled form of economic development that has been "no longer about anything except the creation of suburban sprawl and the furnishing, accessorizing, and financing of it" (ibid.: 222). To bail out the gambling losses of "high-entropy" finance, related to the activities of hedge funds in (mainly suburban) land devel-opment and the innovation of complex derivatives based on mortgages whose value could not ascertained, the US federal reserve reduced interest rates while federal policy encouraged looser mortgage lending. This unleashed the US real estate bubble and encouraged households to extract (fictitious) value from their homes in the form of re-financing, hence keeping levitated a US economy that was now "running on fumes" and no longer producing much real wealth (ibid.: 228–233). Ultimately, therefore, the rise in household debt here derives from state policies that encouraged the public to take on debt as a way of making up for the declining returns from auto-dependent suburbanization and consumption-led development. This argument fits with that of Crouch (2009), who argues that under financialized neoliberalism it is households, rather than the state, whose deficit spending is relied upon to stimulate a moribund economy, although Crouch does not specifically relate this to suburbanization or automobility.

There is a clear relationship between suburban locations and higher debt levels within the metropolitan areas of the most suburbanized, auto-dependent developed nations (discussed in more detail below). However, the relationship is less clear at more macro scales. Implied in the theorizing of housing stress and financial vulnerability put forward under the peak suburbs hypothesis is the idea that those nations most heavily invested in low-density, auto-dependent suburban forms of city-building should be the ones to reveal the highest levels of household indebtedness. In turn, it is the most suburbanized nations that should have suffered the most from the global financial crisis and the "great" recession that followed. Yet, perusal of Table 4.1 indicates that it is not the most suburban nations that have the highest household debt levels, and again as noted above the correlation with vehicle ownership – highly related to auto-dependent suburbanization – is actually *negative*. The archetypal suburban nation – the US – reveals a level of household debt (at 87 per cent of GDP) that is roughly 62 per cent of that of Denmark, and two-thirds of the level of debt in the Netherlands, Europe's two predominant cycling nations (see Chapter 13 in this volume). Similarly, the *rise* in household debt over the 2000s in the US was only about one-third of the increase in debt in these two countries, despite the fact that it was in the US that the housing market collapsed. In fact, among Western developed nations, it is the least auto-mobile suburban nation in Table 4.1 – Denmark – that has the highest household debt burden (140 per cent of GDP).

As well, the relationship between automobility and the experience and depth of financial crisis is at odds with the peak suburbs hypothesis. Outside the US, the suburban nations of Australia, New Zealand, and Canada are placed among the few countries who are seen to have avoided the worst of the financial crisis and subsequent recession, and did not see any formal bank failures (largely due to the ability of their national governments to come to the rescue; see Harvey 2010; Murphy 2011; Randolph *et al.* 2013; Walks 2010a, 2014). Even for the US, the story would appear more complex. The fact that the US was hit earlier and harder by the crisis, and that household debt deleveraging has since that time advanced the most in that country, might even be taken as a potential indication of US strength (Panitch and Gindin 2012). The US has, as a result, been able to adjust to prevailing global economic conditions much more quickly, and going forward will not be as burdened by the high levels of household indebtedness that continue to hamper other nations. The relationship between suburbanization and high average levels of indebtedness remains murky at best.

Of course, while the above three hypotheses are conceptually distinct, they are not mutually exclusive. Peak car use may, for instance, either follow from, or be the major factor driving, peak suburbs, while the triggering of both peak car or peak suburbs may depend on the timing of peak oil. Indeed, there are numerous ways in which all three peaks may be implicated in co-producing each other, and each hypothesis shares some key features (the work of Dodson and Sipe 2008, Kunstler 2005, and Urry 2009 all attempt to incorporate elements of all three explanations). Yet, running through each hypothesis is an unproven assumption that household debt has risen due to the rising costs of the fossil-fuelled auto-dependent pattern of development. In addition to other already-noted limitations, missing from each explanation is any notion of where the money might be coming from that is providing the credit through which households have become indebted. Declining incomes and rising costs should theoretically squeeze the ability of households to consume and access credit, rather than increase it as has occurred. A fuller understanding of the current predicament thus requires some deeper digging.

From auto-industrial complex to financialization

Financialization refers to a process in which profits increasingly accrue through financial channels rather than through production or trade, even among non-financial firms, and is articulated in innovations that transform flows of commodities and profits into tradeable financial assets (Epstein 2005; Krippner 2005). Such financial innovations include, among other things, derivatives such as foreign exchange and interest-rate swaps, oil futures contracts, credit-default swaps, and most importantly the various asset-backed and mortgage-backed securities (ABS and MBS) and the collateralized debt/mortgage obligations (CDOs and CMOs) that since the late 1990s have funnelled enormous additional investment funds into lenders of mortgages and other debt (Engel and McCoy 2011; McNally 2009). Derivatives are financial products – pieces of paper – whose value derives from fluctuations in the values of other commodities or financial products, rather than directly representing such commodities. They are predominantly used for hedging and insuring investments and future transactions, but because of their derivative form, are easily used to gamble on the future values of various assets, currencies, and firms (Steinherr 1998). The development of MBS and CDOs has allowed for the creation of liquidity out of spatially fixed assets such as houses (Gotham 2009) and risky debts such as credit cards (Montgomerie 2006). The ability to package mortgages into MBS and CMOs, and automobile loans and credit-card debt into ABS and CDOs, and sell these to investors, meant financial institutions could move loans off their books, allowing them to continue lending without risk (Engelen *et al.* 2011; Major 2012; Montgomerie 2007). This is the primary reason for the easy access to credit experienced over the 2000s, and in turn the rise of household debt and the ensuing housing bubble that led to the financial crisis (ibid.; McNally 2009).

The financialization of the global economy has rested on three important transformations. First of all, it has required the deregulation of finance enacted since the 1980s by many developed-nation governments under neoliberal policy reforms, allowing for both financial innovation and the pioneering of new markets for financial speculation. Second, it has depended upon disinflation and the low-interest-rate environment that forced lenders to focus on loan volume rather than quality in their business models, and henceforth to innovate with financial products that took advantage of volume. Third, it has depended on the presence of massive surplus funds that could be invested in financial assets, including derivative instruments related to mortgage lending and other credit products.

It might be noted that the latter two transformations contradict mainstream peak oil-debt theories, which assume oil depletion should be articulated in inflation, fewer funds for investment, and credit tightening. However, they are consistent with a view of auto-mobile capitalism as leading to a crisis of over-accumulation (Harvey 1982). The political-economic system of automobility has been highly profitable for firms and investors, providing inordinate surplus funds for investment for more than half a century. For much of this time, profits were ploughed back into expansion and modernization of the system, including new regions within the developed world, and under globalization to developing nations such as China. Cheap petroleum has provided the "lifeblood" for the entire system (Huber 2013). However, in recent years it has become increasingly difficult to identify profitable avenues for new productive investment within the confines of this system – what Harvey (2010) calls the surplus capital absorption problem – as much of the rest of the world has become incorporated into it and the possibilities for expanding automobile consumption and production exhaust themselves. As the surplus funds remaining from past profits search for yield, they become increasingly cheap to borrow for a range of purposes, including for speculation and consumption. As financial innovation arises to mop up the surplus of investable funds, they are lent out at ever lower interest rates on easy credit terms to ever greater proportions of borrowers (Ashton 2009).

Financialization and the rise in household debt are therefore directly related to the historical *success and profitability* of the auto-industrial complex, rather than to its supposed failings or resource depletion. George Soros is thus very much correct in saying that the "current crisis is the culmination of a super-boom that lasted for more than 60 years" (cited in Levitt 2013: 191). It is the inevitable limits to expansion of the system of automobility presented by various global economic and political factors – most of all increasing income inequality (Lysandrou 2011), but also perhaps the peaking of car use in some places or expansion of auto-dependent suburbs – that has reduced expected returns from future investment in the auto-industrial complex in comparison with the expected returns from speculation in financial derivatives. It is precisely during such a crisis of over-accumulation that surplus funds remaining from past investments flow into financial innovations, "creating vast quantities of fictitious capital" in the form of credit that is then used for speculative purposes (Harvey 1982). The new credit is then funnelled into speculative plays, often provoking a "colossal form of gambling and swindling" (Harvey 1982: 304), including recently those in real estate and energy (shale gas, deep-water drilling).

Out of the massive but fading success of the system of automobility has arisen loose credit and predatory finance. The processes at work, and the effects of rising indebtedness, however, operate differently at the international and intraurban scales. Internationally, the massive profits stemming from the auto-industrial complex have given many of the core nations and their multinational firms significant financial power with which to shape global trade and flows of profits, and expand into new foreign markets (Schwartz 2009). Nations who liberally deregulated their financial sectors, allowing for rapid financialization over the 2000s, were the ones to reap much of the benefits, including low interest rates for their sovereign and corporate bonds, in comparison with nations that continued with forms of regulatory "financial repression" (ibid.). This allowed governments and private firms in many of the liberal, financialized auto-mobile nations to borrow cheaply and invest in either new production facilities or (more commonly) new financial products in other nations, often driving up debt levels elsewhere (ibid.). This helps explain the negative correlation between household debt and vehicle ownership among developed nations, and why the US has lower relative levels of debt than many other nations.

At the intra-urban scale, on the other hand, much of the cheap credit accessed by households has been put to work extending the consumption of the key products of the auto-industrial complex, namely new (suburban) houses and automobiles. Indeed, much of the financial power of the core "suburban" nations and their banks, including the US before the financial crisis and both Canada and Australia in its aftermath, has derived from the continued appetite for debt-fuelled consumption on behalf of their urban populations (Schwartz 2009). It is perhaps not a surprise that the key policy responses of national governments in the US and elsewhere to the ensuing recession and crashing housing markets was to bail out not only the banks, but two of the largest automobile manufacturers (General Motors and Chrysler), and actively stimulate automobile consumption via "cash for clunkers" programmes and the easing of credit for automobile purchase (Rattner 2010; Stanford 2010). Since the onset of recession, amortization terms for automobile loans have been loosened considerably from a standard four years to eight years, significantly increasing the affordability of monthly payments and bolstering access to credit in the face of record household debt (Weisbaum 2013). This has led to a recovery in automobile production and sales, and to rising profits for financial institutions making the loans, such that one-quarter of the growth of the US economy since the financial crisis can be attributed to the auto industry (Fletcher 2013). However, this has also meant the reversal of household deleveraging of automobile-based debt in the US that began with the crisis, and rapid growth of such debt in Canada post-crisis (Figure 4.1),

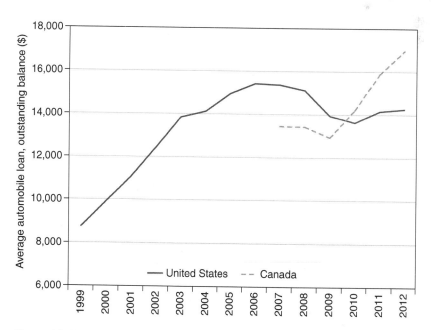

Figure 4.1 Average household outstanding automobile loan balance, United States and Canada, 1999–2012 ($) (source: Federal Reserve Bank of New York (various years), US Census Bureau (various years), and Equifax Canada (2012/2013)).

Notes

Amounts are in local currencies (US$ for the United States and C$ for Canada). Data are only available from 2007 onwards for Canada.

and has come at the cost of doubling the amount of time it will take many car owners to pay off their loans. It has also augmented the shift in financial vulnerability to the auto-dependent suburbs.

From auto-mobile suburbs to financial vampires?

Whereas the relationship between automobility and household debt at the international scale is indirect and highly mediated by flows of capital, financial innovation, and the power of multinational firms and sovereign states, there is a much more direct relationship operating through the geography of the city. Housing and transport costs are the two largest household expenditures, and can take upwards of half of the average household's income in many Anglo cities (Table 4.2). The burdens of high transport costs drive up levels of financial stress, and make affected households particularly vulnerable. It is at the intra-urban scale that this relationship is strongest. In neighbourhoods where, due to the prevailing urban form, households are "driven to drive" (Soron 2009), households are compelled to spend more on transport than those living in denser urban settings more amenable to public transit and active forms of transportation. Coupled with rising housing prices driven higher by the credit bubble over the 2000s, this heightens housing stress and reduces the discretionary income and ability of the residents of auto-dependent areas to save. In their examination of housing stress in Australia's capital cities, Vidyattama *et al.* (2013) found that higher transport costs significantly increased levels of household financial stress. This is particularly true among households with children, who are more likely to require two cars to meet daily travel needs, while the addition of transport costs into the equation significantly elevates the level of financial stress among home-owners *above* that of renters. When transport costs are included, the measured level of financial stress among housing-stressed homeowning families with children, as well as stressed sole-parent renting families, increased by roughly 30 per cent. Among housing-stressed renting couple-families with children, it increased by 44 per cent (ibid.: 1787).

This obviously makes it much more likely that households residing in automobile-dependent neighbourhoods will have to rely on credit for their daily needs, while also potentially preventing them from paying down mortgages and other debt as fast as households with cheaper transport alternatives. Furthermore, higher levels of financial stress, coupled with a dependence on the automobile for travel, makes the residents of auto-dependent suburbs more vulnerable to "shocks" related to either rising oil prices or interest rates (Dodson and Sipe 2008). Dodson and Sipe (2008) construct what they call the VAMPIRE index ("vulnerability assessment for mortgage, petrol and inflation risks and expenditure") to identify neighbourhoods in Australia's cities with the highest levels of financial vulnerability to either oil or interest-rate increases, evoking the image of suburban households being sucked of their vitality. Their research demonstrates that financial vulnerability increases in consistent fashion as one moves out from the CBD to the fringe. Similarly, in Canadian cities it is generally fringe suburban neighbourhoods that reveal the highest debt burdens relative to disposable income, even after controlling for housing tenure, age, family and household formation status, immigration status, and a host of other variables including income (Walks 2013b). Immergluck (2010) has demonstrated that in the US, mortgage foreclosures have generally been concentrated in either poor inner-city neighbourhoods in older deindustrializing cities or in the newer fringe suburbs in growing cities.

The hypothesized relationship between household transportation budgets and levels of household indebtedness holds when analysed among neighbourhoods within Canadian largest cities (Figure 4.2). Here, levels of household debt (vertical axis) are analysed among quintiles of neighbourhoods based on the proportion of household spending going to motor vehicle operation and ownership (horizontal axis). While the degree to which debt levels rise with increasing spending on motor vehicles varies among cities, in virtually all cases there is a strong positive association between the two variables. Despite having only five data points (quintiles of neighbourhoods) per city, the correlations are statistically significant and strong in most cases (between a low $r=0.50$ in Ottawa for consumer debt, and a high of $r=0.97$ in Toronto for total debt), with the only exception the distribution of total household debt in Vancouver (where astronomically high mortgage debt swamps all other factors; see also Walks 2013b). Such a strong correlation at the neighbourhood level implies that the spatial form of the city (which itself is, of course, always socially produced) is a factor linking patterns of consumption within the system/regime of automobility to financial vulnerability. Separate multi-level analysis of Canadian metropolitan areas conducted by Walks (2013b) revealed that, after controlling for a host of different variables at both the metropolitan and neighbourhood levels, for every 1 per cent increase in the neighbourhood proportion of commuters who drive to work, total household debt as a proportion of disposable income increased by 0.18 per cent on average across all cities (statistically significant at $p=0.001$). Other work conducted on Australian cities (Dodson and Sipe 2008; Vidyattama *et al.* 2013) and cities in the US (Immergluck 2010; Schafran 2013) suggests similar relationships operate in these other highly suburban nations. Unfortunately, Canada is the only country for which detailed data pertaining to both household debt and motor vehicle spending is available at sufficiently small spatial scales to test this hypothesis.

The general trends can be seen at the level of census tracts in the Calgary Census Metropolitan Area (CMA) (Figure 4.3). While there is not a one-to-one correspondence between the proportion of household income spent on motor vehicles and consumer debt as a proportion of income, the correlation is fairly strong and positive for total household income ($r=0.59$, which includes mortgages) and very strong for consumer debt ($r=0.81$, not including mortgages). As well, while the relationships with debt are weaker in the centre of the city, having more to do with speculation in housing than automobility, in the suburbs they are strong and consistent. Within the suburbs, it is areas with lower levels of transit use that reveal both higher debt levels and higher spending on motor

Table 4.2 Annual transportation and housing costs as a percentage of income: Canada, US, Australia

Canada 2011 CMA:	Average household income (C$)	Housing cost (C$)	Transport cost (C$)	Housing cost (%)	Transport cost (%)	Total (%)
St John's	82,207	16,849	14,729	20.5	17.9	38.4
Edmonton	97,452	19,297	17,180	19.8	17.6	37.4
Calgary	113,169	22,299	19,947	19.7	17.6	37.3
Saint John	72,189	13,232	13,559	18.3	18.8	37.1
Kitchener–Waterloo	84,823	16,969	14,289	20.0	16.8	36.8
Halifax	76,207	14,504	12,206	19.0	16.0	35.0
Winnipeg	74,675	14,010	11,693	18.8	15.7	34.5
London	74,373	14,818	10,655	19.9	14.3	34.2
Vancouver	83,672	16,541	11,747	19.8	14.0	33.8
Saskatoon	84,812	15,662	12,934	18.5	15.2	33.7
Hamilton	84,230	16,599	11,568	19.7	13.7	33.4
Toronto	95,317	18,951	12,298	19.9	12.9	32.8
Quebec	69,499	11,449	11,089	16.5	16.0	32.5
Ottawa	91,794	16,683	11,952	18.2	13.3	31.5
Montreal	70,277	12,538	8,960	17.8	12.7	30.5

United States 2010 MSA:	Median family income (US$)	Housing cost (US$)	Transport cost (US$)	Housing cost (%)	Transport cost (%)	Total (%)
Miami	55,200	16,326	12,027	29.6	21.8	51.4
Los Angeles	63,000	18,990	12,496	30.1	19.8	49.9
Detroit	55,900	13,166	13,109	23.6	23.5	47.1
New York	65,600	18,962	11,440	28.9	17.4	46.3
San Diego	75,500	19,005	13,605	25.2	18.0	43.2
Phoenix	66,600	14,342	13,187	21.5	19.8	41.3
Atlanta	71,800	15,249	14,348	21.2	20.0	41.2
Dallas	68,900	13,945	13,579	20.2	19.7	39.9

	Average disposable household income (A$)	Housing cost (A$)	Transport cost (A$)	Housing cost (%)	Transport cost (%)	Total (%)
Portland	71,200	15,065	13,083	21.2	18.4	39.6
Chicago	74,700	17,063	12,509	22.8	16.7	39.5
Boston	85,200	19,770	13,305	23.2	15.6	38.8
Denver	75,900	15,788	13,005	20.8	17.1	37.9
St Louis	68,300	11,848	13,514	17.3	19.8	37.1
Baltimore	82,200	16,468	13,579	20.0	16.5	36.5
Seattle	85,600	17,559	13,474	20.5	15.7	36.2
Philadelphia	76,200	15,150	12,353	19.9	16.2	36.1
Washington, DC	101,700	22,122	14,335	21.8	14.1	35.9
Minneapolis	84,800	16,326	13,983	19.3	16.5	35.8
San Francisco	99,400	21,825	12,692	22.0	12.8	34.8
Australia 2003 GCCSA	*Average disposable household income (A$)*	*Housing cost (A$)*	*Transport cost (A$)*	*Housing cost (%)*	*Transport cost (%)*	*Total (%)*
Sydney	57,199	11,486	8,179	20.1	14.3	34.4
Brisbane	44,431	8,028	7,240	18.1	16.3	34.4
Perth	47,729	7,806	7,052	16.4	14.8	31.2
Melbourne	51,488	8,356	7,659	16.2	14.9	31.1
Hobart	45,521	6,491	7,307	14.3	16.1	30.4
Adelaide	46,610	6,961	7,004	14.9	15.0	29.9

Source: Calculated by the author from Census of Canada, 2011; Environics Analytics, special data tabulations, 2012; US Census Bureau, 2010; Hickey *et al.*, 2012; Australian data provided by Yogi Vidyattamma, based on the analysis contained in Vidyattamma *et al.* (2013).

Notes

Amounts are in local currencies.

Census Metropolitan Areas (CMA) in Canada, Greater Capital City Statistical Areas (GCCSA) in Australia, and Metropolitan Statistical Areas (MSA) in the United States are spatial units used in each country to delineate a contiguous metropolitan area, using similar (but not identical) methodologies based on commuting flows. Because the income variables are slightly different, results should be compared among cities within each nation, rather than across nations.

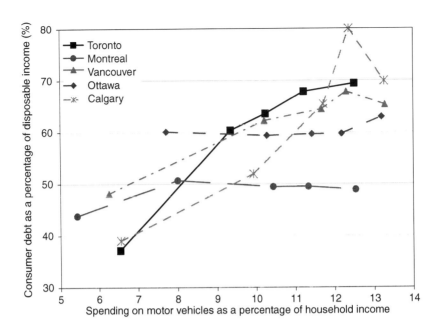

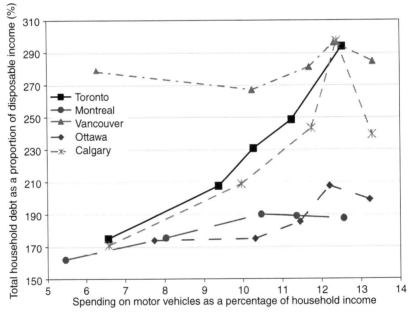

Figure 4.2 Household debt-to-income ratios, and spending on motor vehicles by quintiles, five largest Canadian metropolitan areas, 2011 (source: Calculated by the author from Environics Analytics, special data tabulations, 2012 (data for 2011)).

Notes
Spending on motor vehicles includes all costs related to vehicle purchase, lease, accessories, licensing, driver training, operation, maintenance, repair, tyres, parts, and fuel.

Spending on motor vehicles, % of income   Consumer debt, % of income

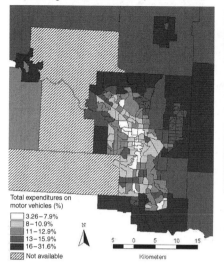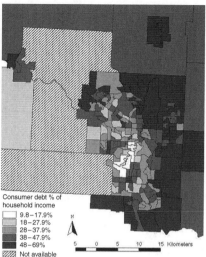

Figure 4.3 Spending on motor vehicles, and household consumer debt, as a proportion of household income, Calgary 2011 (source: Author, calculated from Environics Analytics, 2012, special data tabulations).

Notes

The spatial units being mapped are census tracts. These are spatial units created by Statistics Canada to act as proxies for neighbourhoods. The populations of census tracts average around 4,000 people, although they often vary between 2,000 and 8,000 people.

vehicles, particularly on their operation, maintenance, and fuel (i.e. the costs that vary with distance driven). As a result, the level of automobile dependence has a direct effect on the ability of households to build up their net worth. Across the landscapes of Canadian cities, households that are forced to spend more on basic operation, repair, and gasoline have less left over for vehicle purchase or for housing, and thus face greater challenges in building net worth, while those areas with greater transportation choice reveal higher levels of net worth in relation to income (even after controlling for income).[1] Such findings imply that one easy way to improve the ability of households to reduce debt and build up their savings is to make sufficient public transit accessible and affordable in automobile-dependent locales, and to reconcentrate places of employment in suburban clusters that are accessible via walking and cycling.

Although there is a clear relationship between debt and automobility at the intra-urban scale, there is no such relationship at larger scales of analysis. While the data presented in Table 4.2 suggest that the most sprawled cities have the highest transport costs, if not also the highest housing costs (e.g. Los Angeles, Phoenix, Detroit, Atlanta, and Miami in the US; Edmonton, Calgary, Saint John, and St John's in Canada), the empirical relationships between automobility and debt at the metropolitan scale do not mirror this pattern. Among metropolitan areas in Canada, in fact, there is no correlation between the proportion who drive and levels of consumer debt (r=−0.07), while the relationship with total household debt (including mortgages) is strongly *negative* at r=−0.67 (Figure 4.4). It is actually those metropolitan areas with higher rates of commuting via public transit use, bicycling, and walking – the trendier cities – that reveal the highest average debt-to-income burdens. This is mostly due to much higher average levels of mortgage debt in these places, itself partially resulting from the rapid gentrification of such cities and partially from higher housing demand on behalf of seniors (Walks 2013b). The most automobile-dependent metropolitan areas, meanwhile, are characterized by more production-based employment, often linked to the automotive and related sectors, and lower levels of both housing costs and mortgage debt. Traditional employment bases rooted in the auto-industrial complex are sustaining many such areas. While this may also make them more vulnerable to future economic shocks, it has been less of a factor driving them into debt. Instead, indebtedness is mostly a function of how automobility interacts with the *internal* urban forms and social geographies of the city, and thus how urban space is negotiated and consumed.

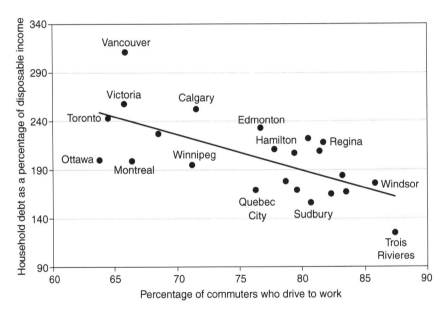

Figure 4.4 Relationship between automobile mode share and total household debt, Canadian metropolitan areas, 2011 (source: Calculated by the author from the 2011 Canada National Household Survey (NHS) (mode of transportation), and special data tabulations provided by Environics Analytics, 2012 (household debt in 2011)).

Notes
Household debt in this figure includes mortgage debt.

Conclusion: driving into debt?

The relationship between automobility and rising indebtedness is complex and multi-scalar. At the macro/global scale, automobility has provided key auto-mobile nations and their firms with significant profits, and significant financial power. Contrary to the potential explanations provided by the peak oil, peak car, or peak suburbs hypotheses, the most auto-mobile, fossil-fuelled nations are neither the most indebted, nor did they see the most rapid rise in indebtedness. Instead, it has been households and governments in nations more marginal to the system, particularly those at the fringe of the group of core developed nations, that either have the highest household debt levels, or that have seen debt levels rise the most (Ireland, Denmark, South Korea, etc.). The profits derived from the significant success of the auto-mobile industrial complex have been used both to expand the system into new territories and markets and to fund the many financial innovations that provided the credit to households to continue with consumption. However, as new profitable avenues for absorbing the surplus capital produced by the auto-industrial complex dry up in correspondence with the limits of its expansion, that surplus capital has gone instead to fund an increasingly predatory financial system. The rise of household debt is thus attributable to the financialization made possible by the vast profits and massive success of the entire system of automobility.

Financialization has, with the help of neoliberal state policies, provided households with easy access to credit for the continued consumption of the products and accoutrements of automobility, as well as for speculating on various asset values. Financialization, and the bailouts and stimulus packages implemented by states to prevent its collapse, have extended the life of the existing system and acted as a source of considerable financial power for both the state and multinational capital in the core auto-mobile, suburban nations. Yet they have come at the cost of rising indebtedness, both among households who are compelled to get into debt by increasingly unaffordable housing costs and among governments who are compelled to socialize financial losses in the face of crisis. However, rising costs of automobility as a whole have not been the cause of rising average debt levels. Just as the most auto-dependent nations are not the most indebted, neither are the most auto-dependent cities. Indeed, automobility has provided substantial benefits – including employment and commodities of real value – that have allowed for the production of significant real wealth and an overall rise in net worth, and this is particularly true among the most auto-mobile nations and cities of the developed world.

Instead, automobility should be thought of as a variable force of *redistribution* at the intra-urban scale, structuring uneven access to its benefits and burdens through the different ways that the geography of the city compels residents to live, work, and travel around. By providing accessibility and transport mode choice in some locales, while compelling people to drive in others, the auto-city affects financial and travel budgets, which are then incorporated into the complex locational decision making of households of different classes and other backgrounds. Those households located in the most auto-dependent neighbour-hoods are compelled to spend more on transport, and as a result often experience heightened levels of financial stress and are more financially vulnerable to changes in oil prices or interest rates, in relation to those living elsewhere in the city (Dodson and Sipe 2008, 2009). This is resulting in rising relative levels of indebtedness, heightened financial and generational risk, and reduced abilities to build net worth in the less-desirable auto-dependent suburbs, while simul-taneously allowing those of significant means to follow a path of successionist automobility to self-segregate into exclusive and protected enclaves (Henderson 2006). The physical and social space of the auto-city itself becomes part of the machine driving certain households into debt, and producing new forms of fin-ancial inequality and vulnerability within the larger context of the dominant system and regime of automobility.

Note

1 Statistically, using a controlled OLS regression model estimating the effect of mode share on the factors producing net worth across all Canadian metropolitan areas, every percentage increase in the proportion who drive to work at the census tract level increases the negative effect of higher fuel/maintenance/operation costs on household net worth in that census tract by approximately 5 per cent, while simultaneously decreasing the positive wealth-building effect of vehicle and housing purchase by a similar rate of 5 per cent (sig.=p<0.01, contact author for details).

5 Driven to school

Social fears and traffic environments

*Ron N. Buliung, Kristian Larsen, Paul Hess,
Guy Faulkner, Caroline Fusco, and
Linda Rothman*

Adults, supported through the "expert and media construction of risk", manufacture all manner of horrific thoughts and images regarding the role of strangers in children's lives (Valentine 1997; Dalley and Ruscoe 2003; Jenkins 2006; Murray 2009: 473), while being seemingly less concerned about a much greater threat – the automobile. People driving cars, typically bearing no malice, present a larger and more persistent child pedestrian safety risk than does intentional harm by strangers abducting children walking along public streets. In 2010, road traffic crashes resulted in approximately 1.24 million people killed and another 20–50 million non-fatal injuries worldwide (WHO 2013). According to Evans (2004), on a monthly basis more Americans are killed in automobile accidents than those who died in the World Trade Center attacks of September 11, 2001. During the course of the twentieth century, more than three million Americans have died in traffic crashes, nearly five times the number of deaths from all American wars from the start of the revolutionary war in 1775 to the 2003 Iraq War (Evans 2004). In 2010, road injury ranked eighth for global death rates, with a 46 per cent increase since 1990 (Lozano *et al.* 2013). It is predicted that road traffic collisions will rise to become the fifth leading cause of death globally, and the seventh leading cause of Disability Adjusted Life Years (DALYs) lost by 2030 (WHO 2008).

Children are especially vulnerable to road traffic injury due to their physical size relative to the motorized vehicle fleet. Moreover, their physical and cognitive skills may not be sufficiently developed for the demands of mixed traffic environments (see Plumert *et al.* 2004); environments designed by an adult technocracy whose *raison d'être* for much of the Fordist era involved sorting out the particularities of building cities for people in cars. Road crashes were consistently one of the top two causes of death for young people aged 5–29 globally in 2002, second only to lower respiratory infections in 5–14-year-olds (WHO 2009). In 2010, road traffic crashes were the fourth leading cause of years of life lost (YLL) for children aged 5–9 years, the second leading cause for children aged 10–14 after HIV/AIDS, and the leading cause for young people aged 15–24 (Lozano *et al.* 2013). Road traffic crashes result in more than 260,000 child fatalities annually and approximately ten million non-fatal injuries, leaving one million children with disabilities (CfGRS 2009).

While much of the planning discourse on children, safety, and automobility focuses on children's active travel (walking, bicycling, etc.) and the issue of pedestrian injury and death, it is actually child automobile-occupant injury that presents the larger problem. In Canada, for instance, vehicle occupant trauma, pedestrian injury, and cycling collisions accounted for 61 fatalities and 9,000 injuries to children under the age of 15 in 2010 (Transport Canada 2010). Car occupants represent the highest proportion of road traffic fatalities (31 per cent) followed by motorcyclists (23 per cent), pedestrians (22 per cent), and cyclists (5 per cent) worldwide (WHO 2009). In Canada, in 2005, the fatality rate of child and youth occupants (aged 0–19) was seven times that of pedestrians (Public Health Agency of Canada 2009). In 2008/09 the hospitalization rate of young automobile occupants (aged 0–24) was almost five times that of pedestrians (ibid.).

Rather than considering the social and environmental conditions that place children in cars in the first place, putting them at risk of violent injury or death, many remain complacent, seeking exogenous material, institutional, legislative solutions to preserve automobility through the enhancement of child occupant safety. When our children's typical body size and habitus can no longer accommodate previous material solutions, we simply "supersize" the car seat, rather than considering, through any thoughtful discourse, the origin of these physical changes – essentially creating a set of material circumstances set on perpetuating a cycle of increasing risk of injury and/or ill health attributable to a generational and primarily adultist automobility.

Injury and loss of life from road traffic events contrast, for example, with the five annual non-familial/friend stranger abductions resulting in injury or death discussed by Dalley and Ruscoe (2003). So why do beliefs regarding threat from strangers contradict the statistical evidence? Why, as Beckmann (2001) and others have asked, are we perhaps unable to be more self-critical about our automobility? Does the presence of the automobile, SUV, truck – an interceding material third party – attenuate our capabilities for risk assessment, or is it that our built environments demand convenience that can only be satisfied, in an immediate sense, through automobility? According to Sheller (2004: 236),

> Cars will not easily be given up just (!) because they are dangerous to health and life ... Too many people [including some children] find them too comfortable, enjoyable, exciting, even enthralling. They are deeply embedded in ways of life, networks of friendship and sociality, and moral commitments to family and care for others.

In this chapter we consider the place of child and parent/adult fear within the context of a common rite of passage of childhood – the school run. Like Murray (2009), the epistemological framing of our work acknowledges the gendering of fear and risk and the agency of children in affecting activity and mobility. We position this chapter around the concept of *fear* rather than safety per se as we assert that it is fear that drives many school travel decisions and the production of safety interventions. We also argue that automobility, in this context, acts to produce objective and perceived environmental risks while also offering some perceived protection from both itself and from other fearsome objects populating a broader landscape of fear. In his landmark work, *Landscapes of Fear*, Yi-Fu Tuan tells us that "Landscape, as the term has been used since the seventeenth century, is a construct of the mind as well as a physical and measurable entity. Landscapes of fear[s] refer both to psychological states and to tangible environments" (Tuan 1979: 6). These landscapes are partly material, containing hostile forces; "People are our greatest source of security, but also the most common cause of our fear" (Tuan 1979: 8). Here we intersect Tuan's landscape concept with automobility to illuminate and discuss objects (such as busy streets) and actions (*crossing* a busy street) that inhabit the school run's landscape of fear. This chapter is also, then, partially about the construction of an automobile childhood – one set within a large North American city. We of course recognize that there is not one child or childhood even here – and that intersections with socio-economic class, gender, mobility, and place produce diversity in childhood. School travel experiences of children vary across neighbourhoods and according to socio-spatial context. In interaction with these social and spatial contexts, automobility plays an important role in structuring the articulation of fear.

Automobility and changing school travel patterns

Changing patterns in school travel over the last half-century provide one example of the effect of a largely adultist auto-mobilization of children's activity and transportation. Declining rates of use in active modes such as walking or bicycling have occurred in tandem with increased automobile use for trips to and from school (Pooley *et al.* 2005; Buliung *et al.* 2009; McDonald *et al.* 2011). This has been a concern in the public health literature because, over the same period, we have seen rising rates of childhood overweight and obesity (Wang 2011). Intervening to reverse the increasing trend in motorized school transport is often seen as one opportunity to insert regular physical activity into the daily activity patterns of children and youth by promoting walking and cycling for the regular school trip (Tudor-Locke *et al.* 2002; Cooper *et al.* 2003). Children who use active modes for school travel have also been found to be more physically active overall (Faulkner *et al.* 2009).

Research into children's school travel safety is often positioned towards understanding and addressing adult concerns and fears, social and otherwise (Joshi and MacLean 1995; Greves *et al.* 2007; Eyler *et al.* 2007; Fesperman *et al.* 2008; Lang *et al.* 2011; Price *et al.* 2011; Zuniga 2012). Assumptions regarding parents' roles in constructing and controlling the activity and mobility of children underpin much of that work. From this adultist context emerges a view of "the" child as something more precious and less capable than "the" adult, requiring protection through surveillance and control (Fotel and Thomsen 2004). Children and adults may have different ways of thinking about safety and risk. In the absence of the child's voice, adults dominate the risk and safety narrative. Policy informed by such work perhaps allays parental concern without proper attention to children's actual social fears and anxieties. The objects and actions children fear during and in relation to school travel may indeed differ from those of adults.

Barriers to active school travel (AST)

Distance between home and school, convenience, and safety are often cited as the most common correlates of AST. The common finding that the likelihood of AST decreases with distance (Ewing *et al.* 2004; McMillan 2007; Larsen *et al.* 2009) needs to be extended with more sophisticated discussions regarding the production of such distances, how to attenuate distance effects, and how distance produces and interacts with other social and environmental barriers. Convenience also associates with mode choice; parents may drive because they perceive that driving allows them to more easily insert their child's school day into other household activities (Faulkner *et al.* 2010). Parents also typically perceive that driving saves time (Schlossberg *et al.* 2006; McDonald and Aalborg 2009; Faulkner *et al.* 2010). There are, however, potentially enormous external and personal costs attached to this quest for speed and time savings, costs that tend to distribute towards the most vulnerable populations (see Chapter 7 by Walks and Tranter in this volume). Indeed, as Tranter argues (2010), there may be little to no time savings when we enumerate all of the time required to produce automobility (i.e. time spent on paid work, commuting, maintenance, etc.).

Concern and even fear about child safety often motivate the decision to drive children to/from school (Collins and Kearns 2002; McMillan 2007; Ahlport *et al.* 2008; Wen *et al.* 2008). The child safety construct incorporates several dimensions of social fear as well as fear of pedestrian injury from traffic. Social fears can include abduction, assault, bullying, or interaction with strangers (National Center for Safe Routes to School 2010). Risk of child pedestrian collision with a motor vehicle develops through several features of the traffic environment. Traffic concerns may relate to both objective and/or perceived vehicle speed, volume, fleet composition, street crossings, availability of sidewalks, separation between traffic and footpaths, presence of crossing guards, traffic lights, and interactions between these dimensions (Roberts 1995; Carlin *et al.* 1997; Posner *et al.* 2002; Retting *et al.* 2003; Garder 2004; LaScala *et al.* 2004). Paradoxically, driving with the intention of reducing risk of child injury or death contributes to the environmental hazard posed by traffic (Collins and Kearns 2002). Epidemiological research into the subject of child pedestrian injury suggests that while there is an uneven geography to the phenomenon, injury is temporally coincident with the periods of travel to and from school (Yiannakoulias *et al.* 2002, 2013).

Ecological studies of school travel typically examine the role of the environment (built and social) in affecting mode choice. Several studies examining child pedestrian safety, AST, and the built environment have found a connection between neighbourhood characteristics and injury risk (Retting *et al.* 2003; Graham and Glaister 2003; Clifton and Kreamer-Fults 2007; Clifton *et al.* 2009). Evidence regarding the relationship between the environment, safety, and travel mode provides little consistent direction, however, with mixed findings reported on associations between intersection density, type of street crossing, and sidewalk availability and quality. For example, some studies indicate that higher intersection density is related to walking rates (Braza *et al.* 2004; Schlossberg *et al.* 2006; Frank *et al.* 2007), while others have not (Timperio *et al.* 2004; Ulfarsson and Shankar 2008). Similar contradictory findings have been produced for road crossings (Wen *et al.* 2008; Schlossberg *et al.* 2006; Larsen *et al.* 2012) and sidewalks (Boarnet *et al.* 2005; Dalton *et al.* 2011; McMillan 2007; Larsen *et al.* 2012). With regard to injury, Harwood *et al.* (2008) and Zegeer and Bushell (2012) found that neighbourhoods with complete sidewalk networks reduce objective risk of pedestrian injury, while other studies have not found a statistically significant relationship (Mueller *et al.* 1990; Stevenson 1997). It is possible that some environments are perceived to be so dangerous that no one ventures into them as a pedestrian, making them "safe" in the sense that there are no pedestrians to be injured. However, uneven findings for specific environmental variables such as street system connectivity are found across the literature that examines relationships between built form and travel mode, and safety findings are more likely related to the complex and potentially locally specific ways that environmental and social variables combine than to lack of pedestrian activity.

Beyond uncertain environmental effects, income and class enter into the school travel literature with many studies reporting that higher-income households or neighbourhoods commonly have lower rates of AST (Frank *et al.* 2007; Larsen *et al.* 2009; Dalton *et al.* 2011; Larsen *et al.* 2012). Low-income households, or more specifically children living in neighbourhoods with higher rates of overall deprivation – along multiple dimensions – appear to engage in AST at higher rates than children living in potentially less socially and economically depressed parts of cities. Moreover, higher rates of AST among lower-income respondents are coupled with higher risk of pedestrian injury in these neighbourhoods (Rivera and Barber 1985; DiMaggio and Li 2012). Income is not only related to mode choice, but pedestrian safety, and in many North American cities, lower-income central-city neighbourhoods have high traffic volumes from influxes of non-local vehicles accessing city centres from higher income areas (Yiannakoulias and Scott 2013).

Separation of socio-economic classes by technologies of mobility, from the horse-drawn carriages of the urban elites of the past to today's automobiles, is a persistent socio-historical theme (e.g. Tuan 1979; Ward 1977). Within such a context, and keeping children in mind, we are reminded that "The children who are most at risk from the car in the city are those who have the least access to its benefits" (Ward 1977: 118). Consider, then, that it is children living in low-income areas who are essentially practising, for reasons that likely speak more to social and economic constraints than to any desire to achieve physical activity targets, what we cannot seem to convince others who do not possess any disabling physical limitations to do – i.e. walk to school. In many North American cities, their reward for such "good behaviour" is increased risk of injury and higher traffic volumes.

Child safety has been a central focus of most qualitative studies of school travel (Joshi and MacLean 1995; Eyler *et al.* 2007; Greves *et al.* 2007; Mitchell *et al.* 2007; Ahlport *et al.* 2008; Fesperman *et al.* 2008; Murray 2009; Kirby and Inchley 2009; Lang *et al.* 2011; Price *et al.* 2011; Zuniga 2012). Social fears and risk of injury from traffic are two consistent themes. Social and environmental barriers to AST identified within these studies include: strangers, perceived threat of abduction, traffic volume, fast-moving vehicles, and inadequate sidewalks (cracked concrete or presence of garbage). Quality sidewalks and safe street crossings are supportive features that may reduce safety risks (Eyler *et al.* 2007; Greves *et al.* 2007; Fesperman *et al.* 2008). Fear of strangers, abductions, fast-moving traffic, and inadequate sidewalks were factors associated with mode choice in one US study (Eyler *et al.* 2007). In Seattle, the fear of traffic and personal security were barriers to active travel in low-income neighbourhoods (Greves *et al.* 2007; Fesperman *et al.* 2008), while supportive environments, including sidewalks and safe street crossings, had positive effects (Ahlport *et al.* 2008). As well, in both the United Kingdom and New Zealand, personal safety and road safety risks are among the most commonly reported barriers (Kirby and Inchley 2009; Lang *et al.* 2011). However, many studies do not consider children's perceptions of their experiences, and thus miss important insights into the power of automobility to frame social fears and vice versa.

Researching child school travel and fear in Toronto

This chapter draws on qualitative empirical research undertaken in the City of Toronto, Canada's largest city. Sampled schools were drawn from four strata: high income and central city (n=1); low income and central city (n=1); high income and inner suburbs (n=1); low income and inner suburbs (n=1). High-income neighbourhoods were defined as census tracts with median household income levels in the upper-income quartile (above the 75th percentile ≥CDN$69,072.50), whereas low-income neighbourhoods were within the lower-income quartile (below the 25th percentile ≤CDN$40,003.50). Figure 5.1 illustrates the central-city and inner suburban areas of the city, along with examples of street design in the neighbourhoods of interest. Such geographical stratification enables exploration of difference in safety perception across distinct built environments. Five motorized travel and five active travel child–parent dyads were recruited from each school. Semi-structured interviews were conducted separately with parents (n=37) and their children (n=37), providing an opportunity for independent thinking about school travel. More details on data collection and sample composition can be found in Faulkner *et al.* (2010).

The sample was gender balanced: boys made up 49 per cent of the sample, while 51 per cent were girls. Median child age was 10 years (range 8–11 years); median parent age was 41 years. The median distance to school was 805.6 m.

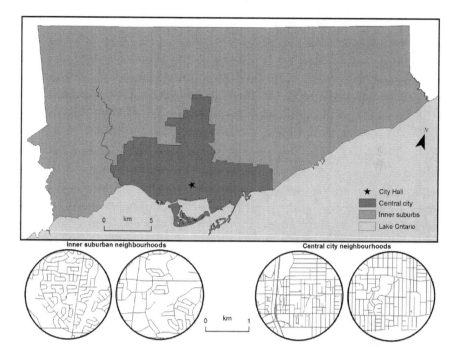

Figure 5.1 Example of neighbourhood types and street design in the city of Toronto.

There was evidence of spatial regularity in route selection: 78 per cent of cases reported use of the same route daily. Most children travelled to school using non-active modes (54 per cent), boys (58 per cent) were more likely to use active transportation than girls (34 per cent). All examples of AST involved walking. Non-active travel was dominated by automobile or minivan, small truck/SUV use, with one student reporting transit as the primary mode. Similar to most studies on children's activity and mobility, the majority of adult respondents self-identified as mothers (81 per cent). There was little evidence of unsupervised school travel in this sample; only four children indicated being allowed (or sometimes allowed) to walk with friends or alone to school, while one child was allowed to take transit alone. Mothers or mothers along with another household (father, sibling) or non-household member (other children) typically exercised responsibility for supervised school travel (70 per cent of cases).

Automobility, social fears, and traffic environments in Toronto

There are several themes related to both social fears and the traffic environment that emerge from our thematic study of the interview transcripts of children and their parents. We have summarized the dominant themes using concept maps to illustrate the diversity of social fears and traffic environmental concerns that spill across parents, children, and socio-spatial contexts (Figures 5.2 and 5.3). Of course, differences in what are perceived to present a safety risk, and indeed what is feared or to be feared, also relate to class, race, and gender, producing a complex web of relationships between the social, the spatial, and school travel (see Figure 5.3 for conceptual mapping of the intersections between these social and spatial articulations of fear). Here, parental perceptions and fears regarding the important barriers to active travel to school are discussed first, followed by those of children.

Fear of automobile accidents that produce child pedestrian and/or occupant injury is one component of the construction of parents' school travel fear in many North American cities. However, social fears, which on first blush may not seem directly related, are nonetheless influenced and reconstituted by and through automobility. An important social fear raised by the majority of parents was kidnapping, accompanying a general fear of strangers. While social fears were a concern at schools located in higher-income neighbourhoods, they were a more dominant theme in lower-income neighbourhoods. Nearly the entire sample of low-income respondents discussed social fears of some kind as presenting a barrier to AST. As might be expected given the gendered construction of parenting, childhood, independent activity, and mobility (see Jenkins 2006), nearly twice as many parents talked about strangers and the risk of abduction when their child was a girl. When one parent was asked why her daughter wasn't allowed to walk to school she responded:

"You know, there's strange people around, I'm *afraid* of that."
(Mother of girl: Inner suburbs: Low income: Auto-based school travel)

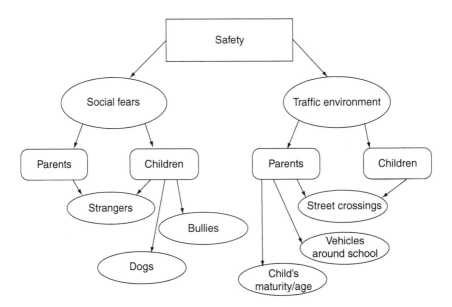

Figure 5.2 Concept map describing differences in child–parent social fears and traffic environmental issues.

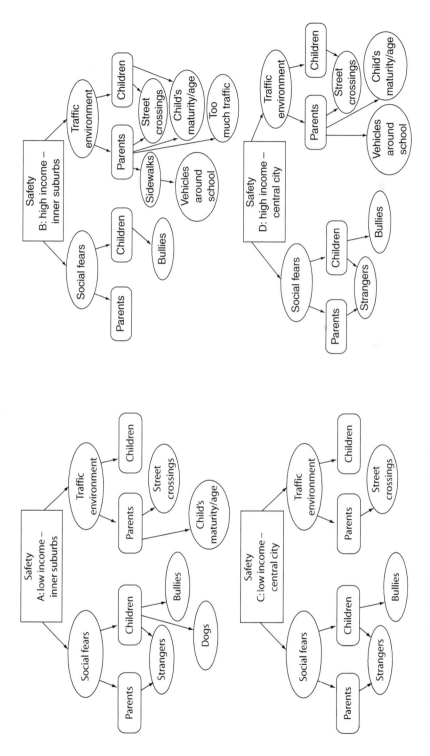

Figure 5.3 Neighbourhood difference concept mapping of social fears and traffic environment.

This was a common response from parents of daughters. Not only were parents of girls, and girls themselves, separately expressing concerns regarding strangers, mothers and daughters had discussed the potential of stranger danger. One parent reported:

> "So obviously everybody's hesitation is, you know, she might meet a stranger on the way to school. We've talked about the scenarios and what you could do."
>
> (Mother of girl: Central city: High income: AST)

We see then the possible internalization of a gendered adult risk narrative by children (Jenkins 2006), and a generational construction of fear and risk perception possibly based upon the risk and mobility experiences of mothers (Murray 2009).

However, even on the topic of abduction, it is notable that automobility looms large. When discussing potential frightening scenarios involving kidnapping or strangers, the presence of an automobile that could trap and take the child away out of sight added urgency and a sense of the sinister. When asked "What are your biggest hesitations or fears about walking to school alone?", one parent responded:

> "Somebody approaching her ... I'd be *afraid* that, you know, a car driving up the hill, some wacko grabbing her and putting her in the back of the truck."
>
> (Mother of girl: Central city: High income: Auto-based school travel)

It is within the context of such fears that the duality of automobility comes to light, as both problem and solution. A tool of the trade of its unwitting ally, the auto-mobile stranger on the one hand – whose culpability in constructing social fears through the occasional perpetration of what Murray (2009) calls "dread events" – contributes to the production of supervised travel both within and without the car.

Yet, within certain socio-spatial contexts, it is traffic safety that above all arouses concern among parents. Indeed, parents in the high-income inner suburbs, where the majority of children are driven to school, have an almost myopic focus on traffic safety risk (Figure 5.3). Among the responses related to traffic, the most common concern or fear, expressed by both children and their parents, was related to crossing busy streets. The following response was common in relation to factors influencing parents' decision to drive their child(ren) to school:

> "I would say up until now, with our kids, has been the volume of traffic and getting across the busy street."
>
> (Father of girl: Inner suburbs: High income: Auto-based school travel)

In the North American urban traffic-engineering regime, the *in situ* or immediate problem of crossing busy streets has been dealt with through mechanistic techno-regulatory and hybridized techno-human solutions: the lighted crosswalk, the stop sign, the zebra crossing, and combinations of one or more of these objects of control with crossing guards. Add to these the corralling of automobiles into kiss-and-ride zones or drop-off areas located some distance from the school site established to reduce traffic volume directly around school locations. These sites of techno-regulatory or techno-human intervention may indeed pull children, parents, and cars en masse into spaces of contestation, the front lines of exposure and conflict between the giving over of the public realm to the child pedestrian and/or to motorized vehicles driven by adults. However, to a degree, the regulation of these zones of conflict has been successful, especially relative to adults, in that child pedestrian injury (particularly for young children) is more likely to occur at mid-block and on lower speed neighbourhood roads (Rothman *et al.* 2012). Moreover, on any school day and in certain neighbourhoods, you can find that many of the parents supervising school travel on foot may walk home only to get in their cars and drive to work – drifting seamlessly between the cultures of walking and automobility (first author, personal observation).

While it might be expected that the various forms of socio-technical interventions described above would allay the fears of parents worried about high levels of traffic, in fact the opposite could be true. Some parents actively avoid crosswalks due to a perception that drivers do not respect such methods of control. But again, and as the work on injury and crossing location suggests, this could be an unwise strategy depending upon crosswalk location (Rothman *et al.* 2012). Regardless, drivers, it would seem, can only be counted on to respect traffic lights, which are meant to regulate in the interests of drivers:

> "There's been a couple of people hit at the crosswalk ... it's easier to explain to kids that a green light is safe to walk than a crosswalk."
> (Mother of boy: Central city: High income: AST)

Another parent said,

> "Crosswalks scare me, they scare me to death. Not all the drivers stop.... And my son, he's not an adult; he's smaller.... People don't always look for kids."
> (Mother of boy: Central city: Low income: Auto-based school travel)

Instead of crosswalks, the preferred option is for another human (crossing guard, etc.) to walk a child across the road, or temporarily disrupt traffic flow, if only for a minute or less, giving over the space of privilege occupied by automobility to the child pedestrian. However, not all parents who drive were convinced:

> "He's not a policeman, cars don't necessarily obey ... they're generally older people, retired, maybe hard of hearing."
> (Mother of girl: Inner suburbs: High income: Auto-based school travel)

Importantly, fears related to traffic, of all kinds, are more evident among parents who drive their children to school (see also Collins and Kearns 2002). Under automobility, keeping children safe from the automobile has become an important component of good parenting (Murray 2009). When asked why they drive to school, one parent who lives in the inner suburbs responded:

> "because we're the parents and we're looking after them ... driving is best because if they were to walk there is a lot of traffic ... I don't think it's safe ... we live really close to an avenue."
>
> (Mother of girl: Inner suburbs: High income: Auto-based school travel)

The main issue among many parents was an inability to trust children to properly perceive the degree of risk that is present. However, trust was constructed in relation to perceptions of the infrastructure in place, and the degree to which the walking environment was made sufficiently safe through some combination of sidewalks and reduced/slower traffic, or adequate parking and road space for the car:

> "In the morning it's very heavy traffic. It is quite hectic and my street is very busy so I wouldn't trust him to walk at all."
>
> (Mother of boy: Inner suburbs: High income: Auto-based school travel)

> "It's insane, it's dangerous, there's not enough parking for the amount of people that drop off their kids ... it's been an issue ever since I've brought my daughter to this school."
>
> (Mother of girl: Inner suburbs: High income: AST)

The issue of trust in the abilities of children to properly perceive the inherent risks associated with the journey to school is aligned with parental judgements of child maturity. Parents often justify the decision to drive their children to school on the assumption that children are not sufficiently mature or responsible to be able to properly assess risk and react quickly:

> "He's not at that stage yet. I don't think he's responsible enough to cross the street by himself. I don't think it's safe."
>
> (Mother of boy: Inner suburbs: Low income: Auto-based school travel)

> "It's maturity, age for sure ... it's a major intersection she would have to cross."
>
> (Mother of girl: Inner suburbs: High income: Auto-based school travel)

Parents reported starting to believe a child is mature enough to travel without adult supervision by the tween years, or about grade 6. One parent who drove their child to school was asked, what is it about grade 6 that makes you feel like he can walk by himself?

"Well he's more mature.... He's moving on to a middle school."
(Mother of boy: Inner suburbs: Low income: Auto-based school travel)

"A child should not ever cross the street alone until they're 10 ... their brain isn't developed to the point where they can ... judge where the car is."
(Mother of boy: Inner suburbs: High income: AST)

It is notable that none of the parents appeared to consider the possibility that the generational construction of fear and risk, coupled with supervised mobility, might actually be infantilizing their children, limiting their opportunities to mature and develop responsibility (Jenkins 2006). However, some parents who drive their children to school for safety reasons did admit their complicity in producing an unsafe situation. For instance, one parent said she drives her child to school because there are too many vehicles on the road and it is not safe to walk, but realizes she is contributing to the problem,

"I think it's just the busyness on the roads ... which we're not helping by driving, but [laughs]."
(Mother of boy: Inner suburbs: High income: Auto-based school travel)

Parents had various approaches for dealing with traffic fears. Supervised travel was the dominant solution, followed by surveillance. Many adults would escort their child to and from school by car or on foot, while others reported "spying" on their children, as well as calling the school secretary to ensure safe arrival had occurred. Others said they "they actually cross on the speed bump on the top" (Mother of girl: Central city: High income: AST). Again, no parent raised the possibility that such actions might be infantilizing their children, sending them a message that they required surveillance to minimize safety risk (Murray 2009).

It is rather easy, of course, to level criticism against parents who are operating from within a complex environment fraught with mixed messages regarding the efficacy of "paranoid" parenting, and the expert- and media-constructed discourse on child safety risk. As Murray (2009) suggests, mothers in particular may find themselves facing such a double-bind, and as our work and the work of others suggests, some reach out to the system of automobility for the answer – with a view to becoming the "good parent" who places their child's immediate physical health at the centre of mobility decisions that produce stealthy surveillance, supervised walking, or automobile use (Jenkins 2006; Murray 2009).

Internalizing automobility

Children present some contrast to the fears of parents regarding school travel. Many children were also concerned about strangers and walking alone; however, these fears were not as evident as they were in parents' transcripts. When a child was asked what makes it difficult to walk to school, he responded:

"People and stuff, sometimes they can like kidnap me and stuff."
                    (Boy: Inner suburbs: Low income: Auto-based school travel)

Many children were actually more concerned about bullies (Figure 5.2). When asked with whom they wanted to walk to school, children often rejected "big kids" or teenagers because they are more likely to be bullies. Yet, bullies were not mentioned in the parents' transcripts at all. Children also have concerns about neighbourhood dogs. One child, when asked what he didn't like about walking to school, responded:

"Like stray dogs ... because when dogs hate you, they could get really scary and they will attack you for no apparent reason."
                    (Boy: Inner suburbs: Low income: AST)

Whereas a number of children talked about bullies and dogs, only one parent expressed a similar concern; her child did not. Overall, shared parent–child concerns about strangers were complemented by children's concerns about bullies, teenagers, and dogs, a finding that was reproduced across neighbourhoods.

The solution for how to deal with potential safety risks among parents and children often differed considerably. Importantly, while for some parents a solution is pursued within the system of automobility, children expressed interest in group-walking, often responding that they would feel safer because if someone tried to hurt them, others could help:

"I would feel safe if I was walking in a big group, but if I was walking alone I wouldn't really."
                    (Girl: Central city: High income: Auto-based school travel)

These responses were common among both active and non-active travellers. A child who was driven to school, and perhaps ironically facing little threat from kidnappers during school travel, but obviously aware of a possible solution, stated that travelling in groups is safer,

"Because if the kidnapper tried to get all of us, he couldn't."
                    (Girl: Central city: High income: Auto-based school travel)

Group travel was also discussed in the context of solving the bullying issue, even by a child whose usual school trip occurs protected from such social threats within the confines of a car:

"If children walked together in a group and big bullies come, it won't be easy for the big bullies to catch them because there will be so many."
                    (Boy: Central city: Low income: Auto-based school travel)

While most parents liked the idea of group travel, reservations remained. Most parents were uncomfortable with unsupervised travel, even unsupervised group travel, particularly at their children's current ages. Parents seemed to favour the idea of allowing group travel over time:

> "Yes, older, probably twelve, thirteen, but with a group of friends, not by herself."
>
> (Mother of girl: Inner suburbs: Low income: Auto-based school travel)

Another solution raised by children involved having trustworthy adults to help. In particular, crossing guards allayed children's concerns about street crossings, irrespective of travel mode:

> "I think it makes everything safe for the children.... And they could also be your friend."
>
> (Boy: Central city: High income: AST)

One child did not like crossing at crosswalks or traffic lights as he did not trust that people would stop; he only felt safe with a crossing guard:

> "Because people don't listen to the lights."
>
> (Boy: Central city: Low income: AST)

Another child said crossing guards make them feel safe,

> "Cause the stop guard always stops ... they tell the car to stop and that's really why I feel safe."
>
> (Boy: Central city: Low income: AST)

These responses from children, which often pointed to human contact, directed walking, and protection from danger through the power of the group, differ from the parental focus on the automobile. This is not to say that none of the children raised being driven as a potential solution to personal safety and fear. However, when asked why she didn't want to walk to school, one child responded,

> "Well, there's a lot of traffic and it would be unsafe when you are crossing the road."
>
> (Girl: Central city: High income: Auto-based school travel)

Yet, here it is unclear whether children are genuinely concerned about street crossings, or if they are merely reporting the outcome of their parents' attempt to inculcate them with a similar set of concerns. As Murray (2009) suggests, a child's perception of risk is based on life experience, social interaction with space, and adopting their parents' views on risk. This child was among the few who pointed to being driven as a primary solution to the safety problem.

Children, however, did not jettison automobility either immediately or entirely, although they were less likely than adults to discuss an automobile solution to their school travel fears. Indeed, many children had clearly internalized the thrill, convenience, and power of driving, or at least of being driven. In the context of automobility and safety, then, we might ask, what role do children play in mobilizing automobility? And how do children perceive their agency in mobility decisions? Consider, for example, the following exchange between interviewer and child:

> "Okay, who decides how you get to school in the morning? Like who makes the decision that you are going to drive?"
>
> (Interviewer)

> "Me."
>
> (Girl: Central city: High income: Auto based school travel)

> "You decide. Okay, so you tell mom that you want to drive?"
>
> (Interviewer)

> "Yeah."
>
> (Girl: Central city: High -income: Auto-based school travel)

> "Okay, so do you ever have a choice to walk?"
>
> (Interviewer)

> "Yeah."
>
> (Girl: Central city: High income: Auto-based school travel)

> "So you chose to drive instead? Okay, so tell me a little bit about that?... How do you choose that decision instead of walking?"
>
> (Interviewer)

> "Because I normally have like practices in the morning..."
>
> (Girl: Central city: High income: Auto-based school travel)

> "Okay, so you are the one that makes the decision. And it's you, so how do you do that? Do you just tell mom?"
>
> (Interviewer)

> "Yup."
>
> (Girl: Central city: High income: Auto-based school travel)

The student's mother held a different view, indicating that while her daughter originally walked to school for a period of time – with her – once she (the mother) went back to work it became more convenient to travel by car. The child may indeed feel that she is in control of the mobility decision, and may well hold considerable influence over the daily travel decision, but it is also the case that the mother's report indicates a choice for automobility based largely on the desire for what is perceived to be a convenient mobility solution to overcome separation between home, school, and work:

> "Things kind of changed ... when I went back to work full-time."
> (Mother of girl: Central city: High income: Auto-based school travel)

> "So that change happened around...?"
>   (Interviewer)

> "Grade one."
> (Mother of girl: Central city: High income: Auto-based school travel)

> "And the main reason was because you had to be at work and driving was more convenient for you?"
>   (Interviewer)

> "Exactly."
> (Mother of girl: Central city: High income: Auto-based school travel)

It is interesting to note the shared experience of time pressures between mother and daughter, and the intersection between a household mobility decision, labour, gender, and the organization of the economy across space. The child's response indicates that she shares the view that the automobile is an adequate solution to overcoming the household's space–time problem.

Some children enjoy automobility, and as the previous case illustrates may "choose to be driven by car" (Fotel and Thomsen 2004: 548), or advocate for automobility. As one child indicated:

> "I would like to drive because it's really fun, because my two brothers and my brother's friend, we all get in the van and it's really squishy. And after we tell funny jokes and stuff. And after, when my dad turns off the radio and puts on the news he says, 'Everybody says hey at the same time' and it's so funny."
> (Girl: Inner suburbs: High income: Auto-based school travel)

Clearly the place of the automobile in childhood is not merely the product of adult experiences or needs. The constraints of time and space thrown up by the automobile and its complementary urban forms, and the scheduling of school and work, cascade across the household, becoming internalized among children who may prefer to be driven if at all possible, and in doing so seek some claim of autonomy, consciously or otherwise, over the school run.

Conclusion

This chapter examined differences in parent and child conceptualizations of fear and safety in relation to school travel. While the automobile and automobility are prominent features of the school transport conversation, and traffic is clearly a barrier, it is important to note the role of social fears. Parents and their children expressed different ideas about social fears. Children complemented stranger danger with specific mention of bullies, teenagers, and dogs. Bullying is largely known for directly impacting mental and physical health. What this chapter makes clear is that it can also influence children's mobility. Bullying is currently one of the most discussed children's health issues, yet while the current bullying conversation remains focused on schools and the Internet, the importance of "the street" as a stage for such interpersonal conflict should not be overlooked.

Parents were more likely to discuss traffic safety, abductions, and strangers, for which being driven to school is then presented as the solution and incorporated into what it means to be a "good parent". Many parents felt compelled to drive their children to school to protect them. This is true, even though many knew that by doing so they were contributing to the problem, increasing the likelihood that others would also feel compelled to drive, raising the overall level of traffic, reducing the number of people on the street, and thus augmenting the safety risks from walking. Others might walk their children to school, only to turn around and drive themselves to work, travelling seamlessly between the worlds of walking and automobility. Such are the complexities of automobility.

Children have a rather sophisticated understanding about the role of the automobile in daily transport, and are able to point to some of the safety risks and reasons for its use. They often have a sophisticated understanding of what is to be gained and lost through automobility (see, for example, Fusco *et al.* 2012). Many children internalize the desires, the power, and the convenience of automobility. Instead of fearing traffic, some children may desire to be part of it. We also see here, as in Barker (2009), that the experience of the child passenger offers a contrasting example of collective modern mobility (as do carpooling and ridesharing), producing mixed paradoxical results. The automobile is a paradoxical material object, and automobility is a paradoxical practice, producing an extinction of experience and safety risk for some (occupant and pedestrian), while attenuating child and parent concerns about time pressures inter alia and, where the child passenger is concerned, operating as a mobile place for socializing with friends and family and play.

Parents, on the other hand, have had to focus on the solutions they have at hand to issues of perceived personal safety of their children. Unlike children, for whom bullies, teenagers, and dogs can be avoided through group walking, some parents reach for the one item most within their control – the car. Yet, by driving their children to school, they not only augment the traffic safety risks associated with walking, they also may impede the ability of children to develop their independence, maturity, and individuality through their own intrinsic auto-mobility via walking and biking, a problem augmented when driving to school also means driving everywhere else. Driving children to or from school may also facilitate the infantilization of children, while simultaneously instilling in them a desire to drive and be driven. It is through such generational processes that automobility has both evolved and become culturally locked-in. This is one way through which, as Sheller and Urry note (2000), automobility self-generates the conditions for its own reproduction. The seeds for the reproduction of automobility may well be planted in childhood. Yet, the processes and relationships that intersects automobility with childhood – that put children in cars in the first place or at risk of injury – are fraught with complexity. They include difficult gendered power relations between children and adults and the construction of built environments through networks of adult and often patriarchally founded and structured institutions and professions that in some households produces demand for automobility to support knowledge acquisition, as in the case of school travel, and the accumulation of material wealth, irrespective of the safety risks.

Part II
Driving Inequality

6 Driving the commute

Getting to work in the restructuring auto-mobile city

Pablo Mendez, Markus Moos, and Rebecca Osolen

The increasing dominance of the automobile in the post-war period has been accompanied by a decentralization of the metropolis and the slow deconcentration of the population. While this pattern is most advanced in the most automobile-dependent, suburban nations – particularly the United States (US) – it is nonetheless a global trend (Beauregard 1993, 2006). The decentralizing auto-mobile city has transformed traditional logics of workplace and residential location, but also the whole process and experience of commuting. The factors shaping metropolitan commuting patterns have important implications for policy makers grappling with how best to deal with increasing congestion and automobile use given their negative effects on the environment, people's health, and inequality (Ewing *et al.* 2008).

Kain (1968) was among the first to articulate some of the problems with the kind of decentralization that was already occurring in the US by the 1960s, and that would in subsequent decades turn many older metropolitan areas effectively inside-out. Whereas low-income and low-skilled workers, particularly blacks, were confined and concentrated in the old inner cities, the kinds of work that had been employing these workers was rapidly moving out to the urban fringes where public transit was infrequent or non-existent, while older factories in the inner cities were closing. Such a "spatial mismatch" in the locations of low-skill workplaces and workers would continue to deepen, and by the 1990s was an entrenched characteristic of the geography of poverty, inequality, and race in the US (Holzer 1991, 1996; Ihlanfeldt and Sjoquist 1998; Ihlanfeldt 1999; Preston and McLafferty 1999; Taylor and Ong 1995; Martin 2004). Those living in the inner cities often did not learn about employment opportunities in the suburbs, and if they did not drive, often they could not get there. Then, even if they were lucky enough to be called to a job interview and could drive to it, they often found that white employers were more likely to hire local white applicants (Ihlanfeldt and Sjoquist 1998). The decentralizing city was now producing new forms of inequality, expressed through mobility, and resulting from unequal access to different forms of mobility.

As the metropolis continued to deconcentrate through the 1980s and 1990s, the new economic and social "frontier" moved to the fringe, producing new kinds of decentred, fragmented "edge cities" characterized by clusters of new office and retail development structured around interstate highway exchanges (Garreau 1992; Barnett 1995). New debates focused on what the auto-mobile city meant for commute distances. Gordon *et al.* (1991) highlighted the commuting "paradox" that had emerged in the US, in which decentralization of the city, coupled with near-universal automobility, was now leading to shorter and briefer commutes on average, as the dispersal of employment into multiple suburban nodes meant that households could now more affordably relocate to be near to work. Beito *et al.* (2002) extended their analysis into a normative promotion of the decentred, low-density "voluntary" city – one that allows individuals and families to express their preference for autonomous mobility in the confines of the free market, with the city expressly expanding to meet their demands. However, Beito *et al.* (2002) pay little attention to those who, for whatever reason, cannot elect to drive and are thus forced into residual forms of mobility in the auto-city.

These two perspectives on the post-war metropolis – spatial mismatch and the voluntary city – provide contrasting pictures of the outcomes and benefits of auto-mobile urban development. Yet, both have some grounding in contemporary realities. Indeed, the tension between the two is one of the defining characteristics of the auto-city. This is perhaps most evident in commuting patterns – the journey to work. While the literature on both spatial mismatch and the voluntary city emerged out of the US context, the basic concepts, and the tensions between them, can be applied to the expanding metropolises of other auto-mobile suburban nations, albeit with differences on the ground. Combining features of both the dispersal of many US cities (in which cities like Houston and Phoenix provide perhaps the most extreme examples) and the more compact patterns evident in other affluent nations, large metropolitan areas in Canada act as informative case studies of how the expansion of automobility has intersected with other, often weakened and residual, mobility systems to produce new social patterns of commuting with their own distinct geographies.

The auto-mobile city has continued to evolve alongside the restructuring of urban economies beset by deindustrialization, financialization, and neoliberalization. The process of inner-city gentrification has, since the 1970s, had an effect on metropolitan housing markets, pushing lower-cost housing away from the inner city and towards the automobile-oriented suburbs that were built during the mid-twentieth century (Skaburskis and Moos 2008). In this way, the residential geographies of lower-income workers and the ways they get to work are also undergoing their own transformations. As noted by Sheller and Urry (2000), automobility is constantly evolving in a non-linear yet auto-poeitic form. As outward dispersal and expansion meet their limits, they provide feedback into the system and a restructuring of the relationships between mobility, work, and inequality.

This chapter uses the three largest Canadian metropolises – Toronto, Montréal, and Vancouver – as case studies with which to explore the structure of the journey to work in the context of the restructuring city. More precisely, we measure patterns and changes in commuting distances between 1996 and 2006 for different kinds of mobilities and different kinds of workers, in order to assess how changes in the distribution of homes and workplaces under ongoing post-Fordist and neoliberal urban restructuring have affected commuting patterns in the auto-mobile city. Although much research has examined the changes in the location of homes and workplaces associated with the transition to post-Fordism, it is unclear how these transformations have affected different kinds of workers, and whether and how they may have spurred new spatial mismatches. This chapter thus builds on the literature on the changing economic and social geographies of Canadian cities (Walks 2001, 2011; Shearmur *et al.* 2007; Shearmur and Coffey 2002; Murdie and Teixeira 2006).

Driving workplace restructuring: the changing metropolitan context

Fordist urbanization largely established the consumption norms and laid down the infrastructure of automobility, expanding car use for daily travel within a low density, dispersed urban form (Freund and Martin 1993, 1996; Bunting and Filion 1999). The private vehicle was a defining consumer product of the Fordist–Keynesian city, wherein redistributive policies and a growing middle class sustained demand for mass-produced, durable products to furnish suburban lifestyles (Aglietta 1979; Paterson 2007). As private vehicles became more common, they became both affordable signifiers of "freedom, individuality, and progress" (Gartman 2004: 180) and increasingly necessary. State investment in large linear infrastructure, such as expressways and highways, facilitated vehicular access to the central city from the rapidly expanding suburbs of the mid-twentieth century North American city (Vojnovic 2000). Growing suburbanization came to be associated with increasing everyday travel for larger segments of the population (Kaufmann *et al.* 2004).

Fordist suburbs were more diffused than their predecessors, and they extended beyond those built around the streetcar lines that facilitated radial urban expansion in the late nineteenth century (Lang *et al.* 2006; see Chapter 3 by Newman and Kenworthy in this volume). In the monocentric metropolitan form of the early Fordist city, commute patterns came to be understood through Alonso's (1964) and Muth's (1969) bid-rent models, which both described and shaped urban development as they formed the basis of transportation planning algorithms (Hanson and Pratt 1988).

These models continue to describe much about today's commuting patterns, as many residents of the suburbs still commute long distances to workplaces in the central cities (Axisa *et al.* 2012). North American cities in particular are dominated by lengthy commute times with significant portions of the population commuting more than one hour in each direction, particularly in the largest metropolitan areas (Table 6.1). Yet there is a strong negative correlation between average commute duration and proportion of automobile-driving commuters, which can be linked to the briefer commutes of car drivers in the auto city (as per Gordon *et al.* 1991) combined with long commute times for

Table 6.1 Commuting in the US and Canada (selected metropolitan areas)

USA – Metropolitan Area	Population, 2012	Land area, in km² (2010)	Population density, per km²	Mode: drive alone 2012 (%)	Average commute in minutes, 2012	Commuters travelling > 60 minutes, 2000 (%)
New York	19,831,858	17,319	1,145	49.8	35.2	18.4
Washington, DC	5,860,342	14,500	404	65.8	34.0	12.8
Chicago	9,522,434	18,640	511	70.9	30.6	13.2
San Francisco	4,455,560	6,399	696	60.4	30.4	11.8
Atlanta	5,457,831	21,597	253	78.0	30.0	11.8
Boston	4,640,802	9,032	514	68.6	29.5	9.9
Los Angeles	13,052,921	12,557	1,039	74.1	28.9	11.2
Houston	6,177,035	22,863	270	79.6	28.6	9.2
Philadelphia	6,018,800	11,919	505	73.3	28.6	9.5
Seattle	3,552,157	15,209	234	69.6	28.5	9.1
Miami	5,762,717	13,150	438	77.6	28.0	8.8
Dallas	6,700,991	23,122	290	80.9	27.1	7.8
Denver	2,645,209	21,616	122	75.6	26.9	6.1
Detroit	4,292,060	10,071	426	83.7	26.2	6.6
Phoenix	4,329,534	37,725	115	77.3	25.8	6.3
Portland	2,289,800	17,311	132	70.8	25.1	5.7
Minneapolis	3,422,264	15,610	219	78.2	24.9	4.2
Cincinnati	2,128,603	11,375	187	83.5	24.2	4.6
Las Vegas	2,000,759	20,439	98	78.5	24.1	5.2
Oklahoma City	1,296,565	14,275	91	82.9	22.0	3.7
Buffalo	1,134,210	4,053	280	82.9	20.9	3.2

Canada – Census Metropolitan Area	Population, 2011	Land area, in per km² (2011)	Population density, per per km²	Mode: auto driver, 2010 (%)	Average commute in minutes, 2010	Commuters travelling >60 minutes, 2010 (%)
Toronto	5,583,064	5,906	945	64.5	32.8	15.8
Montreal	3,824,221	4,258	898	66.4	29.7	11.9
Vancouver	2,313,328	2,883	803	65.9	28.4	9.9
Calgary	1,214,839	5,108	238	71.3	27.0	6.9
Hamilton	721,053	1,372	526	77.8	26.9	11.0
Ottawa	1,236,324	6,287	197	63.8	26.3	6.3
Edmonton	1,159,869	9,427	123	76.7	25.6	6.1
Halifax	390,328	5,496	71	68.7	23.7	5.0
Winnipeg	730,018	5,303	138	71.0	23.3	3.9
Quebec	765,706	3,349	229	76.4	22.0	3.7
Victoria	344,615	696	495	65.8	21.8	4.3
Kitchener–Waterloo*	477,160	827	577	81.4	21.7	6.2
London	474,786	2,666	178	78.6	21.1	4.3
Sudbury	160,770	3,411	47	80.7	20.1	3.1
Saskatoon	260,600	5,215	50	80.5	19.9	3.4
Windsor	319,246	1,022	312	85.9	18.8	2.2
St John's	196,966	805	245	79.7	17.9	2.8
Regina	210,556	3,408	62	81.7	17.3	2.1

Sources: Statistics Canada Census of Population, 2011; Statistics Canada National Household Survey, 2011; US Census Bureau Annual Estimates of the Population, 2012, Table CBSA-EST2012-01; US Census Bureau Decennial Census, 2010, Table GCT-PH1; US Census Bureau ACS (American Community Survey), 2012, Tables GCT0801, GCT0802; Pisarski (2006).

Notes
Commuting data for Canada is from the 2011 National Household Survey (NHS). The accuracy of the results of the voluntary, self-administered NHS may be particularly affected by non-response bias, which occurs when the survey respondents differ from non-respondents. Twenty-one per cent of Canadians participated in the NHS in 2011, and Statistics Canada estimates the response rate to be 68.6 per cent.
The Pearson correlation of data column 5 (drive mode) with 6 (avg. mins.): US = −0.788; Canada = −0.608.
* Includes Cambridge, ON.

those travelling by public transit (Ong and Blumenberg 1998), which in the US have grown faster than those for automobile commuters despite shorter trips for transit users (Figure 6.1). In 2009, trip lengths in the US averaged 12.09 miles for private vehicles and 10.18 minutes for public transit (Santos *et al.* 2011: 48). Taylor and Ong (1995) refer to this diverging pattern between modes – briefer commutes via auto coupled with longer transit commuting – as "auto-mobile mismatch".

Of course, many standard commuting models are predicated on the behaviour of white, middle-class men, and even the models of spatial mismatch have been less able to account for much of the diversity of commuting behaviour and preferences in cities (Hanson and Pratt 1988; Bauder 2000). This includes the travel behaviour of women, who face important gendered constraints augmented by the spatial patterning of the automobile-dependent city (Fava 1980; England 1991; Preston and McLafferty 1993), and are especially entrenched where gender-based barriers intersect with race and class (Wyly 1996; Preston and McLafferty 1999; Blumenberg 2004a; Parks 2004).

Post-Fordism and decentralization of metropolitan labour markets

The post-Fordist transformation of production and consumption over the past 40 years has forged new employment and residential geographies (Gospodini 2006; Marcuse and van Kempen 2000), and led to new profiles of intra-metropolitan mobility. There are various ways of interpreting post-Fordism (Amin 1994), and

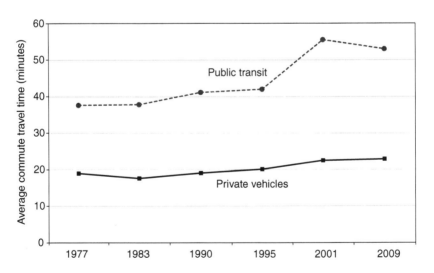

Figure 6.1 Time spent commuting, United States, 1977–2009 (source: Santos *et al.* (2011: table 27)).

Notes
Public transit includes local bus, commuter bus, commuter train, subway, trolley, and streetcar.
Private vehicles include automobiles, motorcycles, and trucks.

its associated processes are contingent on a number of local conditions that in many places continue to support Fordist economic sectors (Marcuse and Van Kempen 2000). Nonetheless, shifts in production and consumption have generally been linked with automobile-oriented metropolitan expansion and the emergence of new economic spaces in cities around the world. In addition, the high-growth sectors of the post-Fordist urban economy are associated with the reconfiguration and redevelopment of urban production and consumption spaces. The vertical disintegration of firms and the globalization of production chains have led to a decline in manufacturing employment in Western cities. Although industry is still an important urban sector, the US lost roughly one-third of its manufacturing workforce between 2000 and 2009 alone (Walks 2010a). Services, particularly producer services, personal services, and "cognitive-cultural" activities (especially information production and management), have become the leading sector and source of employment growth (Scott 2008, 2011).

One of the most salient features of metropolitan restructuring is the growing decentralization of employment and the rise in the proportion of jobs in suburban areas (Garcia-López and Muñiz 2010; Glaeser and Kahn 2001; Kneebone 2009; Kolenda and Liu 2012; Shearmur *et al.* 2007; Weitz and Crawford 2012). While producer services – many tied to the financial sector – and cognitive-cultural activities tend to concentrate in central cities or "suburban downtowns", manufacturing locates in suburban and, increasingly, in exurban locations (Shearmur *et al.* 2007). High-technology industry districts, often combining research and development with production, are typically located on greenfield sites, the classic example of which is Silicon Valley (McNeill and While 2001). Processes of employment decentralization have increased the proportion of metropolitan jobs located in suburban areas, often entailing faster job growth in the suburbs – particularly in manufacturing and retail jobs – and in some cases a decline in the absolute number of jobs in the central city (Shearmur and Coffey 2002; Shearmur *et al.* 2007; Heisz and LaRochelle-Côté 2005).

A diversity of metropolitan structures has arisen among Western cities, wherein employment may be concentrated in the downtown, dispersed in a highly diffuse pattern across the suburbs, or re-concentrated in suburban centres (Garcia-López and Muñiz 2010). So, while traditional commutes towards the central city for work have persisted over time (particularly in cities where the central city makes up a large proportion of the metropolitan region, as is the case with a number of Canadian cities), the fastest-growing form of commute is that between one suburb and another (Figure 6.2). Spatial mismatches are not only a problem of low-income or racialized groups residing in the central cities, but also among suburban populations (Preston and McLafferty 1993, 1999; Gottlieb and Lentnek 2001; Houston 2005), creating a large diversity of potential effects across metropolitan areas, even in the US (Cooke 1996). While it may be difficult to disentangle racial segregation from processes of spatial mismatch in the US (Preston and McLafferty 1999; Bauder 2000; Stoll and Covington 2012), research from Britain demonstrates that spatial mismatches there are largely class-based (Houston 2005).

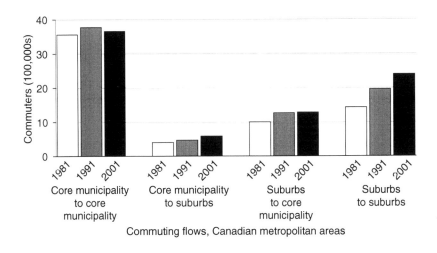

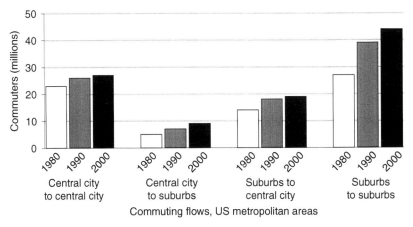

Figure 6.2 Growth trends in metropolitan flows, Canada and US (source: adapted from Statistics Canada (2003: 24); Pisarski (2006: 52)).

As the location of workplaces has changed with the restructuring of urban economic sectors and sites of production, so has the depth and configuration of inequality. Occupational polarization involving the increase in both low-skill, low-income jobs and high-skill, high-income jobs has been associated with deepening income inequality among occupations and neighbourhoods (Walks 2001, 2011, 2013c; Marcuse and van Kempen 2000; Sassen 2006). Neoliberal restructuring that reduced the size of the welfare state and heightened labour market insecurity has exacerbated inequality. Unemployed workers and those in manufacturing and lower-status consumer service jobs have largely shifted their residential location towards the inner suburbs. Findings from both the US and Canada show a high level of demographic, socio-economic, and cultural diversity in suburban areas (which have become places of immigrant settlement and increasingly diverse household types) but also a mix of affluent enclaves and dilapidated housing (Frey and Berube 2002; Walks 2001, 2011; Walks and Bourne 2006; Hiebert *et al*. 2006; Hanlon 2008, 2009).

Coupled with the restructuring of work, the uneven distribution of transportation infrastructure has led to a distinct geography of accessibility patterns, and hence differences in what Kaufmann (2002) and Urry (2007) call "motility capital" – all of those attributes that enhance the capacity to be mobile. As urban planning was able to influence accessibility gradients through the location of public transit corridors to the expressway development of the mid-twentieth century, suburban growth largely entrenched an urban form where private vehicle use became the norm. Today, despite the congestion, pollution, and health risks it entails, the automobile continues to be the prevalent mode of daily commuting, and built environments are still being constructed to accommodate its use (Filion 2003). More recently, developers and politicians have cast a favourable eye on the development of transit-oriented neighbourhoods and the renewal of downtowns to promote walking, cycling, and transit use. The work of Newman and Kenworthy (1989, 1999), which linked commuting and the density of the built form, helped create a mandate for the planning profession to promote more compact and mixed-use development as one means of attaining more sustainable transportation patterns (Hall 1996; Cervero 1998; Low and Gleeson 2003; Ryan and Turton 2007). Mees (2001, 2009) has shown that integrated and networked transit systems operated by public agencies do the best job at capturing the dispersed trip origins and destinations typical of the auto-city.

Given metropolitan restructuring, however, high-income earners are now outcompeting other households for access to central neighbourhoods, driving the gentrification of inner cities and their increasing densification and privatization (Lees *et al*. 2007; Quastel *et al*. 2012; Rosen and Walks 2013; Walks and Maaranen 2008). This has important effects on the ability to appropriate different forms of mobility. There are two logics that construe central neighbourhoods as sites of locational advantages and support the accessibility of gentrifiers, in turn enhancing their motility capital: the first relates to the aesthetic and lifestyle distinctions of urbanism, and the second to high levels of proximity and accessibility to sites of employment and consumption in central-city neighbourhoods (Rérat and Lees 2011). Central-city homes in Canada, for instance, are attractive to dual-income households and professional women, as female labour force participation in centrally concentrated professional occupations has increased (Rose and Villeneuve 1994, 1998). Alternatively, an affinity for urbane culture and heterogeneity and a rejection of suburban lifestyles are integral to the desirability of the gentrifying neighbourhoods for households belonging to what Ley (1996) has termed the "new middle class", whose members are typically employed in high-skill service-sector jobs. In such neighbourhoods, residents are more likely to commute by active transportation modes, although car use is also high in gentrifying areas (Danyluk and Ley 2007). Gentrification may thus induce higher car use downtown. Research from the US suggests that compact urban form, such as that of the central city, does not necessarily diminish commute distance or car use, particularly for dual-income households (Jarvis 2003, 2005). For households that have access to a vehicle, motility capital is increased by the option of driving within neighbourhoods characterized by high accessibility and multi-modality.

Driving the commute in Canada's global cities

Commuting patterns are shaped by an array of household- and individual-level characteristics such as income, gender, occupation, and immigration status, in addition to the location of transportation infrastructure, places of work, and affordable housing (Hanson and Giuliano 2004; Horner 2004; Plaut 2006; Shearmur 2006). As affordable housing is increasingly more likely to be located in the suburbs than in the central city, questions remain about the mobility patterns of low-income residents. The US literature on the spatial mismatch hypothesis investigates how decentralization of manufacturing employment, combined with racial segregation of ethno-racial minorities, hinders labour market participation and earnings (Ihlanfeldt and Sjoquist 1998, Kain 1968). Much less is known about spatial mismatch in the Canadian context, but less pronounced inner-city decline than in the US may reduce the conditions leading to spatial mismatch (McLafferty and Preston 1996). In the US, low-skilled workers may face protracted commutes to suburban manufacturing jobs under conditions of spatial mismatch – but evidence of short commutes does not necessarily translate into equitable mobility, particularly when workers do not gain better wages or working conditions from working locally (Wyly 1996). Nonetheless, a shorter commute should be beneficial, other things being equal, if it leaves more time for other pursuits, including work, and enhances the range of choices in everyday life.

US scholars have also highlighted the mobility challenges of lower-income job seekers and workers in automobile-oriented urban contexts. Sanchez (1999) finds a correlation between transit access and local labour participation rates, but cautions that this may not be a causal relationship. Low-income workers often choose to live near transit to gain mobility, but they may also base their residential choices on numerous other household and place characteristics (Boschmann 2011). Transit route configurations and schedules may not be appropriate for shift workers, or for accessing spatially dispersed, low-status service employment (Sanchez 1999). Unsurprisingly, jobs are more accessible by private vehicle in cities such as Detroit and Buffalo, where transit access is relatively poor (Hess 2005; Grengs 2010). But the logics of the metropolitan autoscape mean that even where transit systems are relatively extensive and roads are commonly congested, such as in the San Francisco Bay Area, commute times by private vehicle are relatively shorter, and less variable, particularly in suburban places (Kawabata and Shen 2007).

In this section, we explore the links between changes in the geography of these characteristics as an outcome of contemporary metropolitan restructuring, using as the main case studies Canada's three global metropolises: the Toronto, Montréal, and Vancouver Census Metropolitan Areas (CMAs). Toronto, it might be noted, has the longest commuting distances of Canada's three largest metropolitan areas according to Statistics Canada 1996 and 2006 census data, likely due to metropolitan size but also to the characteristics of the city's workforce (Turcotte 2011). As Canada's financial centre, Toronto has a higher share of workers in high-order services, with university education, and higher average incomes – factors traditionally associated with longer commuting distances (Shearmur 2006) – as well as higher shares of households with children present. The social geography of the city still sees families with children residing mostly in suburban areas (Murdie and Teixeira 2006), which tend to be further from jobs (Shearmur and Coffey 2002).

In terms of mode-split, the automobile remains the dominant mode of transport, albeit with some variation related to the era during which development occurred in each place (either before or after the Second World War) (Table 6.2). Vancouver, which is often celebrated for its planning policies that promote walking, cycling, and transit (Quastel *et al.* 2012), saw the largest increase in transit use and largest decrease in automobile use between 1996 and 2006, but at about 2–3 per cent the changes are overall modest. Changes in density associated with the transformation of inner-city landscapes from manufacturing and industrial spaces to residential and recreational ones are visible in all three metropolitan areas (see Ley 1996). Yet these changes, which particularly in Vancouver have most intensely contributed to reduced car use and shorter commute distances in the core (City of Vancouver 2012), are muted by the large suburban populations that continue to rely mostly on their cars. Longer and more automobile-dependent commutes are related to the cost of housing (Alonso 1964; Gurran 2008) – highest in Vancouver (Table 6.2) – and contribute to suburbanization.

To understand the relationships between urban form and commuting patterns, we analysed customized data from the 1996 and 2006 census for Toronto, Montréal, and Vancouver at the level of census tracts.[1] We define an inner-city urban form, and the post-war suburbs, based on the age of the housing stock within census tracts, following Walks (2001) and Filion *et al.* (1996). The inner city's housing stock was predominantly built before 1946, when walking and transit, particularly streetcars, were the predominant means of transportation. The inner suburbs were built between 1946 and 1970, when automobile use was becoming more common and accessibility became less dependent on propinquity. The outer suburbs correspond to tracts built after 1970, and a density threshold of less than 1,000 people per square kilometre was used to identify the urban fringe. Finally, three occupational groups – professional, manufacturing/industrial, and services/sales occupations – were created following the method used by Walks (2001).

Consistent with what is already known about commuting patterns in large North American cities, the distance travelled to work tends to be longer for people who work in higher-order occupations, have higher incomes, and live further from employment clusters, such as the CBD of a metropolitan area

Table 6.2 Summary descriptive statistics (percentage shares and change)

	Toronto		Montréal		Vancouver	
	2006	Change 1996–2006	2006	Change 1996–2006	2006	Change 1996–2006
Distance:						
< 5 km	28.5	−0.5	33.9	1.1	35.4	2.0
5–9.9 km	23.2	−0.5	24.7	−0.3	26.3	−0.3
10–14.9 km	16.3	0.4	16.1	0.0	14.9	0.1
15–19.9 km	10.8	0.0	10.1	−0.1	9.6	−0.6
20–24.9 km	7.2	0.0	6.5	0.0	5.9	−0.3
25–29.9 km	4.7	0.4	3.4	−0.1	3.3	−0.2
> 30 km	9.3	0.1	5.3	−0.6	4.6	−0.6
Mode:						
Car	70.2	−0.8	69.3	−1.2	74.5	−2.8
Public transit	23.6	0.5	22.8	0.8	17.4	2.3
Walking	5.2	0.1	6.1	−0.1	6.5	0.5
Bicycle	1.0	0.2	1.8	0.6	1.6	0.0
High-order service workers	31.1	3.7	28.7	3.1	29.5	3.3
Intermediate service workers	19.9	0.6	21.7	1.3	21.0	0.4
Low-order service workers	30.7	−4.3	32.4	−2.1	32.2	−3.2
Manual workers	18.3	0.0	17.3	−2.3	17.2	−0.6
Income (2005 dollars, in '000s)	48.259	9.325	40.134	5.491	42.060	4.762
Immigrant	50.6	4.6	21.4	2.9	42.3	5.2
Children present	67.5	20.7	58.4	13.3	59.9	17.9
No high school diploma	12.1	−10.2	12.9	−8.1	10.9	−10.2
High school diploma	25.5	0.3	22.1	−5.1	27.0	1.6
Trades/Non-univ. cert.	24.6	−1.2	33.6	5.9	27.2	−1.6
University degree	37.8	11.0	31.5	7.3	34.9	10.2
Female	48.3	0.8	48.6	2.0	48.8	1.4
Non-family household	12.6	−1.1	18.7	0.8	16.5	−2.3
Dwelling value/room (2005 dollars, in '000s)	58.388	19.242	37.148	10.004	72.083	29.428

Source: Calculated using Statistics Canada census data (1996, 2006).

(Shearmur 2006; Maoh and Tang 2012). Occupation is related to commute distance, both because the higher earnings of high-status occupations allow workers to afford the costs of a longer commute, and because workers are willing to travel further to earn higher salaries, although there are also occupational differences in commuting propensities that are independent of income (Cubukgil and Miller 1982). Controlling for gender and income, commute distances are longer for managers and professionals than for blue-collar workers, while service and sales workers have the shortest commutes (Shearmur 2006). With a few notable exceptions, this ordering of occupational groups characterizes most of the findings in Tables 6.3 and 6.4.

Table 6.3 shows the average commutes by occupation and mode in the urban zones of each city. In the inner city, industrial workers have longer average commute distances than professionals by both private vehicle and transit. This is due to the decentralized locations of industrial jobs, coupled with the centralized location of many professional jobs. Industrial workers who live in the inner city are located in an area where manufacturing employment has been declining since the 1960s. The long commutes by industrial workers in the inner city has also been influenced by planning policies that encourage mixed-use redevelopment of industrial districts where employment is declining, because such policies can discourage industrial uses in central areas (Green Leigh and Hoelzel 2012). The finding of disproportionately long commutes among manufacturing workers in Canada's inner cities provides evidence of spatial mismatch among manufacturing workers, despite largely lacking the legacy of racialized city–suburban segregation found in the US city.

In contrast, in the suburbs and urban fringe, industrial workers have shorter average commute distances than professionals by both private vehicle and public transit. Industrial workers who live in the suburbs and urban fringe are commuting shorter distances to suburban employment centres where manufacturing employment is increasing. Manufacturing employment continues to grow in employment centres outside the CBD, even when deindustrialization characterizes the overall shift in employment at the metropolitan scale (see also Shearmur and Coffey 2002). In the inner city of all three metropolitan areas, professionals commuting by private vehicle have longer average commute distances than those travelling by transit, while the reverse is true in many of the suburbs (Table 6.3). A number of professionals who commute by private vehicle from homes in the inner city are now travelling to suburban employment locations, using inner-city locations to manage the spatially expansive "everyday time-space co-ordination" linking commuting to work, shopping, and transporting children to school (Jarvis 2003: 592). The relatively long transit trips compared to automobile trips in the suburbs are particular to professional workers whose jobs are clustered in the CBD, including those working in the FIRE sector. In the auto-mobile city, it is largely for these often high-paid workers that the state has subsidized rapid commuter rail networks into the central city, where full-scale automobile commuting is made unworkable by the very concentrated nature of financial and producer service jobs in the CBD.

Table 6.3 Average commute distance of occupational groups within urban zones (km)

Urban zone		Total		Professional			Industrial			Service and sales		
		2006	1996	2006	1996	Relative change	2006	1996	Reiative change	2006	1996	Relative change
Montreal												
Inner city (n = 235)	All modes	7.81	8.21	8.06	8.44	0.03	9.60	9.31	0.69	6.17	6.80	0.23
Inner suburb (n = 342)		10.47	10.94	11.32	11.69	0.10	10.20	10.55	-0.12	8.30	9.29	0.52
Outer suburb (n = 92)		14.96	15.66	15.98	16.28	0.40	13.90	14.89	0.29	11.54	13.65	1.42
Urban fringe (n = 100)		17.63	18.38	18.41	18.93	0.24	16.81	17.89	0.33	14.52	15.87	0.60
Inner city	Private vehicle	10.06	10.33	10.17	10.43	0.01	10.93	10.90	0.30	8.40	9.40	0.74
Inner suburb		11.40	11.90	12.01	12.40	0.11	11.04	11.48	-0.07	9.48	10.44	0.47
Outer suburb		15.36	16.07	16.24	16.54	0.40	14.34	15.19	0.15	12.26	14.40	1.43
Urban fringe		17.91	18.72	18.54	19.19	0.17	17.23	18.20	0.16	15.05	16.57	0.71
Inner city	Transit	6.47	6.71	6.46	6.59	-0.10	8.44	7.66	1.02	5.91	6.14	-0.01
Inner suburb		8.95	8.91	9.68	9.47	0.18	7.99	8.07	0.12	7.70	8.14	0.48
Outer suburb		13.82	13.00	15.31	13.92	0.57	11.81	11.81	0.82	10.84	11.18	1.16
Urban fringe		18.29	16.48	19.74	17.71	0.21	15.05	14.40	1.16	12.59	13.86	3.08
Toronto												
Inner city (n = 235)	All modes	9.78	10.54	10.26	10.82	0.20	11.76	12.33	0.18	7.72	8.94	0.47
Inner suburb (n = 342)		13.06	13.38	13.99	14.37	-0.06	12.35	13.00	0.34	10.21	10.86	0.32
Outer suburb (n = 92)		17.05	17.70	18.04	18.74	-0.05	14.98	15.85	0.22	13.27	14.78	0.86
Urban fringe (n = 100)		19.51	20.60	20.60	21.92	-0.24	17.41	19.31	0.82	16.05	18.04	0.91
Inner city	Private vehicle	12.51	12.85	12.66	12.81	0.11	13.40	13.99	-0.25	10.56	11.50	0.59
Inner suburb		14.11	14.29	14.91	15.06	0.03	13.04	13.62	0.39	11.65	12.21	0.38
Outer suburb		17.33	17.97	18.22	18.88	-0.03	15.37	16.14	0.13	13.92	15.56	1.00
Urban fringe		19.66	20.87	20.66	22.03	-0.16	17.66	19.74	0.87	16.71	18.70	0.78

		1	2	3	4	5	6	7	8	9	10	11
Transit	Inner city	7.88	8.29	7.89	8.01	−0.27	10.39	10.24	0.56	7.09	7.76	0.26
	Inner suburb	11.46	11.46	12.28	12.37	−0.09	10.31	11.03	0.72	9.49	9.91	0.41
	Outer suburb	17.75	17.69	19.51	19.72	−0.27	12.18	13.76	1.64	12.58	12.09	−0.43
	Urban fringe	20.80	20.41	23.10	23.24	−0.53	13.79	17.87	4.47	14.41	12.35	−1.67

Vancouver

		1	2	3	4	5	6	7	8	9	10	11
All modes	Inner city (n = 175)	8.19	8.60	8.49	8.98	−0.07	9.34	9.25	0.51	6.63	7.18	0.13
	Inner suburb (n = 366)	11.03	11.55	11.57	12.05	0.04	11.23	11.86	−0.11	8.70	9.54	0.32
	Outer suburb (n = 198)	15.42	16.44	16.35	16.93	0.44	14.63	15.89	0.25	12.28	14.50	1.20
	Urban fringe (n = 73)	15.61	16.11	15.91	16.42	−0.02	15.98	16.63	−0.15	12.86	14.15	0.80
Private vehicle	Inner city	9.42	9.66	9.66	9.87	0.03	10.24	9.95	0.52	8.10	8.61	0.27
	Inner suburb	11.85	12.32	12.23	12.65	0.05	11.91	12.41	−0.04	9.89	10.56	0.20
	Outer suburb	15.62	16.61	16.36	16.99	0.36	14.96	16.19	0.24	12.81	14.91	1.11
	Urban fringe	16.01	16.48	16.37	16.83	0.00	16.36	16.83	0.00	13.49	14.55	0.59
Transit	Inner city	7.49	7.27	7.56	7.55	−0.21	8.14	7.98	−0.07	6.53	6.28	−0.03
	Inner suburb	10.21	10.34	10.82	10.49	0.45	9.68	10.32	0.52	8.72	9.40	0.56
	Outer suburb	16.85	18.52	18.81	19.19	1.28	13.91	15.40	−0.18	14.00	16.02	0.36
	Urban fringe	16.75	16.01	17.63	15.85	0.72	12.95	14.57	2.35	11.68	14.32	3.37

Source: Calculated using Statistics Canada Census data for 1996 and 2006

Notes

Occupational categories based on the National Occupational Classification (NOC-S 2006) and 1991 Standard Occupational Classification (1991 SOC), The Professional Occupations category includes workers in Management, Natural and Applied Sciences and Engineering, Health, and Social Science, Education, Government Services, and Religion Occupations. The Industrial Occupations category includes Occupations Unique to Manufacturing, Processing and Utilities, and Trades, Transport and Equipment Operators Occupations. Sales and Services Occupations are a distinct category that captures non-professional service-sector work.

Table 6.4 Average commute distance of occupational groups by median income (km)

Tract median household income to CMA median		Total		Professional			Industrial			Service and sales		
		2006	1996	2006	1996	Relative change	2006	1996	Relative change	2006	1996	Relative change
Montreal												
<0.7500 (n = 218)	All modes	7.99	8.33	8.40	8.56	0.18	9.01	8.94	0.42	6.59	7.19	0.25
0.7500–0.8999 (n = 159)		9.37	9.73	9.74	9.98	0.12	9.94	10.16	0.14	7.67	8.29	0.26
0.9000–1.0999 (n = 127)		11.32	11.85	11.73	12.08	0.18	11.65	12.36	−0.17	9.12	10.21	0.56
1.1000–1.4999 (n = 178)		15.38	15.28	15.85	15.35	0.40	15.34	15.75	−0.42	12.33	13.25	1.02
≥1.5000 (n = 74)		15.74	15.93	16.09	15.93	0.34	15.39	16.02	0.27	12.64	14.19	1.36
<0.7500	Private vehicle	9.90	10.20	10.35	10.45	0.20	10.10	10.27	0.14	8.37	9.39	0.71
0.7500–0.8999		10.52	10.87	10.88	11.23	0.00	10.82	10.94	0.24	9.01	9.51	0.14
0.9000–1.0999		12.27	12.74	12.55	12.78	0.24	12.46	13.02	−0.08	10.46	11.31	0.38
1.1000–1.4999		15.90	15.91	16.33	15.92	0.41	15.73	16.15	−0.07	13.20	14.17	0.97
≥1.5000		16.04	16.31	16.27	16.26	0.19	15.68	16.34	0.32	13.23	14.98	1.49
<0.7500	Transit	6.87	6.96	6.93	6.81	−0.09	7.83	7.53	0.39	6.37	6.58	0.12
0.7500–0.8999		8.07	8.17	8.00	8.00	−0.10	7.97	7.77	−0.30	7.26	7.49	0.13
0.9000–1.0999		9.45	9.66	9.56	9.28	−0.28	9.04	8.88	−0.37	8.15	9.26	0.90
1.1000–1.4999		13.67	11.85	14.11	11.87	0.42	12.55	11.19	0.46	10.67	10.57	1.72
≥1.5000		14.79	13.16	15.70	13.05	0.80	12.79	15.72	−0.56	9.72	10.06	1.97
Toronto												
<0.7500 (n = 174)	All modes	11.03	11.16	11.46	11.46	0.13	11.97	12.38	−0.28	8.73	9.35	0.49
0.7500–0.8999 (n = 158)		12.08	12.21	12.66	12.58	0.21	12.09	12.70	−0.48	9.71	10.13	0.30
0.9000–1.0999 (n = 162)		13.53	14.05	13.82	14.64	−0.30	13.80	14.05	0.27	10.67	11.76	0.57
1.1000–1.4999 (n = 237)		17.14	16.77	17.81	17.39	0.05	15.72	15.96	0.61	13.42	14.18	1.12
≥1.5000 (n = 82)		17.73	18.62	18.21	19.06	0.05	16.47	17.51	0.14	14.70	16.24	0.64

Private vehicle		<0.7500	13.16	13.13	13.70	13.35	0.32	13.00	13.49	−0.20	11.03	11.46	0.45
		0.7500–0.8999	13.57	13.62	14.26	13.95	0.36	13.06	13.61	0.51	11.25	11.79	0.49
		0.9000–1.0999	14.90	14.93	15.29	15.45	−0.13	14.44	14.58	0.12	12.19	13.04	0.82
		1.1000–1.4999	17.53	17.25	18.11	17.84	−0.01	15.98	16.28	0.57	14.37	15.16	1.06
		≥1.5000	17.95	18.79	18.30	19.20	−0.06	16.76	17.67	0.08	15.58	16.69	0.27
Transit		<0.7500	9.64	9.49	9.57	9.25	−0.17	10.48	10.70	−0.37	8.49	8.78	0.45
		0.7500–0.8999	10.20	9.95	10.56	9.57	−0.02	10.03	10.10	0.02	8.83	8.64	0.06
		0.9000–1.0999	11.02	11.60	11.02	11.96	−0.35	10.25	12.05	0.32	9.36	10.00	0.06
		1.1000–1.4999	15.97	14.44	16.83	14.76	0.53	13.47	12.72	0.78	10.93	11.28	1.88
		≥1.5000	17.97	18.62	18.73	19.21	0.17	12.89	18.46	4.93	11.68	12.72	0.40

Vancouver

All modes	<0.7500	(n = 35)	8.56	9.49	8.73	9.70	−0.04	10.14	10.86	0.22	6.76	7.91	0.20
	0.7500–0.8999	(n = 46)	10.18	10.26	10.46	10.54	0.00	10.69	10.73	0.04	8.40	8.62	0.13
	0.9000–1.0999	(n = 91)	11.52	11.77	11.68	11.96	−0.02	12.36	12.29	0.32	9.44	10.10	0.41
	1.1000–1.4999	(n = 93)	14.54	15.37	15.02	15.64	0.21	14.60	15.58	−0.15	11.57	13.29	0.89
	≥1.5000	(n = 33)	14.07	14.60	14.42	14.66	0.28	14.26	15.47	−0.68	11.54	13.00	0.94
Private vehicle	<0.7500		10.38	11.31	10.46	11.22	−0.01	11.41	11.95	0.39	8.68	10.20	0.59
	0.7500–0.8999		11.02	11.17	11.42	11.40	0.17	11.28	11.35	0.08	9.28	9.49	0.06
	0.9000–1.0999		12.38	12.41	12.50	12.54	−0.01	13.09	12.79	0.33	10.56	11.00	0.42
	1.1000–1.4999		14.98	15.71	15.32	15.90	0.15	14.87	15.92	−0.10	12.48	13.92	0.71
	≥1.5000		14.23	14.76	14.52	14.80	0.25	14.66	15.72	−0.53	12.22	13.45	0.70
Transit	<0.7500		8.72	9.06	8.86	9.51	−0.31	9.36	10.16	−0.45	7.49	7.89	0.06
	0.7500–0.8999		9.34	8.68	9.14	8.46	−0.02	9.29	7.87	−0.77	8.00	8.19	0.85
	0.9000–1.0999		10.34	9.94	10.59	9.94	0.24	9.74	9.52	0.18	8.76	8.53	0.17
	1.1000–1.4999		13.69	14.64	14.55	15.27	0.24	12.74	14.16	0.47	10.89	11.50	−0.34
	≥1.5000		14.08	14.09	15.84	13.63	1.29	12.94	16.41	−1.17	8.91	10.33	1.42

Source: Calculated using Statistics Canada Census data for 1996 and 2006.

The restructuring of the city raises the question of how commuting distances have changed over time (Table 6.4).[2] Professionals tend to commute longer distances, but as the residential location of professionals becomes concentrated in the inner city (where many professional jobs are also concentrated), one would expect that the commute distance of this group would decline over time. As more industrial and manufacturing activities are located in suburban locations further from the downtown, one might expect industrial workers and service and sales workers to travel further to work, on average.

While we found some stability in the relative commute distances of workers within tracts of varying income levels and among occupations (with the average commute distances generally higher among tracts with above-average median incomes), the average commute distance generally increases as one moves from the inner city towards the urban fringe in both years. The small changes in the average commute distances that have characterized different occupational groups have not altered the general pattern, in which professional workers tend to have the longest average commutes, followed by industrial workers, and then service and sales workers. But despite the small magnitudes of change, there are some clear patterns. First, the average commute distance of service and sales workers declines in absolute terms, and in virtually all cases, in relation to the average for all commuters and income levels. In contrast, in most cases professionals have longer average commutes, and the distance travelled more often deviates from the average for all commuters. The exception is middle- and high-income tracts in Toronto, where commutes among automobile drivers converge with the average.

Meanwhile, there are divergent trends in the average commute distances of industrial workers who live in tracts with lower- and higher-income levels. With only minor exceptions, industrial workers in lower-income tracts in particular tend to have longer-than-average commutes, even when changes over the study period have led to shorter relative commutes among other workers. Even though commuters who live in low-income neighbourhoods tend to have shorter commutes, industrial workers who live in low-income tracts often have fewer housing opportunities, particularly in those parts of the metropolis where industrial employment is growing. By contrast, in Montréal and Toronto industrial workers in tracts with higher median household incomes (at least 110 per cent of the CMA median) tend to have lower commute distances than other workers in the same neighbourhoods. Furthermore, changes in the average commute distance of these industrial workers led to shorter relative commutes over time as manufacturing workers who live in higher-income suburban areas are located relatively close to those places near the fringe where manufacturing jobs have been increasing. Metropolitan restructuring in the context of the auto-mobile city has introduced new divergent realities among production workers, articulated in greater inequality in their capacities to be mobile.

In the auto-city, mobility is largely determined in relation to access to private vehicles. However, even among lower-income neighbourhoods, commutes by private vehicle tend to be further than commutes by public transit.

Since low-income commuters use transit more often than higher-income commuters, this finding demonstrates that regional and suburban transit systems are not adequately serving commuters in low-income areas. Even though transit is a more affordable mode than a private vehicle, patterns of development have isolated many employment locations from where they might be reached in a reasonable time by transit. These findings are congruent with studies in US cities showing that many suburban jobs are accessible from low-income neighbourhoods by private vehicle, but not by transit (Wachs and Taylor 1998; Ong and Blumenberg 1998). In an urban landscape that supports travel by private vehicle, all else being equal, the car allows access to a much wider range of employment opportunities.

Modelling commuting patterns under metropolitan restructuring

Regression analysis allows for explicit determination of the relative importance of different factors shaping commute patterns. We model commute mode and distance as functions of workers' occupation, income, immigration status, presence of children, education, gender, household type, and value of dwelling (on a per room basis) in the Toronto, Montréal, and Vancouver CMAs. Occupations are grouped into four categories: high-order services, intermediary services, low-order services, and manual work.[3] Adapting a previous model (see Moos and Skaburskis 2010; Moos 2012), we include data for two census years, 1996 and 2006 – representing a period during which substantial changes occurred in Canadian cities in terms of downtown revitalization, rising property values, and continuing suburbanization. Various methods appropriate to the analysis of data arranged in this way are employed, with interaction terms created by multiplying the variable indicating census year with other explanatory variables, enabling us to estimate how the effect of each explanatory variable changes over time (Vandersmissen *et al.* 2003).[4]

As might be expected, the analysis points to a great deal of complexity in the explanatory variables of commuting patterns. There are four key elements of metropolitan restructuring reflected in our analysis (Tables 6.5 and 6.6). First, there are the shorter commute distances travelled by immigrants, who reveal higher propensities to travel by public transit in comparison with non-immigrants (Heisz and Schellenberg 2004). However, the effect of the interaction term suggests that immigrants in Toronto are actually travelling further than non-immigrants by 2006, and furthermore that their transit use has declined over time, a pattern also visible in Montréal. Immigrants are less likely to walk or bicycle to work in Toronto and Vancouver. The trends speak to the increasingly suburban location patterns of immigrants, at least in these two cities (Hiebert *et al.* 2006), and to changing relationships between accessibility and immigrant location in the auto-mobile city (see Chapter 9 by Hess *et al.* in this volume).

The second notable trend relates to the impact of household restructuring and downtown revitalization on mobility patterns. Downtown revitalization is in part driven by the growth of non-family households, and these tend to have shorter

Table 6.5 Generalized ordered logit estimates – workers' commuting distance

	Toronto coefficients	Montréal coefficients	Vancouver coefficients
Year 2006	-0.142 **	-0.386 ***	-0.181 **
High-order service occupations	0.173 ***	0.161 ***	0.053
High-order service occupations* Year 2006	0.303 ***	0.125 **	0.227 ***
Intermediate service occupations	0.200 ***	0.233 ***	0.155 ***
Intermediate service occupations* Year 2006	0.127 ***	0.007	0.085
Manual workers	0.222 ***	0.239 ***	0.354 ***
Manual workers *Year 2006	0.119 ***	0.176 ***	0.157 ***
Income (2005 dollars, in '000s)	0.005 ***	0.005 ***	0.005 ***
Income (2005 dollars, in '000s)* Year 2006	-0.004 ***	-0.002 ***	-0.002 ***
Immigrant	-0.049 **	-0.304 ***	-0.145 ***
Immigrant* Year 2006	0.052 *	0.037	0.006
Children present	0.109 ***	0.199 ***	0.154 ***
Children present* Year 2006	-0.109 ***	-0.161 ***	-0.158 ***
High school diploma	0.229 ***	0.154 ***	0.209 ***
High school diploma* Year 2006	0.161 ***	0.138 **	0.086
Trades/Non-Univ. Cert.	0.411 ***	0.238 ***	0.344 ***
Trades/Non-Univ. Cert.* Year 2006	0.225 ***	0.266 ***	0.087
University degree	0.290 ***	0.117 ***	0.084 *
University degree* Year 2006	0.229 ***	0.298 ***	0.236 ***
Female	-0.304 ***	-0.258 ***	-0.235 ***

Female* Year 2006	0.007	0.040	0.023
Non-family household	−0.331 ***	−0.259 ***	−0.302 ***
Non-family household* Year 2006	−0.133 ***	−0.173 ***	−0.215 ***
Dwelling value/room (2005 dollars, in '000s)	−0.006 ***	−0.023 ***	−0.007 ***
Dwelling value/room (2005 dollars, in '000s)*Year2006	0.002 ***	0.011 ***	0.003 ***
Constant – Commuting distance: <5 km	0.799 ***	1.049 ***	0.724 ***
Constant – Commuting distance: 5–9.9 km	−0.236 ***	−0.026	−0.405 ***
Constant – Commuting distance: 10–14.9 km	−0.938 ***	−0.796 ***	−1.126 ***
Constant – Commuting distance: 15–19.9 km	−1.516 ***	−1.458 ***	−1.794 ***
Constant – Commuting distance: 20–24.9 km	−2.037 ***	−2.097 ***	−2.418 ***
Constant – Commuting distance: 25–29.9 km	−2.496 ***	−2.631 ***	−2.999 ***
Log likelihood	−210,626.240	−146,441.580	−78,874.875
Wald chi2(25)	5,908.930 ***	5,760.840 ***	2,250.720 ***
Number of observations	119,797	88,112	48,082

Source: Calculated using Statistics Canada census data (1996, 2006).

Notes

Base for the models are: year 1996, low-order service workers, non-immigrants, no children present, less than high school diploma or certificate, males, in family households.

*** $p < 0.0001$, ** $p < 0.01$, * $p < 0.05$.

Table 6.6 Multinomial logistic regressions – workers' commute mode

	Toronto			Montréal			Vancouver		
	Public transit	Walking	Bicycle	Public transit	Walking	Bicycle	Public transit	Walking	Bicycle
Year 2006	0.491 ***	0.570 ***	0.575 **	0.372 ***	0.434 ***	0.628 **	0.727 ***	0.996 ***	0.512
High-order service workers	-0.339 ***	-0.067	0.038	-0.306 ***	-0.234 ***	-0.460 **	-0.408 ***	-0.279 **	-0.398 **
High-order service workers* Year 2006	-0.117 **	-0.418 ***	-0.187	-0.157 **	-0.111	0.295	-0.001	0.069	0.447 *
Intermediate service workers	-0.289 ***	-0.273 ***	0.146	-0.260 ***	-0.336 ***	0.084	-0.244 ***	-0.190 *	-0.013
Intermediate service workers* Year 2006	-0.090 *	-0.063	-0.108	-0.108 *	0.047	0.056	0.053	0.081	0.233
Manual worker	-0.603 ***	-1.109 ***	-0.159	-0.504 ***	-0.885 ***	-0.062	-0.695 ***	-0.933 ***	-0.453 **
Manual worker* Year 2006	-0.307 ***	-0.519 ***	-0.391 *	-0.427 ***	-0.370 ***	-0.705 ***	-0.184 *	-0.780 ***	0.028
Total income	-0.008 ***	-0.019 ***	-0.019 ***	-0.017 ***	-0.022 ***	-0.039 ***	-0.016 ***	-0.015 ***	-0.015 ***
Total income* Year 2006	0.005 ***	0.010 ***	0.013 ***	0.006 ***	0.004 *	0.016 ***	0.004 **	0.000	0.010 ***
Immigrant	0.557 ***	-0.169 ***	-0.733 ***	0.887 ***	0.208 ***	-0.609 ***	0.278 ***	-0.220 ***	-0.810 ***
Immigrant* Year 2006	-0.142 ***	-0.098	-0.066	-0.163 ***	-0.109	0.057	-0.010	0.103	-0.111
Children present	-0.459 ***	-0.363 ***	-0.670 ***	-0.459 ***	-0.273 ***	-0.746 ***	-0.321 ***	-0.388 ***	-0.170
Children present* Year 2006	0.272 ***	0.047	0.364 *	0.332 ***	0.070	0.613 ***	0.046	-0.112	-0.226
High school diploma	-0.002	-0.240 ***	0.244	-0.009	-0.249 ***	0.203	-0.063	-0.296 ***	0.215
High school diploma* Year 2006	-0.019	-0.207 **	-0.452 *	-0.063	-0.144	-0.683 ***	-0.045	-0.116	-0.649 **

Trades/Non-Univ. Cert.	-0.185 ***	-0.670 ***	-0.259	-0.100 **	-0.514 ***	0.079	-0.129 *	-0.453 ***	-0.082
Trades/Non-Univ. Cert.* Year 2006	-0.103 *	-0.279 **	-0.569 **	-0.059	-0.166	-0.786 ***	-0.050	-0.130	-0.440
University degree	0.184 ***	-0.212 ***	0.650 ***	0.113 **	-0.247 ***	0.774 ***	0.033	-0.228 *	1.085 ***
University degree* Year 2006	-0.039	-0.215 **	-0.502 **	0.209 ***	-0.014	-0.730 ***	0.094	0.000	-0.708 **
Female	0.621 ***	0.360 ***	-0.841 ***	0.564 ***	0.293 ***	-0.826 ***	0.374 ***	0.349 ***	-0.837 ***
Female* Year 2006	-0.151 ***	-0.131 *	0.145	-0.222 ***	-0.067	0.030	-0.143 **	-0.225 **	0.067
Non-family household	0.540 ***	0.692 ***	0.895 ***	0.488 ***	0.580 ***	0.728 ***	0.545 ***	0.743 ***	0.687 ***
Non-family household* Year 2006	0.208 ***	0.244 **	-0.120	0.172 ***	0.118	0.001	0.031	0.087	-0.168
Dwelling value per room	0.014 ***	0.017 ***	0.012 ***	0.020 ***	0.025 ***	0.006	0.010 ***	0.016 ***	0.009 ***
Dwelling value per room* Year 2006	-0.012 ***	-0.010 ***	-0.007 **	-0.015 ***	-0.014 ***	-0.003	-0.011 ***	-0.012 ***	-0.008 ***
Constant	-1.628 ***	-2.318 ***	-3.877 ***	-1.375 ***	-2.142 ***	-2.897 ***	-1.547 ***	-2.495 ***	-3.498 ***
LR chi2(75)	-95,477.579			-71,328.849			-37,723.031		
Log likelihood	14,952.170 ***			12,932.470 ***			5,880.950 ***		
Number of observations	132,633			95,443			55,041		

Source: Calculated using Statistics Canada census data (1996, 2006).

Notes

Base for the models are: year 1996, low-order service workers, non-immigrants, no children present, less than high school diploma or certificate, males, in family households, commuting by car.

*** p < 0.0001, ** p < 0.01, * p < 0.05.

commute distances and are less likely to travel by automobile in all three metropolitan areas. In a context of rising land values (Skaburskis and Moos 2008; Walks 2012), non-family households are more likely than larger households to reside in the ever smaller apartments being built in the high-density areas that are generally better served by transit. The variable measuring dwelling value per room also demonstrates that residence in more valuable properties is associated with shorter commutes and higher propensity to travel by transit, walking, or cycling. However, this effect is reduced in 2006, following from the continued deindustrialization of the city – forcing remaining blue-collar workers to commute to the suburbs – and gentrification – which brings in more affluent households and their cars (and greater capacity for commuting to professional jobs in the suburbs). The presence of children is associated with longer commutes and a higher likelihood of travelling by automobile, but again this effect is becoming smaller over time as the geography of employment continues to shift and ever wealthier households appropriate the most accessible locations.

The third finding relates to the relationship between income, education, and commuting patterns, with higher-income earners continuing to travel further to work and more likely to travel by automobile. But again, this relationship demonstrates some weakening over time, as alternative modes of travel (walking, cycling) have become popular among some high-income earners in the central cities (Quastel *et al.* 2012; Garrett and Taylor 1999; Plaut 2004). Notably, although university education alone is associated with further travel to work, it does not display the same changing spatio-temporal pattern as high income in that it remains associated with even longer commutes in 2006 than in 1996. This is related to the declining abilities of young people, who are more likely to be educated, to afford housing near places of employment when their household size increases.

Finally, these findings have implications for the restructuring of blue-collar and low-skill work. Low-order service workers are found generally to have the shortest commutes and highest rates of transit use and walking to work, partly because of the proliferation of retail establishments and other places of employment, across the metropolis. Blue-collar workers, on the other hand, have experienced the most dramatic decreases in employment opportunities due to the deindustrialization of North American metropolitan areas and the internationalization of the production process (Bourne *et al.* 2011). These workers are less likely to travel to work by transit or walking in all three metropolitan areas, and are also less likely to bicycle in Vancouver. In terms of residential location, manual workers have become increasingly decentralized as an outcome of downtown revitalization and gentrification (Walks 2001, 2011). This has relegated them to areas where residents are more reliant on the automobile. However, remaining manual jobs have also become more decentralized as downtowns have become more specialized in new economy sectors (Bourne *et al.* 2011). The result is that blue-collar workers travel further to work, and their commuting distances have increased over time.

Conclusions

Post-Fordist processes of metropolitan restructuring are typically associated with an increased presence in inner-city neighbourhoods of residents in managerial and professional occupations, while workers in consumer services and manufacturing increasingly tend to reside in the suburbs. In Canadian cities, the location of employment in producer services, a sector that employs many managers and professionals, remains highly centralized despite an increase in the number of suburban employment centres. The inner city also has mobility characteristics and transportation infrastructure that support greater "motility capital" – capacities to be mobile – and that have increasingly become appropriated by affluent households as the auto-mobile city evolves. Manufacturing employment, however, continues to decentralize into the suburbs, even as deindustrialization has decreased its overall share of metropolitan employment. These processes affect commute patterns. Although higher-status occupations still tend to commute longer distances, and changes over time are slow and incremental, the auto-city restructures the relationship to mobility, creating new forms of mobility-based inequality. These trends are emergent in Canada's global cities.

In gentrifying inner-city neighbourhoods, there has been an increase in the presence of managers and professionals, who benefit from proximity to downtown jobs and services and are able to out-compete other households for housing in a context of increasing real estate values. Among professionals, those who commute by private vehicle commute longer distances on average than those who commute by transit, and commute longer distances from suburban locations on average. The relationship between income and commute distance is in decline. But high education levels are still associated with ever lengthier commutes, and low-income blue-collar workers are compelled to commute farther in the face of deindustrialization. Blue-collar workers are less likely to commute by transit and walking than workers in other occupations, and the distance that these workers travel to their jobs has increased. The auto-city is one that produces such spatial mismatches, even as it allows higher-income workers with access to private vehicles to reduce their commutes.

High levels of road congestion have become a common experience in Canada's largest metropolitan areas, especially during rush hour when a large share of the labour force travels to and from work. The automobile, commonly depicted as a symbol of Fordism, has not declined in importance in the post-Fordist, neoliberal city. However, its social function and meaning are slowly undergoing transformation. Despite the potential peaking of car use (see Chapter 3 by Newman and Kenworthy in this volume), metropolitan commuting patterns continue to be dominated by the car, and in some areas commuting by car and the distances travelled have even increased in recent years. This is a real concern because there are serious environmental, social, and economic implications associated with automobile dependency (as discussed in Chapters 2, 3, and 4 in this volume). While the restructuring of the auto-city has brought about more walking, cycling, and public transit, in part through the work of planners who promote higher densities and mixed-use (Ewing *et al.* 2008), these initiatives have not yet brought about significant aggregate changes in commuting behaviour primarily because the auto-city compels residents to drive. In places where these strategies have been successful, the beneficiaries have been higher-income earners in professional occupations, who are most able to marshal their capacities to be mobile through the accoutrements of automobility or the appropriation of accessible space in the central city.

Notes

1 Census tracts are areal units defined by Statistics Canada to have relatively stable boundaries and populations between 2,000 and 8,000 residents on average. We calculated tract income levels as the ratio of the median household income of the tract to the CMA median household income. Commute distances are the distance from home to place of work, which are grouped in five mutually exclusive ranges: less than 5 km, 5 to 9.99 km, 10 to 14.99 km, 15 to 24.99 km, and 25 km and above. Average commuting distances were estimated by assigning the mid-point distance to the intervals defined by Statistics Canada and multiplying by the count of commuters in each category. Commuting distance data in Canada is recorded for a household's primary maintainer only, and does not include households where no one commutes from home to work, commuters who work outside Canada, and those who do not work at a fixed address.

2 The relative change column of Tables 6.3 and 6.4 shows the change in the difference between the average commute per household and the average commute for each occupational category in tracts of an urban zone or a given income level. A negative value in the relative change column indicates that the average commute distance of an occupational group is converging with the average per household, while a positive relative change indicates that the average commute distance is diverging from it. It is possible for the absolute change in the average commute to decrease while the relative change increases, and vice versa.

3 Occupational definitions based on groupings of Statistics Canada (2009) census data. *High-order service workers* include: senior managers, middle and other managers, and professionals. *Intermediate service workers* include: semi-professionals and technicians, supervisors (clerical, sales, and service), supervisors (crafts and trades), administrative and senior clerical personnel, and skilled sales and services personnel. *Low-order service workers* include: clerical personnel, intermediate sales and service personnel, and other sales and service personnel. *Manual workers* include: skilled crafts and trades workers, semi-skilled manual workers, and other manual workers.

4 While Statistics Canada's census data provide limited information on commuting, they are the only consistent dataset across entire metropolitan areas. Publicly available data on commute distances are made available in categorical form. We therefore estimate commute distance through a generalized ordered logistic model, which measures the likelihood of higher values of an ordered categorical variable. Ordered logistic models for commuting distance were found to violate the proportional odds assumption, so we opted for generalized ordered logistic models in which the independent variables were constrained to meet this assumption (see Williams 2006). We then analyse commute mode using multinomial logistic regression, which measures the likelihood for different kinds of workers of taking public transit, walking, or bicycling to work, compared to driving.

7 Driving mobility, slowing down the poor

Effective speed and unequal motility

Alan Walks and Paul Tranter

It has become a truism that the automobile and automobility have been responsible for significantly enhancing and enlarging the mobility of modern societies. Underlying this truism is the assumption that automobility confers "enormous benefits on many millions of drivers and their passengers" (Taylor 2006: 279). Because of its flexibility and high speed, the car extends the range and number of travel destinations that can be accessed, and widens the potential activity spaces pertaining to individuals and households (Farber and Paez 2011). Availability of an automobile is seemingly a key factor augmenting access to nutritious food and other retail amenities (Covenay and O'Dwyer 2009; Paez *et al.* 2009), broadening the extent of social networks and new social connections (Urry 2007, 2012), enhancing exposure to varied arts and leisure activities (Schlich *et al.* 2004), providing parents with greater choice for their children's schools (Fyhri *et al.* 2011), extending the range of out-of-town vacation possibilities, and perhaps most importantly, considerably facilitating accessibility to jobs and expanding the spatial reach of labour markets (Kawabata and Shen 2007). Thus, the automobile not only facilitates increased travel, but appears to enlarge the capacities for mobility. The automobile significantly enhances what Kaufmann (2002) terms "motility" – the capacity to be mobile (see also Urry 2007).

Of course, automobiles, car drivers, and destinations do not exist in isolation or on some abstract limitless plain, but within a social, cultural, political, economic, and physical context characterized by many drivers, cars, destinations, established routes of varying capacities, other forms of transport and their users, and land uses of varying density, functional mix, and levels of spatial concentration. The city, as a complex non-linear multi-dimensional system, is constantly and reflexively evolving in response to changes in all these parameters, while simultaneously influencing changes in their intensity and direction. As the city grows in a dispersed fashion and travel becomes ever more mono-modal, the benefits of automobile-based mobility are eroded and transformed. What once conferred special advantages and options becomes a necessity that constrains choice. In the modern dispersed city, many are "driven to drive" by long trip distances, the lack of alternatives, and the political and cultural "lock in" created by the weight of so many interests vested in the maintenance of automobility (Sheller and Urry 2000; Soron 2009). As noted by Urry (2004: 28), automobility often compels people to restructure their daily activities around the car in order to deal with the time–space constraints that it generates.

The form of mobility afforded by driving enlarges not only the real activity spaces but also the perceived activity sets (PAS) of drivers and their families (Farber and Paez 2011; Le Vine *et al.* 2013). However, paradoxically the increasing distances and time spent driving between activities reduces the amount of time available for socializing outside the family and for other non-work activities (Farber and Paez 2011; Tranter 2010). Automobility thus impacts the overall social interaction potential (SIP) inherent in the modern metropolis, albeit with differential effects among different social groups (Farber and Paez 2009; Farber *et al.* 2013). Long commutes reduce neighbourhood social satisfaction (Delmelle *et al.* 2013) and have negative implications for health, including those related to lack of exercise, unhealthy meals, and higher stress levels (Tranter 2010). While accessibility to various destinations is theoretically enhanced, accessibility to many activities in practice is often much lower than expected because the time available for travel is limited. It is partly for this reason that free-flowing fast-moving routes (highways and rail) become seen as a necessity, as higher trip speeds and faster commuting modes are demanded as a way to offset longer travel distances (Ma and Kang 2011). This is particularly true if it is not feasible for households to reorient residential locations to be closer to work – for instance, when jobs remain highly concentrated (downtown, etc.) leading to high land values among nearby residential neighbourhoods (Levinson and Wu 2005). In this way, the dispersed modern metropolis can be considered a machine compelling ever greater movement and velocity.

Despite criticisms of the coercive and constrained nature of mobility facilitated by driving (Freund and Martin 1993; see also Banister 2011), the assumption that automobility has provided for a greater quantity of mobility, exemplified by more distant travel and ever faster speeds, is rarely questioned. If measured only via in-vehicle trip velocities, such an assumption is often valid (except when an excess of congestion reduces mobility), and current trends can be described as promoting hyper-mobility (Freund and Martin 2007; see also Chapter 2 in this volume). However, it is not clear that it is only the trip time spent in-vehicle that should count when considering capacities for mobility. Indeed, there are a host of practices and activities, including those involved in the production, purchase, and maintenance of various modes of transport, that are necessary before a traveller can even take a trip or access a vehicle. Any analysis of mobility and motility needs to take into account the range of such activities for a more complete understanding of the underlying capacity for mobility.

The concept of effective speed incorporates such out-of-vehicle practices and activities. In doing so, it provides a holistic perspective on the capacities for mobility that are afforded by different modes. Seen through the lens of effective speed, a different story emerges concerning the reality of mobility in the modern metropolis than typically understood. Among other things, when mobility is defined in terms of effective speed, commuting via the automobile – and by implication the whole system of automobility – is shown to confer no greater levels of mobility than that provided by cycling, for example. Automobility also affects the distribution of mobility benefits. This chapter first develops the concepts of mobility and motility in relation to the auto-mobile city. It outlines the concept of effective speed, and discusses its parameters and application to contemporary cities across the developed and developing world. It then compares effective speeds among different transport modes, using Canadian metropolitan areas as case studies, and develops from this a metric for measuring effective motility. The analysis is then extended to low-income households, and the way that inequalities are articulated in the realm of mobility. The chapter ends by interrogating the implications of this analysis for understanding the meaning and importance of mobility in the auto-city.

Mobility and motility

Spatial mobility is a multi-faceted concept that implies not only physical and geographic, but also social and metaphorical, forms of movement (Kaufmann 2002; Urry 2007). There are at least four distinct forms of mobility examined in the literature, defined by their length of duration and proximity to one's place of residence (Kaufmann 2002). Mobilities of long duration include migration (between regions, nations) and moving house within the same city (residential mobility). Travel for business, visiting family, and tourism is another form of mobility, one that is increasing as both transportation and communications technologies facilitate cheaper travel (Urry 2007). Daily mobility – particularly the commute to work – is the most common form of movement and the most studied form of mobility. Commuting has been a core topic for analysts of the geography of cities and urban change. Effective mobility in this case refers to the actualized capacities for daily movement both to and from work, and elsewhere within the city. The measuring of mobility has involved the "trilogy of distance, speed, and time" (Banister 2011). In its geographic sense, the clear implication is that more travel equates with greater mobility. However, scholars have mostly focused on trip times when assessing the social benefits and disadvantages of different kinds of mobility (see Hine 2011; Farber *et al.* 2013). Representing the amount of distance for any given unit of time, speed incorporates both distance and time and thus provides a unifying concept for measuring mobility.

Motility is a concept related to yet distinct from mobility. Motility is defined as "the capacity of a person to be mobile … the way in which an individual appropriates what is possible in the domain of mobility and puts this potential to use for his or her activities" (Kaufmann 2002: 37). Motility involves not just actualized mobility, but the propensity for mobility as determined through the interaction of three particular elements: (1) the possible choices available for mobility and access to particular transport and social networks facilitating mobility, including those encouraged by prevailing settlement patterns and urban forms; (2) skills of those involved in training, attainment of certification, understanding rules and signage, and organizational skills involved in navigating routes and other information, not to mention the physical skills involved in movement (ability to walk, bike, drive, etc.); and (3) appropriation of information, skills, and resources in light of various life aspirations (of where to work, live, visit, etc.) and what kinds of mobility are appropriate in different circumstances for different kinds of individuals, given prevailing attitudes, standards, and norms (ibid.: 38–39). Motility is thus a broader term, and is dependent on the choices available within a given social, economic, cultural, and political context. Although thus far posited as a property of individuals, the effective motility provided by the mix of transport options, social skill levels, concentration of local infrastructure, and settlement patterns means the concept can also be applied to different communities, social groups, and cities.

As noted in a number of chapters in this volume, the low-density auto-city encouraged decentralization and deconcentration of destinations – places of work, shopping, institutions, education – and allowed wealthier households who could afford a car to live far from places of work. The contemporary dispersed city compels a majority of workers to be mobile, and to commute by automobile. The compulsion to drive that characterizes the auto-city thus brings into question the assumption of the link between automobility and enhanced mobility, in quality if not quantity. While the distances travelled are more extensive, the level of motility may be imperilled, both because of lack of choices, as well as contradictions between the ideology of autonomous individuality and the actual experience of driving in congested, regimented traffic (Sheller and Urry 2000; Sheller 2004; Soron 2009). At the macro scale, the freedom and independence assumed to be provided by the automobile often ends up being illusory: as noted in the introduction, the mobility of car drivers is not free but entirely dependent on goods and services provided by a multitude of others, including the manufacturing of vehicles, the building of roads, repairs and other services, traffic management and policing, the gasoline and other fuel industries, as well as a massive energy subsidy (Ker and Tranter 2003).

The collective impact of automobility could even be reducing individuals' access and autonomy through a series of feedback impacts. As car ownership and use increases, distances to shops, schools, and services also increase (Farber and Paez 2011). Data from Melbourne, Australia, indicate that between 1951 and 2005, the number of land uses accessible within 800 m of an individual's home declined dramatically. In 1951 over 70 per cent houses were within 800 m of five or more land uses. By 2005 only 40 per cent had this level of accessibility (Kelly *et al.* 2012). The supposed freedom that automobility provides is thus eroded over time by the adaption of urban land uses to automobility. Another important factor that is largely overlooked by both drivers and policy makers is the considerable time costs involved in owning and operating automobiles. These costs are taken into account in the holistic concept of effective speed.

Effective speed

The concept of effective speed is relatively simple to grasp and its lineage can be traced back to the musings of Henry David Thoreau in *Walden*, first published in

1854, regarding the benefits of walking over train travel, given the additional time that would need to be spent working to pay for the train ticket (see Tranter 2011). Illich (1974: 18–19) adopted the same reasoning in arguing that once all the time spent in traffic courts, hospitals, garages, watching car commercials, and, of course, working to pay the monthly loan instalments are taken into account, the "model American" (driver) "puts in 1,600 hours to get 7,500 miles: less than five miles per hour". Effective speed is a holistic measure, in that it takes into account the real monetary and time costs involved in the reproduction and performance of mobility. It can be calculated for any mode of transport, or any combination of modes. Furthermore, it can be applied at any spatial scale – individual, household, neighbourhood, even whole metropolitan areas and nations. Since it can be calculated separately by transport modes, it can be used to compare the real underlying mobility benefits associated with different modes of travel. When the combined effective speed of all modes of transport is analysed for a given collective, it provides one representation of the overall level of mobility associated with that place. Effective speeds can also be calculated and compared among different social groups, and can be used as a tool for policy analysis.

Effective speed is also a concept with radical social and political potential, and its use constitutes an example of what Wyly (2009, 2011) calls strategic positivism. Unlike most economic approaches, which seek to express all forms of value in terms of unbounded nominal money units, under effective speed all costs are converted to units of time. In contrast to money, which is unequally distributed, unbounded at the top end, and for which it is claimed there is no technical or physical limit (albeit perhaps a socio-political one), the amount of time available in each day is limited, constant, and is distributed exactly equally. Time is the great equalizer (Adam 2004). As a concept, the intent and philosophical underpinnings of effective speed dovetail with Marx's theory of value, in which the true units of value involve socially necessary labour time (Marx 1972). Monetary units make comparisons difficult between currencies, between different eras, and even between different occupations. Unlike money, which Marx persuasively argued is fetishized and used to justify and mask inequalities in underlying power differentials, time is a concept that is instantly recognizable and understood, and is directly rooted in universal human experience. This is one reason for the existence of time-travel budgets, in which individuals and households resist extending the amount of time spent commuting (Marchetti 1994, discussed in Chapter 2 in this volume). Much of the goal of transportation policy, including massive highway programmes, put in place over the last 50 years primarily concerns the reduction of trip times (given fixed distances), instead of the shortening of people's travel distances. This is precisely why speed is so central to improving the value of mobility through space–time compression (Harvey 1989), and why it is under capitalist labour relations (in which time is often falsely equated with money) that "faster is better" (Adam 2004).

Although the concept is quite simple, the data required for an accurate assessment of effective speed is often difficult to attain. In addition to trip distances and times, data is required for all expenditures related to mobility, including costs related to vehicle purchase, maintenance, fuel, parts, parking, insurance, licence fees, as well as transit and taxi fares. To convert such costs into units of labour time, data for hourly wages is necessary. Even this information can only get at the direct out-of-pocket costs associated with different mobilities. In the spirit of Illich's (1974) discussion of the real costs of energy use, a holistic accounting of mobility also needs to take into account the indirect/external costs related to effects on the health care system and other health costs/benefits, environmental resource externalities and pollution, road and transit subsidies, and other costs associated with producing and maintaining the entire transport system in its various elements. When the totality of such direct and indirect/ external costs are factored into the equation, the resulting metric is considered an indicator of the "social" effective speed (Seifried, cited in Whitelegg 1993; Tranter and Ker 2007). Of course, data on such external costs is even more diffi-cult to ascertain.

While analysis of effective speed as an indicator of mobility has been growing over time, its application has been limited, in part due to these data limitations. Seifried (cited in Whitelegg 1993) applied such an analysis to German cities using data for the 1980s, finding that when both direct and indirect (social and environmental) costs are factored into the analysis, the social effective speed of bicycling could be higher (provide more effective mobility) than a car. Kifer (2002) examined the effective speed for selected cities in the United States (US) (without factoring in indirect costs of a social/environmental nature), estimating the direct effective speed of all US motorists at roughly 9.7 miles/hour. Macer (2006: 75) similarly examined the direct effective speed of drivers in two Japa-nese cities, concluding that once the time spent earning money to pay for a car is considered, the effective speed is roughly "equivalent to vigorous bicycling speed". More systematic analysis of Australian cities (Tranter 2004; Tranter and Ker 2007) and, more recently, international cities (Tranter 2012) has included estimates of some indirect/external costs, finding that the social effective speeds related to driving vary considerably among cities.

The calculation of social effective speed relies on a series of disparate data sources, as evidenced by this example from Vancouver, Canada (Table 7.1). In the Canadian case, quality data for average expenditures for virtually all costs related to vehicle purchase, operation, licensing, and maintenance is available at the level of the household (only) from the Survey of Household Spending (SHS) and made available by Environics Analytics (2012b), necessitating analysis at the level of households. These expenditures are then compared to custom census data estimates of the average hourly household incomes of households commut-ing via each different transport mode, as well as custom data regarding commute distances by mode.[1] Average trip speeds by each mode of transport are here modelled using data from the 2006 census of Canada and a popular online route-mapping tool (see Tranter 2012 for additional methodological details). Because the data for expenditures in the SHS pertain to the average for all vehicles used by Canadian households, they reflect the real expenditures of the average household related to owning/leasing, operating, and maintaining the average car in the city, making it unnecessary to calculate the effective speeds of any specific makes or models of automobile here. However, it should be noted that the social effective speeds of different makes of vehicles can vary dramatically (Tranter 2004, 2010).

Most difficult to ascertain are the indirect environmental and social costs of driving and other modes of transport. Litman and the Victoria Transport Policy Institute (VTPI) have estimated per-mile and per-vehicle costs associated with a host of such indirect but nonetheless real expenditures for urban areas across the US (2009). Some of these costs are not expected to vary among cities, or even nations, such as pollution per kilometre and resource externalities, and thus can be assumed to apply in similar urban contexts in Canadian cities (but must be made applicable to these contexts, for instance by converting to local currencies and distance metrics, e.g. CDN$/km). Other costs have a highly uneven expression across space, related to land-use mix, population density, barrier effects related to the extension of non-motorized modes of transportation, and transport diversity, among other things. In the absence of separate spatially dependent measures for such latter costs in each city, the local extent to which public transit must be subsidized is taken to provide a good proxy, under the assumption that more sprawled cities require greater per-passenger transit subsidies (Pucher 1981; Newman and Kenworthy 1999). The combined transit fare/subsidy is, for the sake of simplicity, here assumed to cover all transit maintenance and operating costs, with the subsidy paid through local taxes.

Calculated in this way, it is households whose members are able to catch a ride as a passenger in other automobiles that reveal, on average, the greatest effective speeds in Vancouver (mainly because they are assumed to not bear any of the direct costs), followed by households who depend on public transit to get to work (Table 7.1). Notably, the total average weighted social effective speed of all commuting households (averaged across all modes) in Vancouver, at 10.66 km/hour, is almost identical to the social effective speed of cycling commuters (10.98 km/hour). And once direct and indirect costs are taken into account, owning a car and driving are shown to provide even less effective speed (10.09 km/hour) in comparison with cycling, despite the fact that auto-driving households have the highest incomes and can thus best afford the costs associated with owning a car (which is taken into account in this analysis). Driving one's own automobile likewise fares poorly in comparison with public transit in Vancouver, regardless of whether any indirect "social" costs and transit subsidies are included in the analysis.

Driving effective speeds internationally

While data availability is variable and uneven, it is possible to compare the effective speeds for automobile drivers in cities of both the developed and developing world (Table 7.2). Due to the differences and limitations in data availability for cities in different nations, a simple comparative international

Table 7.1 Effective speed by transport mode, Vancouver Census Metropolitan Area 2006

	Auto driver			Auto passenger			Public transit			Bicycle			Walk		
	Km	Hours	C$	Km	Hours	C$	Km	Hours	C$	Km	Hours	C$	Km	Hours	C$
Number of commuting households (#)	426,240			16,980			78,270			9,750			33,195		
Distance and speed of commuting households:															
Average distance commuted per commuting day[1]	12.93			10.31			10.13			4.93			2.50		
Average trip commute speed (km per hour)[2]	43.00			39.67			24.14			14.00			4.90		
Average trip distance commuted per household per year (assuming 240 commuting days per year)[1]	6,205.01			4,949.26			4,863.78			2,356.15			1,200.94		
Average annual commuting hours in motion		144.30			124.77			201.51			169.01			245.09	
Walk from parking/transit station to work = 5 mins/day * 240 days		20.00			20.00			20.00			20.00				
Gas fill/check tyres/windshield (2 min/day) + wash car (30 min/month = 1 min/day) = 3 min/day * 240 days		12.00													
Total annual time spent on commute trip		176.30			144.77			221.51			189.01			245.09	
Costs of transport ownership and operation:															
Average hourly income (annual household income ÷ 50 weeks ÷ average hours worked per week)[1]			45.47			32.92			27.27			31.40			28.16
Average annual direct costs per household:[3]															
Vehicle purchase (auto/truck/motorcycle, bicycle)		140.48	6,387.42		6.16	202.87		131.17	3,576.80		6.32	198.55			
Vehicle leasing/renting (driver), and taxi fare (driver and passenger), or public transit fares (transit)		33.72	1,533.29												
Insurance premiums		54.29	2,468.66												
Driving lessons, licence and registration costs, indirect insurance, fines, tolls, and other services		15.89	722.61												
Parking, garage rent, storage		5.53	251.24												
Vehicle accessories		2.00	91.07								0.60	18.87			
Vehicle maintenance and repair		29.05	1,321.08								3.45	108.45			
Tyres, batteries, other automobile parts		10.24	465.41								0.84	26.32			
Gasoline, diesel, and other fuels		93.65	4,258.01												
Total annual direct costs per household		384.85	17,498.79		6.16	202.87		131.17	3,576.80		11.22	352.19		0	0
Average indirect/external ("social") costs per household (C$):[4]															
Injury, death, health costs/benefits (per km C$3.0858 auto, $0.0049 tran, $-0.056 bk, $-0.096 wk)		11.70	532.08		16.16	532.08		0.89	24.18		-0.42	-13.23		-4.08	-114.92
Road facilities, maintenance, traffic services (per km C$0.0286 drive, $0.00704 tran $0 0025 bk/wk)		3.90	177.36					1.26	34.25		0.19	5.88		0.11	2.98

	(1)	(2)	(3)	(4)	(5)	(6)	(7)	(8)	(9)	(10)
Air pollution/greenhouse-gas emissions (ghg) (per km C$0.0503 drive, $0.00288 tran, $0 bk/wk)	312.31	6.87			140.53	5.15				
Congestion costs (per km C$0.0808 drive, $0.0279 transit, $0.0062 bike, $0.00186 walk)	501.23	11.02			136.00	4.99	14.70	0.47	2.24	0.08
Resource externalities (C$0.0285 drive, $0.0240 transit, $0 bike/walk)	177.36	3.90			116.86	4.29				
Water pollution (per km C$0.0087 drive, $0.0014 transit, $0 bike/walk)	53.98	1.19			7.05	0.26				
Average public transit subsidy per household (paid through taxes, rate based on hourly wage)[5]	685.61	15.08	496.43	15.08	411.18	15.08	473.47	15.08	424.58	15.08
Total annual indirect ("social") costs per household	2,439.92	53.66	1,028.50	31.24	870.06	31.91	480.82	15.31	314.88	11.18
Total annual direct and indirect costs per household	19,938.71	438.51	1,231.37	37.40	4,446.86	163.07	833.01	26.53	314.88	11.18
Total annual number of hours required for direct commuting trips/costs (before indirect/social costs)	561.16		150.93		352.67		200.23		245.09	
Total annual number of hours required for all commuting trips/costs (including indirect/social costs)	614.82		182.17		384.58		215.54		256.27	
Direct (only) effective speed of commuting households (in km per hour), before indirect costs	11.06		32.79		13.79		11.82		4.90	
Social effective speed of commuting households (in km per hour), all inclusive	10.09		27.17		12.65		10.98		4.69	

Source: [1] Custom tabulations from the 2006 Census of Canada, provided by Statistics Canada; [2] Calculated and modelled using common route-planning software, according to the method proposed by Rietveld et al. (1999); [3] Calculated (and inflation-adjusted back to 2006) from the Survey of Household Spending, Statistics Canada (Environics Analytics, 2012b); [4] per mile per commuter peak costs from Litman/VTPI (2009), converted to C$/km (US$1/mile = C$1.609344/km); [5] per household transit subsidy costs, calculated from farebox recovery estimates in Hollingworth et al.,/Transportation Association of Canada (2010).

Notes

Public transit subsidy costs, which vary among CMAs, substitute here for the pan-US estimates of land-use impacts, barrier effects, transport diversity effects, and land value effects provided by Litman/VTPI (2009). It is assumed here that the combined fare/subsidy costs for transit cover all maintenance, fuels, and parts related to the servicing of public transit vehicles (and thus, that separate measures for these items are not required). For simplicity, in the determination of how to represent the transit subsidy it has been assumed here that on balance the taxes levied to fund such subsidies (via property, income, and other taxes) are neither progressive nor regressive (i.e. everyone pays according to their means). It is assumed here that auto passengers do not bear the costs associated with auto travel, except for those passengers commuting via taxi, of which there are very few.

Table 7.2 Effective speed estimates for automobiles, selected cities

	Los Angeles	Sydney	Tokyo	New York	London	Singapore	Delhi	Nairobi
Direct/internal and indirect/external costs of motor vehicle:								
Average annual income (most recent year available) (US$)	43,056	42,409	58,296	43,056	30,002	28,368	1,450	1,200
Est. average hourly income (annual inc. ÷ 50 wks ÷ 38 hrs/wk) (US$)	22.66	22.32	30.68	22.66	15.79	14.93	0.76	0.63
Vehicle operating costs, including purchase, fuel, repair, tyres (US$)	6,393	4,402	10,350	6,405	7,111	15,271	1,434	2,958
Parking, storage (US$)	1,135	4,114	920	1,700	6,469	1,574	200	200
Tolls, licensing, fines, other services (US$)	2,865	2,718	1,200	3,600	2,701	2,600	50	50
Total estimated direct costs (US$)	10,392.80	11,233.93	12,470.00	11,705.40	16,280.85	19,445.00	1,684.00	3,208.00
Indirect/external costs ($0.845 per vehicle mile, 240 days/year) (US$)	1,912.31	1,797.74	1,602.63	1,700.90	1,531.15	2,119.43	842.00	1,604.00
Total estimated direct and indirect/external costs (US$)	12,305.11	13,031.67	14,072.63	13,406.30	17,812.00	21,564.43	2,526.00	4,812.00
Hours devoted to direct transport expenses/costs	458.62	503.30	406.43	516.54	1,031.05	1,302.37	2,206.62	5,079.33
Hours devoted to all transport expenses/costs	543.01	583.84	458.66	591.60	1,128.02	1,444.32	3,309.93	7,619.00
Average trip speeds (miles/hour)	26.2	23.9	19.2	22	20	31.3	27.00	29.60
Average annual aggregate trips distance (miles)	13,577.34	13,702.23	13,875.84	13,545.40	13,914.00	13,300.41	13,545.90	13,392.72
Annual hours in motion in vehicle	518.22	573.32	722.70	615.70	695.70	424.93	501.70	452.46
Other time devoted to the car (hrs/year):								
Walking from car to work (5 mins per day* 240 days)	20	20	20	20	20	20	20	20
Wash car (30 mins per month, 1 min average per day)	12	12	12	12	12	12	12	12

Repairs and servicing time (2 hours per 6 months)	4	4	4	4	4	4	4	4
Buy fuel, check tyres, windscreen (14 min per week)	11.7	11.7	11.7	11.7	11.7	11.7	11.7	11.7
Total annual vehicle hours	565.89	620.98	770.37	663.37	743.37	472.60	549.37	500.12
Total hours for transport (incl. direct costs only)	1,024.51	1,124.28	1,176.79	1,179.91	1,774.42	1,774.97	2,755.99	5,579.46
Total hours for transport (direct and indirect/ external costs)	1,108.89	1,204.82	1,229.03	1,254.97	1,871.38	1,916.92	3,859.30	8,119.12
Direct (only) effective speed	13.25	12.19	11.79	11.48	7.84	7.49	4.92	2.40
Social effective speed	12.24	11.37	11.29	10.79	7.44	6.94	3.51	1.65

Sources: For data on average incomes, the following sources were used: for US Cities, the Bureau of Labor Statistics (http://ycharts.com/indicators/sources/bls); for Sydney, the Australian Bureau of Statistics (www.abs.gov.au/ausstats/abs@.nsf/Lookup/6302.0main+features7Nov%202012); for Tokyo, the Statistics Bureau, Japan, (www.tradingeconomics.com/japan/disposable-personal-income); for London, HM Revenue and Customs (www.hmrc.gov.uk/statistics/personal-incomes/tables3-1_3-10.pdf); for Singapore, the Ministry of Manpower, (www.mom.gov.sg/statistics-publications/national-labour-market-information/statistics/Pages/earnings-wages.aspx); for Delhi, International Monetary Fund (www.imf.org/external/pubs/ft/weo/2012/01/weodata/index.aspx); for Nairobi, Central Bureau of Statistics. For data on car-operating costs, the following sources were used: for US cities, the Auto Channel (http://db.theautochannel.com/db/newcarbuyersguide/toc_results.php); for Sydney, the NRMA (formerly National Roads and Motorists Association), (www.mynrma.com.au/motoring/buy-sell/buying-advice/car-operating-costs/class-winners.htm); for Tokyo, The Costs of Buying and Owning a Car (www.supermelf.com/japan/ajetdrivingbook/chap1.html#c); for London, the Automobile Association, UK, (www.theaa.com/resources/Documents/pdf/motoring-advice/running-costs/petrol2012.pdf); for Singapore, One Motoring, (www.onemotoring.com.sg/publish/onemotoring/en/lta_information_guidelines/buy_a_new_vehicle/car_cost.MainPar.30963.File.tmp/Car_Cost_Update.pdf); and for Nairobi, the Automobile Association of Kenya. Figures are in US$ using exchange rates as at July 2013.

Notes

Data on trip speed is based on areas within 15 km from the centre of each city, using averages for six routes from the city centre (three under 5 km and three around 15 km) during peak hours. Indirect/external costs based on estimates provided by Litman/VTPI (2009) of the average car during the morning peak.

methodology to estimate effective speeds (easily applicable to any city) is needed when average direct costs are not available. The information on trip speed in Table 7.2 is in turn based on areas within 15 km of the centre of each city, using averages for six routes from the city centre, three under 5 km and three around 15 km. Speed data are based on current conditions during peak hours, using the same online route-mapping tool used for the Canadian cities. Data on operating costs is sourced from motoring organizations and incomes from official published statistics. The data are based on the operating costs of the "fastest" new cars available in any region (i.e. those with the lowest operating costs). For example, in Australia, the data are based on the Suzuki Alto, which according to the motoring organization NRMA (formerly known as the National Roads and Motorists Association) has the lowest overall operating cost of cars currently sold in Australia.

Table 7.2 indicates that the effective speeds of cars vary hugely from city to city. The effective speed of cars is affected mainly by three variables: the costs of driving in any city (which is dependent on costs of tolls and parking, as well as the running costs of individual cars); average incomes (which influence the time required to earn the money to pay for cars); and the level of traffic congestion (which influences trip speeds). The lowest effective speeds are in low-income nations, whereas the highest effective speeds are in cities with high average incomes, low parking costs, and minimal toll charges. Singapore provides an example of a city where traffic congestion is kept under control through fiscal policies that make operating a car very difficult for people on low incomes. In addition to a congestion tax known as Electronic Road Pricing, registration costs, road taxes, and insurance, Singapore drivers must bid for a Certificate of Entitlement to drive. Vehicle operating costs in Singapore (in US$) are almost two-and-a-half times those in Los Angeles. If other cities emulate the strategies of Singapore, effective speeds are likely to fall in these cities. Even in the cities with the highest effective speeds for automobiles, cycling represents a viable and competitive alternative to driving. In cities with the lowest effective speeds, meanwhile, even walking is evidently a "faster" mode than the car for the average resident.

The implications of the data in Table 7.2 for the future viability of car use are profound. If developed nations are unable to maintain their access to cheap energy, or if concerns about climate change lead to increases in the price of fossil fuels, then the effective speeds of cars in cities throughout the world will likely fall. This means that any investment in car-based transport systems is an increasingly high-risk venture. Given the identification of the concept of "peak car" (Newman and Kenworthy 2011b; Metz 2013) in many cities in developed nations, further investment in active modes of transport may see the effective speeds of cars fall further in comparison with other modes. From an effective speed perspective, such investment would be much better spent expanding public transit and active transportation possibilities.

Effective speeds by mode and the effective place motility of cities

As a holistic measure of mobility, social effective speed encapsulates varied information not only about trip times and distances, but the resources required to support different modes of transportation and their resource intensity, different levels of access to such resources, and the physical, economic, and social/human externalities resulting from both the act of travelling and the way production is organized around each mode. Higher social effective speeds thus represent real higher overall levels of mobility. Extending the analysis in Table 7.1 to all Canadian metropolitan areas, Table 7.3 indicates that in over 70 per cent of cities the bicycle provides greater effective mobility than driving one's own automobile (first seven columns of Table 7.3). Similarly, and mirroring both the findings from Table 7.1 and the literature on the topic from other contexts (Tranter 2012), in most cities the overall weighted social effective speed of all modes taken together is not statistically significantly different from that provided by cycling. Only in those metropolitan areas that have subsequently become linked into the Greater Toronto Area (GTA) via high-speed expressways, and that now partially function largely as commuter suburbs (Barrie, Hamilton, Oshawa), does the overall effective speed, as well as that for driving, surpass that of cycling in any meaningful way, and even then only by 2 to 3 km/hour. Of course, for long-distance commuters in such places, while the (slightly) higher trip speeds provide an advantage, this higher speed comes at the cost of more time spent in cars over longer distances. On average, across all Canadian metropolitan regions, inclusive of the above three areas, the bicycle is "faster" by 1.2 km/hour.

As noted above, the concept of motility goes beyond actual travel to include ideas regarding the potential capacities for mobility. A central assumption concerns access to different modes of transport and the freedom of choice of different households to employ alternative modes if they so desire. Effective motility involves incorporating, among other things, information regarding mode choices available to commuters, information that can be applied at the scale of the entire city. Taking this into account, a simple measure of effective "place" motility can be calculated for cities from the previous analysis by multiplying the overall weighted social effective speed for each metropolitan area by a measure of the level of mix of different commute modes in each place (last two columns of Table 7.3). Those urban regions containing both a high social effective speed and higher levels of transport mix rank highly in terms of effective place motility – they provide not only greater effective mobility (in relation to both the resources available to support travel and those needed to absorb its externalities), but also greater choice in transport modes. Such an exercise works to elevate in relative terms those cities with fewer fast expressways but more transit users and cyclists (including Victoria, Halifax, and Ottawa–Gatineau in the Canadian context), while simultaneously dampening the motility of places well connected by expressways but largely mono-functional, dispersed, and limiting with respect to transport mode (including Barrie,

Table 7.3 Social effective speed (SES), and effective place motility, all households, Canadian Census Metropolitan Areas (CMAs), 2006

CMA	Auto driver	Auto passenger	Public transit	Bicycle	Walk	Weighted total SES (a)	Transport mix (1−H) (b)	Effective motility (a × b)
Abbotsford	11.13	27.06	10.12	10.78	4.73	11.58	0.28	3.24
Barrie	13.95	28.41	11.17	11.57	4.88	14.17	0.32	4.51
Brantford	11.31	22.67	8.05	10.27	4.79	11.40	0.33	3.78
Calgary	9.58	28.02	12.93	11.39	4.72	10.29	0.48	4.94
Edmonton	9.31	27.62	9.58	11.17	4.75	9.73	0.41	3.95
Fredericton	8.16	25.66	6.45	10.94	4.84	8.84	0.41	3.64
Greater Sudbury	9.72	28.52	8.23	11.20	4.88	10.24	0.37	3.82
Guelph	11.08	24.71	8.79	10.29	4.75	11.12	0.40	4.41
Halifax	9.23	26.89	10.62	11.09	4.81	9.88	0.53	5.23
Hamilton	12.96	28.02	15.08	11.29	4.83	13.34	0.40	5.27
Kamloops	8.24	29.69	6.57	11.43	4.84	9.04	0.34	3.03
Kelowna	7.28	25.83	7.47	11.15	4.81	7.83	0.31	2.41
Kingston	9.46	24.75	7.13	11.16	4.80	9.77	0.43	4.24
Kitchener	10.28	23.80	7.08	10.89	4.76	10.40	0.36	3.79
London	9.76	25.72	7.00	10.80	4.84	10.00	0.40	4.03
Moncton	7.90	25.07	5.47	11.02	4.83	8.63	0.41	3.51
Montréal	11.54	25.45	12.02	10.47	4.58	11.62	0.51	5.97
Oshawa	14.11	29.85	18.68	11.50	4.79	14.71	0.36	5.23
Ottawa–Gatineau	10.42	26.85	14.33	10.91	4.56	11.27	0.55	6.19
Peterborough	10.50	29.05	7.50	10.68	4.88	11.11	0.39	4.32
Québec	9.46	25.93	8.68	10.68	4.69	9.47	0.41	3.91
Regina	6.65	20.88	5.32	9.84	4.59	6.99	0.34	2.40

Saguenay	9.39	27.27	5.65	11.67	4.82	9.66	0.25	2.46
Saint John	10.19	25.52	6.89	11.35	4.89	10.47	0.40	4.22
Saskatoon	7.50	25.01	4.88	10.82	4.77	7.91	0.35	2.77
Sherbrooke	9.09	28.39	7.04	10.88	4.81	9.27	0.34	3.11
St Catharines–Niagara	11.06	25.70	7.43	10.98	4.89	11.32	0.32	3.62
St John's	8.04	23.53	6.56	11.56	4.86	8.80	0.40	3.55
Thunder Bay	7.98	26.37	6.34	11.21	4.85	8.44	0.34	2.86
Toronto	12.66	28.51	14.28	10.85	4.62	13.10	0.53	6.95
Trois-Rivières	9.23	27.80	5.74	10.81	4.83	9.30	0.26	2.45
Vancouver	10.09	27.17	12.65	10.98	4.69	10.66	0.50	5.32
Victoria	8.02	25.42	11.32	10.89	4.74	8.63	0.53	4.56
Windsor	9.57	26.03	5.60	11.37	4.89	9.83	0.29	2.87
Winnipeg	8.22	23.80	10.05	10.83	4.82	8.85	0.48	4.22
AVERAGE – All CMAs	9.80	26.31	9.05	10.99	4.78	10.22	0.39	4.01

Source: Effective speeds calculated by the author as in Table 7.1. Transport mix calculated using Census of Canada, 2006.

Notes

Cities with populations over one million are in bold face. Effective motility is calculated as the social effective speed (SES) × the level of transport mix. Transport mix is calculated by subtracting from 1 the Herfindahl index (H) of the concentration of transport among modes. Values of the Herfindahl index range from 0 to 1: if 1 one single mode dominates, while as H approaches 0 there is greater mix and competition among different modes. Subtracting H from 1 flips this relationship, such that 1 represents the maximum level of mix and competition among modes, while 0 represents absolute dominance of only one single mode. The formula for the Herfindahl index is the following (where x = share of transport mode i):

$$H = \sum_{i=1}^{N} x_i^2$$

Oshawa, Abbotsford, and St. Catharines). Using this simple metric, the level of effective place motility of the highest-motility Canadian metropolitan area (Toronto) is almost three times that of the lowest-motility urban region (Regina).

A number of conceptual and policy implications flow from analysis of the effective mobility and motility of cities when measured through social effective speeds. First, any policies aimed at increasing trip speeds for automobiles will have only minor effects on the effective mobility of urban residents. This is because the greater time component for households is not the time spent in the vehicle, but that spent working to earn the money to pay for the various costs of the automobile, as well as building and maintaining the infrastructure on which transport must flow and the costs of the externalities that result. Using the above Vancouver example, even if average trip speeds for cars could somehow be increased by 10 km/hour at no additional cost, this would only result in an increase in the average effective speed of 0.5 km/hour. If public policy is aimed at building new roads, this may even reduce overall effective speeds, as it could mean more time collectively spent on transport as a proportion of the working day, instead of less. Of course, to actually increase trip speeds by any significant amount would require large investments in new expressways, the repurposing of land uses away from residential and employment lands over to transportation uses, and perhaps even the demolition of large areas of existing cities. Not only would such efforts incur substantial costs on their own (including the opportunity costs associated with the removal of productive lands), the amount of pollution from such activities and the traffic it induced, as well as other negative resource and health externalities, would easily nullify any benefits derived from increasing trip speeds. Such a reinvigorated expansionist vision for roadways would have an even greater negative impact on effective place motility, as it would produce a city even more dispersed and automobile dependent, reducing the ability to take advantage of active non-motorized transport modes and compelling more households to travel by automobile while forking over more of their income (time) for this privilege (for increased fuel and repair costs, etc.). Perhaps counter-intuitively, one of the best methods of increasing effective speeds, and thus overall effective mobility, is to limit actual trip speeds, as this reduces not only the per-kilometre fuel and maintenance costs, but significantly ameliorates the negative health and safety impacts of motoring, not to mention the opportunities this might present for retaining local shops, schools, and services (Aarts and van Schagen 2006; Soole *et al.* 2013).

Second, if trip speeds among cyclists could be increased, virtually all of the additional benefit would be experienced in increased effective mobility. This is because for cyclists the main time component of travel involves the trip itself. Furthermore, this could be attained at very little cost – through the implementation of on-street cycling lanes, for instance (see Chapter 13 in this volume for discussion of the political difficulties and opportunities around this). Similarly, the benefits of increased trip speeds for public transit riders would mostly be experienced in increased effective mobility, although the exact amount depends on the costs of the infrastructure and any repurposing of existing land uses that might be required. If more lanes on existing streets could be utilized for buses and light rail vehicles with priority movement, the overall social effective speed of all commuters could be significantly improved for little additional cost.

Third, even disregarding cycling and transit, there is room for improving the effective mobility of those who by necessity must commute via automobile. If auto passengers who, as we noted above, reveal the highest effective speeds could utilize more of the seats in the existing automobile fleet, the effective mobility of the entire transport system could be vastly increased (see Buliung *et al.* 2009 for more discussion). Again, using the Vancouver example from Table 7.1, if just one-third of drivers could be enticed to give up their car and commute as a passenger in the car of a friend or colleague, the combined weighted social effective speed of all commuters would increase by over 3.5 km/hour, a 33 per cent increase (and likely more, since pollution, congestion, and resource externality costs can also be expected to decline with fewer cars on the road). A shift away from driving and into other transport modes would have an even greater effect on city measures of effective motility, since the diversity and mix of transport choices available to the average commuter would then be widened. In our Vancouver example, if one-third of drivers gave up the car and shifted into other modes, this would lead to a roughly 77 per cent increase in the effective motility measure.

Slowing down the poor: unequal mobility and motility

International comparisons indicate huge differences in the effective speeds of drivers in cities between rich and poor nations, and even within a rich country large differences in effective speeds between high- and low-income groups are evident. It is perhaps not surprising, given the extensive literature concerning social exclusion and barriers to mobility among the poor and disadvantaged (Hine 2011; Lucas 2004), that the latter also suffer from much lower effective speeds. But it may be surprising that when drivers are compared with other modes, being poor has the greatest negative effect on mobility. Table 7.4 provides measures of social effective speed for Canadian metropolitan areas (CMAs), but in this case for (only) low-income commuters – those with incomes half of the CMA median or less. Across all transport modes, social effective speeds are considerably lower for poor households, by roughly 35 per cent on average across all CMAs, in comparison with all households in Table 7.3. But among drivers, low-income households suffer from effective speeds 44 per cent lower on average, double the 22 per cent gap found for low-income public transit commuting households, and roughly triple the gap experienced by poor automobile passengers and cyclists. In fact, the average effective speed of low-income driving households (5.51 km/hour) is very similar to that of walking (4.69 km/hour). Among those who walk to work, of course, being poor has virtually no impact on effective mobility. Driving, it could thus be said, slows down the poor more than any other mode. Automobility can be understood as a distinct factor not only articulating income inequality, but also producing unequal mobility, both between and within each mode.

Table 7.4 Social effective speed (SES) and effective place motility, low-income households (only), Canadian Census Metropolitan Areas (CMAs)

CMA	Auto driver	Auto passenger	Public transit	Bicycle	Walk	Weighted total SES (a)	Transport mix (1−H) (b)	Effective motility (a × b)
Abbotsford	6.22	23.29	5.11	9.02	4.77	7.25	0.22	1.57
Barrie	7.92	24.89	6.50	9.99	4.91	8.70	0.23	2.01
Brantford	6.41	19.05	3.16	8.10	4.76	7.02	0.30	2.10
Calgary	5.17	24.59	11.09	9.60	4.54	6.76	0.45	3.02
Edmonton	5.45	24.32	8.42	9.49	4.57	6.59	0.39	2.60
Fredericton	4.89	22.82	3.51	10.31	4.92	6.30	0.37	2.33
Greater Sudbury	5.74	24.82	7.14	8.85	4.83	6.91	0.36	2.48
Guelph	5.50	22.24	5.72	8.77	4.64	6.45	0.33	2.12
Halifax	5.43	24.43	8.58	9.36	4.71	7.16	0.58	4.13
Hamilton	7.11	25.38	11.26	9.88	4.72	8.39	0.38	3.23
Kamloops	4.39	25.10	5.63	8.65	4.77	5.68	0.30	1.73
Kelowna	4.61	24.58	4.67	9.51	4.76	5.50	0.27	1.50
Kingston	5.03	20.97	6.60	9.78	4.71	6.13	0.43	2.64
Kitchener	5.90	20.89	5.13	9.52	4.61	6.61	0.30	2.00
London	5.35	22.63	8.15	9.08	4.73	6.57	0.39	2.56
Moncton	4.05	19.29	5.36	9.74	4.88	5.52	0.39	2.14
Montréal	7.02	22.50	9.88	8.95	4.40	8.17	0.63	5.15
Oshawa	8.41	24.71	16.71	10.09	4.64	9.60	0.26	2.51
Ottawa–Gatineau	6.09	22.98	12.01	9.52	4.37	7.92	0.56	4.40

Peterborough	5.57	21.38	4.55	9.15	4.94	6.70	0.38	2.52
Québec	5.75	22.53	7.34	9.05	4.56	6.34	0.44	2.80
Regina	3.09	18.47	4.47	7.89	4.45	4.03	0.32	1.30
Saguenay	5.57	22.24	4.68	10.85	4.83	6.27	0.22	1.41
Saint John	5.62	20.46	6.11	9.28	4.80	6.59	0.44	2.93
Saskatoon	3.90	21.60	4.11	8.78	4.63	4.89	0.33	1.63
Sherbrooke	5.54	25.24	6.10	9.38	4.72	6.33	0.36	2.25
St Catharines–Niagara	6.05	23.64	6.02	8.70	4.82	7.12	0.31	2.18
St John's	4.36	19.94	4.43	10.44	4.76	5.67	0.36	2.02
Thunder Bay	4.18	23.82	3.23	8.87	4.70	5.20	0.34	1.75
Toronto	7.21	24.84	11.51	9.20	4.38	8.82	0.58	5.08
Trois-Rivières	4.95	25.02	4.70	8.31	4.78	5.74	0.30	1.70
Vancouver	6.31	24.34	11.12	9.38	4.50	7.86	0.56	4.40
Victoria	4.80	22.43	9.07	9.18	4.61	6.12	0.57	3.46
Windsor	4.73	22.94	5.06	9.48	4.81	5.65	0.28	1.56
Winnipeg	4.58	21.19	8.64	9.08	4.70	6.10	0.51	3.11
AVERAGE – All CMAs	5.51	22.84	7.02	9.29	4.69	6.65	0.38	2.58

Source: As in Table 7.3.

Notes
Low-Income Households are defined here as households with annual incomes 50 per cent or less of the overall CMA median household income (this is the standard 'Low-Income Measure', or 'LIM' method used to define low-income households).

Note as well that the gap between the effective speed of cycling and that of driving widens considerably when it is only low-income households in the analysis. Cycling provides poor commuters with higher effective speeds than driving in every single urban region, by a 69 per cent margin on average, while public transit delivers higher levels of effective speed than driving to such households in all but nine (out of 34) Canadian metropolitan regions. Simply put, in no Canadian metropolitan area does it make sense – from an effective speed perspective – for the average low-income household to drive. While the reasons why the poor continue to drive are as multi-faceted as the reasons for poverty, in many cases it is increasingly because they are concentrated in places where driving is the only viable option.

As with the analysis of effective speeds, poor households also experience lower overall levels of effective motility, by 37 per cent on average across all the cities in our Canadian example. This is largely due to the lower effective speeds of poor drivers, as well as, to a lesser extent, public transit users. This is one reason why levels of effective motility among the poor are even lower than average in the more automobile-dependent urban regions and those cities where post-war patterns of urban settlement make it more difficult to extend transit lines or take advantage of active forms of transport. The more mono-functional the local transport system and the fewer opportunities there are for commuters to walk or bike, the more the poor – many of whom cannot afford to drive – suffer from low levels of effective motility.

There is a clear spatial dichotomy involved in analysis of the motility of the poor. It is in places where the transit system is extensive and where opportunities for cycling and walking are proficient, foremost Montreal, but also Toronto, Vancouver, and Ottawa-Gatineau, and a few older cities such as Halifax, that the poor continue to enjoy relatively decent levels of effective motility on the whole, and where the gap in the effective motility of low-income households is relatively marginal. Meanwhile, in those urban areas linked to the larger cities via high-speed expressways and/or expensive rapid train lines, and functioning largely as commuter suburbs, including Barrie and Oshawa in the Toronto region and Abbotsford in the Vancouver region, low-income households suffer from much lower levels of effective motility than other households (around 50–55 per cent lower). This analysis suggests that the provision of a viable public transit system, a denser and mixed-use urban pattern, and a greater diversity of transport choices are key factors for improving the mobility of disadvantaged households in any given city, provided of course that the poor are allowed to live in such areas. As the tendency under capitalism is for wealthier households to usurp the most accessible spaces and displace the poor, it is necessary to also limit the gentrification of dense, mixed-use, accessible urban spaces from which the poor derive significant mobility benefits (Lees *et al.* 2007; Walks and Maaranen 2008).

Conclusion: the effective speed of automobility

Analysis of effective speed provides for a holistic examination of mobility and motility. Effective speed takes into account the costs involved in fueling, provisioning, accessorizing, and storing the vehicles pertinent to each form of mobility, as well as the indirect costs related to air pollution, traffic accidents and a host of other externalities. The effective place motility of cities takes into account not only these factors, but also the mix of transport modes and thus the choices available to local residents. Comparing the effective speeds of different transport modes, it is clear that the automobile, and automobility, do not provide significantly greater social levels of mobility over and above that provided by the bicycle. Despite higher trip speeds, once all the time spent working to pay the additional costs related to driving are taken into account, driving commuters in the auto-mobile city as a whole go no "faster" than the average cyclist. Furthermore, the current system/regime of automobility, in which those with higher incomes are the ones able to pay for both more accessible locations *and* the faster trip speeds of the automobile, significantly inhibits the relative effective speed of – and thus *slows* down – the poor (both those who drive and those who cannot afford to do so).

The significance of this finding warrants elaboration. It means that the automobile, and automobility, has *not* significantly improved real urban levels of effective mobility beyond that achievable with non-motorized transport. If all urban dwellers were to bicycle to work, they would achieve roughly the same, and in some cases enhanced, levels of effective mobility than is currently attained using other modes, particularly in comparison to driving their own automobiles. One implication of this is that if the physical and social forms of cities could be reformed such that a much greater proportion of the population were to cycle, this would have little to no negative effect on the overall levels of mobility measured in effective speed terms, and in fact would provide for a more equitable distribution of mobility across society. A similar positive result would transpire if more car drivers could be shifted into ride-sharing, public transit, and other modes. The socio-economic distribution of effective mobility, and motility, would be likewise significantly enhanced by such a shift.

Of course, in many cities (not least of all those in Canada) inclement weather makes cycling a less desirable mode of transport during certain seasons. But non-motorized, pedal-based vehicles protected from the weather (with roofs, etc.) do exist, and could easily be put into general circulation. And of course, there are many demonstrated models of successful networked public transit systems that could be purposefully planned to integrate with cycling. It is not only or even primarily the desire of the automobile manufacturers to use their power to quash any incipient competition or other threats to their profits that has prevented alternative methods and designs from being implemented (see Paine 2006). It is the design and governance of modern, auto-mobile cities that act as the main barrier to the promotion of sustainable transport. It would be very difficult for all commuters to now switch over to the bicycle, even if everyone moved to be as close to work as possible: places of residence are on average simply too distant from most places of employment. It will instead be necessary to depend on well-placed public transit system improvements that augment the effectiveness of non-motorized transport modes (see Mees 2009).

But our findings concerning the social effective speeds of different modes are indicative of additional factors at work, with real significance for an understanding of the import of automobility. As noted in the introductory chapter and by other contributors to this volume, automobility involves far more than just a mode of transport or a flexible form of mobility. It is a system central to the organization and distribution of production and consumption. For its maintenance and expansion, the system of automobility requires considerable resource extraction, manufacturing, trade, finance, and sales efforts, and the continuation of considerable natural energy subsidies, not least in the form of fossil fuels. A key production, distribution, trade, and consumption complex at the core of the capitalist global economy, the auto-industrial complex supports the livelihoods of a huge proportion of the population and influences the shape and direction of key social, economic, and political networks.

The analysis of effective speeds presented herein shows that automobility is *not* primarily a system providing enhanced mobility and motility overall – since non-motorized forms of transport could provide similar or higher levels of effective mobility than we have at present. Automobility is instead a system by which wealth and value – and this includes the value of mobility and "motility capital" – is first produced, and then *socially distributed* across space. Automobile capitalism works to limit the underlying value and hence power of many members of the poor – not least through slower effective mobilities – while bolstering the power of key multinationals and nation states at the core of the auto-industrial system, as well as those individuals working in key decision-making roles within it and their families, and some well-connected members of the working classes who then have a vested interest in maintenance of the system. Automobility under capitalism is thus as much, or even more, about social redistribution and power, as it is about mobility, and mobility is as much or more about power and social redistribution as it is about trip speed.

Note

1 Of course, in reality, it could be that different members of a household use different modes of transport for commuting. The question then becomes how to classify households, such that the expenditures related to mobility (re. driving versus transit, cycling, etc.) can be accurately assigned to them. In the Canadian data, households are assigned to a commute mode in a hierarchical fashion: if anyone in the household commutes as an automobile/truck driver or motorcycle rider, the household is classified in the auto driver category. Only households containing a driver are assumed to incur expenses related to the purchase, operation, and maintenance of a car. If not in this first category, then if anyone in the household commutes as a passenger in an automobile, the household is classified as such. If still not, then if anyone in the household commutes via

public transit, the household is classified in this category, etc. In this way, households are classified according to modes based on the hierarchy of costs associated with each transport mode, and in such a way as to produce the most conservative estimate of total costs (and thus higher estimated average effective speeds). Note that the Canadian census only includes data for commute trips and not other trips.

8 Automobility and non-motorized transport in the global South

China, India, and the rickshaws of Dhaka

Ron N. Buliung, Annya C. Shimi, and Raktim Mitra

Non-motorized transport (NMT), also called active transport, human-powered transport, and fuel-free transport, includes travel modes such as walking and all non-motorized vehicles (NMVs), for instance bicycles, cycle-rickshaws, carts, working tricycles (World Bank 1995; Litman 2004; Bari and Efroymson 2005a; Norcliffe 2011). NMT plays a critical role in transportation systems, and remains an essential means of transport in many developing cities in the global South, particularly for the socially and economically disadvantaged (Denmark 1998; Pucher *et al.* 2007). More than 20 years ago Replogle (1992) conducted an extensive survey of NMT modes across Indian cities demonstrating the dominance of walking, bicycles, and cycle-rickshaws over motorized vehicular traffic, and the concentration of NMT use within lower-income households. Since that time, however, we have seen rising auto ownership and use in many cities of the global South, and while taken together, NMT modes tend to continue to dominate the traffic stream, their overall share has been in decline over several decades in some places (BTRC 2011; Singh 2005; MOUD 2008; Tiwari and Jain 2008). Of course, the mix and intensity of use of mobile machines varies with city size, the supply of and access to transit options and automobiles, and other factors including terrain, transport policy, and planning, and individual and household characteristics (Singh 2005; Pucher *et al.* 2007; MOUD 2008).

Contemporary transportation planning in the developing world is biased towards supporting a vision of modernity centred on motorized transport (MT) and automobility (Khisty 2003; Whitelegg and Williams 2000). Policy makers and planners in some developing countries have placed the construction of the system of automobility at the centre of the project to modernize the city, under the assumption that it is vital for economic development (Khisty 2003). The push to adapt, build, or rebuild socio-technical systems, built environments, and economies to support and sustain automobility places NMT modes and users at odds with an emerging automobility paradigm. Lack of access to convenient and affordable transport for low- and middle-income people creates difficulties in getting to schools, accessing health care, moving goods to market, and finding and maintaining suitable employment. The rising cost of transport and outright exclusion of NMT through transport policy, coupled with enforcement and harassment, contribute to maintaining and reproducing the cycle of poverty.

Given the geographical origin of much of the automobility discourse, from within a largely Western academic tradition (reacting to a particular Western and Northern system of automobility), it is useful to ask what automobility looks like in the global South. Such a question necessitates discarding the a priori assumption that what has happened "here" will happen "there", and consider that some form of policy transfer, oriented towards correcting the pitfalls of automobility, may fall flat in the developing world.

This chapter considers the tension between a contemporary vision for automobility and the legacy of NMT, with regard to the production of unequal mobilities, forms of land development, and uses of cities in the global South. We begin with an examination of the status of NMT modes with the context of the rising dominance of automobility in the global South, particularly in large and fast-growing regions and nations. We then take aim at the situation in Dhaka, Bangladesh – the city with the largest cycle-rickshaw fleet on the planet. The Dhaka case study examines intersections between NMT policy and the plight of those for whom Dhaka's cycle-rickshaws are essential to the conduct of everyday life.

Automobility and non-motorized transport in the global South

The rise of automobility within the global South is not a geographically and temporally even process. The overarching issue has less to do with per capita auto ownership (which is still relatively low on the global scale, particularly when compared with the United States (US)) than with the *pace* of transformational motorization, and auto-centred urbanization, which in some places is increasingly similar to the rate of automobile consumption in the US during the roaring 1920s (Ng *et al.* 2010). The effects of the rise of a "southern" system or regime of automobility are piling up quickly.

The automobility "project" exists at different stages of completion in the developing world. Consider the People's Republic of China (PRC), the so-called "Kingdom of the Bicycle", where the result of the rapid take-up of the values of automobility is articulated highly unevenly across its cities. Among them are Kangbashi, in Inner Mongolia (Hitchens 2011), where the rise of an auto-city populated by a mesh of ghost roads, awaits cars and drivers who have yet to materialize within the urban landscape. Then there is Hangzhou's Public Bicycle System – the world's largest bike-sharing operation (Shaheen *et al.* 2012). Yet from Beijing, the PRC's capital city, we observe the automobile project in full swing, articulated in a "de-bikification" of that city (Bruno 2012), and a reduction in bicycle mode share between 1986 and 2010 of approximately 46 per cent (Table 8.1) (BTRC 2011). This lamentable (yet perhaps not so lamented by a rising and liberalizing middle class) loss of cycles and cycling within parts of urban China has even become the subject of a contemporary artistic criticism

Table 8.1 Mode share, Beijing, China (1986–2010)

Mode	1986	2000	2005	2007	2008	2009	2010
Car	5.0	23.2	29.8	32.6	33.6	34.0	34.2
Bus	26.5	22.9	24.1	27.5	28.8	28.9	28.2
Subway	1.7	3.6	5.7	7.0	8.0	10.0	11.5
Bicycle	62.7	38.5	30.3	23.0	20.3	18.1	16.4
Taxi	0.3	8.8	7.6	7.7	7.4	7.1	6.6
Other	3.8	3.0	2.5	2.2	1.9	1.9	3.1

Source: BTRC (2011).

levelled against a state-sponsored auto-oriented vision of modernity (Ai Wei Wei 2012). The latter is a criticism deemed so egregious by the state that its author(s) face a steady stream of harassment, constant surveillance, and seemingly endless and absolute mobility restrictions, including, for a time, house arrest.

Yet even within Beijing, the automobile emerges as both problem and solution – with the introduction of the licence-plate lottery, and new vehicle quotas, the expansion of automobility continues, but at a slower rate than before. Notably, the decline in bicycle use occurred parallel to, and perhaps enabled by, enormous investments in the motorized transport infrastructure during the last decade (transit and cars). Between 2006 and 2010 government spending on transport infrastructure in Beijing alone rose by 92.5 per cent, from 26.7 billion yuan to 51.4 billion, while car mode share increased by 4.4 per cent and subway by nearly 6 per cent (BTRC 2011).

In sharp contrast to the Inner Mongolian ghost city, large cities of India and Bangladesh hum under the weight of so much activity and mobility – a mobility that "by design" is, in some places increasingly, oriented towards the system of automobility. Many Indian cities have historically been walking and transit cities when compared with the dominance of cycles and cycling in Chinese cities, although NMT remains the most important means of transport for the transport disadvantaged (Table 8.2) (Replogle 1992; Pucher *et al.* 2007). While many Indian households own cycles (35–65 per cent own one or more bicycles in the medium to large cities) (Tiwari and Jain 2013), as the primary mode for daily trips, work, and otherwise they are used less often than walking or public transit (Pucher *et al.* 2005).

NMT in many, but not all, Indian cities has been in decline since the 1980s, although the results vary considerably across urban regions and by size of conurbation (Table 8.3). In Jaipur, for example, bicycle mode share for all trips would decline by 58 per cent from the 1980s into the mid-1990s (Tiwari and Jain 2013). Recently, Kolkata has enacted a ban on bicycles, non-motorized rickshaws, carts, and cycle vans on 174 roads during business hours, with significant potential implications for those whose economic livelihood depends upon such modes (Gowen 2013). But exceptions can be located, including Chennai and Patna where marginal increases in NMT mode share have been reported (Tiwari

Table 8.2 Mode share and income in three Indian cities

City/household income	Share of total trips (%)			
	Walk	*Bicycle*	*Bus*	*Other*
Vadodara				
Low income	57.8	15.4	25.1	1.7
Mid income	40.3	16.2	33.0	10.5
High income	23.9	13.0	36.7	26.4
Jaipur				
Low income	47.0	24.6	14.9	13.5
Mid income	38.9	25.2	22.3	13.6
High income	27.3	19.6	24.2	28.9
Patna				
Low income	58.0	14.5	19.4	8.1
Mid income	40.3	13.6	33.9	12.2
High income	22.8	10.8	34.0	34.2

Source: Replogle (1992).

Table 8.3 Mode share in Indian cities classified by population size, 2006 (%)

City population (in millions)*	Walk	Cycle	Two-wheeler**	Public transport	Car	IPT***
< 0.50 (Plain terrain)	34	3	26	5	27	5
< 0.50 (Hilly terrain)	57	1	6	8	28	0
0.50–1.0	32	20	24	9	12	3
1.0–2.0	24	19	24	13	12	8
2.0–4.0	25	18	29	10	12	6
4.0–8.0	25	11	26	21	10	7
>8.0	22	8	9	44	10	7

Source: MOUD (2008).

Notes
*   The following Indian cities can be found in different rows in this table:
       Patna (row 4); Jaipur (row 5); Chennai (row 6); Delhi, Kolkata (Calcutta), and Mumbai (final row).
**   Motorized two-wheel vehicles registered with government.
*** IPT: Intermediate public transit/para-transit, auto rickshaw.

and Jain 2013). In the large Indian cities where walking as primary mode has been in decline, public transit still moves more people on a daily basis than automobiles. Moreover, and with the rise of rail transit projects in cities such as Bangalore (Namma Metro), Mumbai (Mumbai Metro), and now Chennai, some travel demand may be siphoned away from automobile traffic and absorbed by these emerging rail transit systems (e.g. Phase I of Bangalore's Namma Metro opened in October 2011).

There are problems in accurately determining mode share for NMT modes, such as cycle-rickshaws in many cities, because planners, transport policies, and policy makers often do not recognize the contribution of these modes to the transportation system (Tiwari and Jain 2013). Geographically and temporally uneven tensions exist between automobility and NMT in India, with examples of movement backwards and forwards regarding the plying of cycle-rickshaws in Delhi, for example, because this mode serves well to connect a growing and liberalizing mobile middle class with transit services (Nair 2013). One irony is that as many of India's large cities struggle with congestion, solutions often turn towards the exclusion of NMT, and cycle-rickshaws in particular, from major roads (Whitelegg and Williams 2000; Tiwari and Jain 2013). The system of automobility, it could be said, is largely incompatible with a built environment tied to pre-automobile and, in many instances, colonial-era land uses and transport systems.

India's vertical urbanism offers perhaps the most striking example of how the system of automobility may be scaled up in the developing world context. Consider the Maharashtra State Road Development Corporation 50 Flyovers Mumbai Traffic Improvement Project. The relatively short span and limited number of elevated road spaces in cities like San Francisco (excluding the bridges) or Toronto (see Chapters 12 and 14 in this volume) contrasts with the flyovers constructed or scheduled for construction in Mumbai. NMT modes and heavy vehicles are excluded from using the flyovers – a space of privilege for automobile elites drifting across a vertical city without being troubled by the life, poverty, and congestion beneath. Kothari (1993) and Fernandes (2004) wrote about a growing amnesia of a liberalizing middle class, setting "into motion a politics of forgetting with regard to social groups that are marginalised" (Fernandes 2004: 2416). The flyover symbolizes a particular politics and economy of mobility in the modern city, contributing to or contained within a much broader politics of forgetting the subaltern, who find themselves dwelling beneath and around concrete and steel projections constructed to support a still emerging system of automobility for a liberalizing middle class.

Undoubtedly, the automobile system and its complex of social and technological inter-linkages, including the fixed capital infrastructures that facilitate its use, hold iconic status at the global scale (Freund and Martin 1993; Sheller and Urry 2000). Their presence communicates something about modernity to others who would take a look at cities and regions within the global South. The global advance of automobility, and the system of automobility, however, bears significant external effects and costs that are well known, including: traffic congestion, widespread air and noise pollution, increased fuel consumption, higher infrastructure costs, and higher accident rates, and the specific and widespread human and environmental impact of these phenomena (Khisty and Zeitler 2001; Khisty and Ayvalik 2003). In contrast, NMT holds many potential advantages over MT in terms of energy conservation, congestion reduction, environmental impact, social equity, economy, and convenience (Khisty 2003). In the remainder of this chapter we examine further the tension between NMT and MT policy and use in Dhaka, the capital city of Bangladesh, a city with an informal economy of mobility comprising the largest fleet of cycle-rickshaws on the planet (Hasan 2013; Replogle 1992).

Driving mobile modernity in Dhaka, Bangladesh

Since gaining independence in 1971, large cities in Bangladesh have experienced spontaneous and imbalanced population growth and rising travel demand, resulting in severe congestion and poor levels of service provision. Issues surrounding inclusive mobility, traffic congestion and safety, and the impact of transport on the environment have become increasingly critical (Hasan and Dávila 2012; Hoque *et al.* 2005). More than one-third of the population of Bangladesh lives in cities. It is projected that by the middle of this century more than half of the population of Bangladesh will be urban. Much of the Bangladeshi population is concentrated in and around Dhaka (World Bank 2009). Since 1971, Dhaka's population has increased eightfold. The city now has a population of 15 million people distributed over an area of approximately 1,529 km^2 (9,810 people per km^2), making it the eighth largest city in the world. It is projected to become the third largest city by 2020 (World Bank 2010). This rapid urban population growth has begun to outstrip the capacity of existing urban infrastructure, leading to low levels of service across utilities and particularly within the transport system (World Bank 2009). In Dhaka, contemporary transport planning and investment has focused principally on motorization and auto-mobility, largely ignoring NMT modes and users.

Using Dhaka as a case study, the rest of this chapter explores the tension between NMT and MT use and policy.[1] First, we provide an overview of Dhaka's current transport system with a focus on NMT modes and facilities. Second, we consider the NMT policy context, offering some critical discussion of the impacts of state-sponsored restrictions on NMT infrastructure and use. Finally, we discuss the economic importance of NMT, the impact of Dhaka's urban fabric on NMT use, the role of NMT in combating traffic congestion, and the importance of NMT in maintaining mobility for the transport disadvantaged.

Dhaka's transport system

Dhaka's transport system includes a mix of MT modes such as buses, minibuses, trucks, cars, auto-rickshaws, taxis, motorcycles, and NMT modes like walking, bicycles, cycle-rickshaws, hand-carts, and rickshaw-vans (Strategic Transport Plan (STP) 2005). According to data from the STP (2005) and the Dhaka Urban Transport Network Development Study (DHUTS) (2009), NMT modes (mainly walking and cycle-rickshaw) account for between 48 and 58 per cent of all trips, respectively (Table 8.4). Overall, private automobile ownership is low in Bangladesh, and in Dhaka. This is likely explained by government policies such as high import cost for vehicles and parts, and relatively high fuel costs. For example, sellers currently have to pay a minimum of 131 per cent import duty on cars (*Financial Express* 2013), and although Bangladesh produces its own fossil fuel, a litre of petroleum costs more than $1, which is relatively high compared to an average monthly household income of Tk 33,600 (equivalent to $420) (DHUTS 2009). Recently, the government has promoted the use of

Table 8.4 Modal share in Dhaka

Travel mode	Trip share (%)	
	STP (2005)	DHUTS (2009)
Walk	14	20
Cycle-rickshaw	34	38
Public transport	44	30
Motorized transport	8	12

Source: Derived from STP (2005), DHUTS (2009).

Notes
Public transport = bus; Motorized transport = car, auto-rickshaw (CNG), taxi, motorcycle.

concentrated nitrogen gas (CNG) as an auto-fuel as part of the Asian Development Bank (ADB) supported "Dhaka Clean Fuel Project" (ADB 2013), and this new and highly subsidized form of fuel costs five times less than petroleum. Affordable fuel, combined with a recent trend toward easily available bank loans for purchasing private automobiles, has triggered a sharp increase in automobile ownership in Dhaka in recent years. According to the Bangladesh Road Transport Authority (BRTA 2013), the number of registered vehicles for household use increased by 30 per cent between 2009 and 2013. Here we see, then, a technical adaptation towards CNG with a view to removing a specific barrier to the rise of Dhaka's system of automobility. But what about other forms of motorized passenger transport?

Heavy and rapid public transit (e.g. rail-based urban transit, bus rapid transit) is non-existent in Dhaka. Buses are the primary form of public transit in the city. Since most urban roads are narrow with inadequate capacity to support bus services, bus routes are mainly limited to major roads. A wide variety of para-transit services, most of which use eight to 12 passenger human haulers/converted mini trucks, which are by no means comfortable rides, provide connections to major destinations within relatively short distances and bus transit services. Other privately operated motorized passenger transportation modes include auto-rickshaws and taxis, both of which are inadequate in supply and expensive.

The average trip length in Dhaka is 5.4 km, and 76 per cent of trips are less than 5 km, ideally suited to NMT modes (STP 2005). NMT plays an important role in enabling the mobility of the city's residents (Bari 2007). NMTs (particularly cycle-rickshaws and walking) remain the most commonly used modes across all income groups (DHUTS 2009). Despite such patronage, the future of NMT in Dhaka is threatened by growing motorization, loss of road space for NMT facilities due to state-sponsored appropriation of road space for motorized travel, and policy intervention favouring automobility and MT modes more generally.

Current transport policy tends to favour motorized modes for the movement of people and goods, despite token acknowledgement of the importance of walking and cycle-rickshaws (DHUTS 2009). Planned provision and incorporation of NMT modes in the motorized traffic stream is often neglected. Recent policy documents have not explored NMT as an alternative to public transport and private automobiles. Instead, these modes are viewed as transit supportive, deployed to transport people to and from their houses and transit stops (Fjellstrom 2004; Rahman *et al.* 2009).

NMT fleet characteristics

The NMT fleet includes bicycles, cycle-rickshaws, flat-topped rickshaw vans, and hand-carts (known locally as "thela-garis" or pulled carts), and, on rare occasions, cow-drawn carriages. In Dhaka the cycle-rickshaw accounts for over 75 per cent of the NMT fleet (UN ESCAP 1997). Historically, Dhaka had only 100 cycle-rickshaws in 1944 and 575 in 1948 (Gallagher 1992). The number increased during the 1970s and 1980s (UN ESCAP 1997). The official number of cycle-rickshaws in 1972–1973 had increased to 14,667, which then doubled to 28,703 by 1982–1983, thereafter increasing rapidly to reach more than 88,159 by the end of 1986–1987 (Gallagher 1992). After that, the Dhaka City Corporation (DCC – the city government) established regulations to control the number of existing cycle-rickshaws and the manufacture of new ones, setting 89,000 as the official legal number of allowable cycle-rickshaws in service (Bari and Efroymson 2005a). Despite this attempt at regulation, cycle-rickshaw numbers in Dhaka have continued to grow. The estimated number of cycle-rickshaws currently in circulation varies between 500,000 (STP 2005) and 1.1 million (Hasan 2013); the largest such fleet of any city on the planet, with roughly one rickshaw per 30 residents.

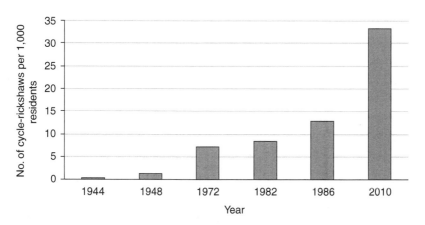

Figure 8.1 Per capita growth of the cycle-rickshaw fleet, Dhaka, 1944–2010 (source: Authors 2010 (Data from Gallagher 1992; STP 2005; Hossain 2008)).

Note

To calculate number of cycle-rickshaws per 1,000 population, population data from 1941, 1951, 1974, 1981, 1991, and 2010 are used for the cycle-rickshaw population data of 1944, 1948, 1972, 1982, 1986 and 2010, respectively. The official number of registered cycle-rickshaws for the considered years other than 2010 is used; for 2010, the number of cycle-rickshaws actually plying the roads is used.

NMT facilities

Though a significant share of total person trips (58 per cent of total person trips) is made by NMT modes (DHUTS 2009), NMT-enabling infrastructure is largely non-existent. The interests of pedestrians, cyclists, and cycle-rickshaw users are sacrificed for the planning of roads to accommodate motorized vehicles.

Walking

Over 70 per cent of Dhaka's roads have no sidewalks and those that exist are not continuous and are often host to vendors, parked motorized vehicles, bus ticket counters, construction materials, garbage, electricity and telephone poles, open sewers, and drainage ditches (DOE 2009; STP 2005). According to the STP (2005), about 40 per cent of the sidewalks are occupied illegally, primarily by vendors. The presence of commercial activity on the sidewalks reduces effective sidewalk width, forcing pedestrians onto the road and putting them at greater risk of injury or death from vehicular traffic. On most busy streets, controlled intersections are either absent or are not enforced. Pedestrian bridges offer some opportunity to pass over busy roads, but they are often not co-located with pedestrian flows and needs in mind. Overpasses can be excessive in height (nearly equivalent to a two-storey building), making them inconvenient particularly for the elderly, children, pregnant women, and physically disadvantaged people. The underpass alternative (underground tunnel) tends to play host to various antisocial activities and crime. The rise of MT, coupled with pedestrian mobility within an inadequately designed and controlled traffic environment, contributes to a situation where nearly half of all traffic collisions and 72 per cent of all fatalities in Dhaka involve pedestrians (Mahmud *et al.* 2006; Maniruzzaman and Mitra 2005).

Bicycle

Although a popular mode in villages and smaller towns, bicycles account for only 2 per cent of vehicular traffic in Dhaka (STP 2005). Low bicycle mode share is partially the result of the absence of cycling facilities, cycle lanes, and secure places to store bicycles, facilities that support urban cycling in other parts of the world (Furth 2012). Other barriers include the high price of bicycles relative to income, high risk of theft, and the low social status ascribed to bicycle users (STP 2005). Despite the barriers, the popularity of cycling is on the rise, particularly among youth and young adults. Cycling is increasingly being perceived as an economical, time-saving, and healthy alternative to other transportation options in Dhaka (*Khabar South Asia* 2013). Largely due to this increased awareness, and fuelled by several online movements such as the one by Bangladesh Cycling Community (www.bdcyclists.com), the rate of cycling for both commuting and recreational purposes has increased in recent years; however, the ratio of bicycles to overall traffic remains very low.

Cycle-rickshaw

Despite one-third of person trips being made by cycle-rickshaws (Table 8.4), and despite this mode's many environmental and economic benefits, many municipal and national government policies have been instituted to reduce the size of the cycle-rickshaw fleet. These include: drastic measures such as the destruction of unlicensed vehicles, banning their use of major roads, and preventing them from legally parking in many areas (Bari and Efroymson 2005b). Few facilities for cycle-rickshaws are provided, and those that exist are often implemented with restriction in mind. Cycle-rickshaw lanes have been created along several major sections of roads, but they are quite narrow and discontinuous, causing inconvenience and costs to cycle-rickshaw commuters both in terms of increased fares and travel time (Bari and Efroymson 2005b). Fares for cycle-rickshaw travel are negotiated between commuters and pullers based on travel time and distance. Following state policies of limiting the use of cycle-rickshaws, fares have increased. Commuters now experience longer journeys because of discontinuous and obstructed travel routes. Where provided, narrow cycle-rickshaw lanes may be obstructed by long queues of cycle-rickshaws, as there is only space for one cycle-rickshaw travelling in each direction (Bari and Efroymson 2005c). Sometimes parked or turning motor vehicles within cycle-rickshaw lanes make them unusable (Bari and Efroymson 2005a). Overall, commuting by cycle-rickshaw is discouraged through government policies that ban them on major roads and/or direct them to narrow and discontinuous cycle-rickshaw lanes (Figure 8.2).

NMT policy

The popular argument against unregulated/unrestricted cycle-rickshaw use in Dhaka, underlying restrictive policies, is that cycle-rickshaws obstruct motorized vehicles, contributing to slower speeds and congestion on urban arterial roads, while offering little in terms of efficient passenger transport. This is the same case made by other auto-centric experts, technocrats, and politicians in a growing number of cities located across the global South (Tiwari and Jain 2013; Whitelegg and Williams 2000; Gowen 2013; Norcliffe 2011). However, unlike in Kolkata (see Gowen 2013) and parts of China (Norcliffe 2011), other NMT modes such as bicycles and hand-carts are basically left alone (UN ESCAP 1997). There is an interesting historical and geographical link between what we are calling the Northern and Southern instances of automobility. During the inter-war period in continental Europe, a similar modernist vision of the city advocated the inevitable supplanting of cycles with automobility and motorized passenger transport more broadly. Hence an emerging class of urban professionals (planners, traffic engineers) also argued a similar case against NMT at that time (see Oldenziel and de la Bruhèze 2011).

In 1986, cycle-rickshaws were banned from one of the most important roads in Dhaka, the Airport Road (Gallagher 1992). This intervention was not entirely

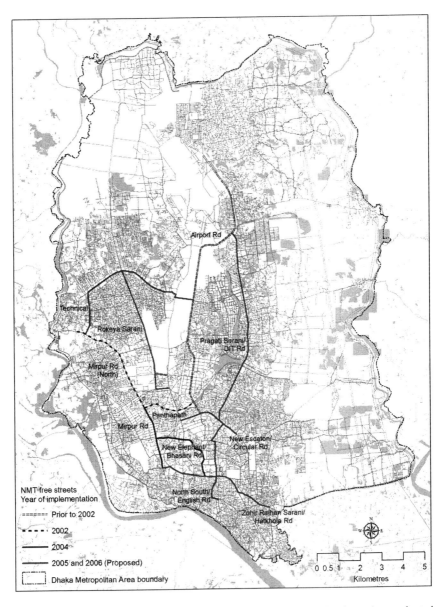

Figure 8.2 Cycle-rickshaw street bans in Dhaka (source: Created by the authors, adapted from Efroymson and Bari (2005) and DUTP (2005)).

unreasonable because the Airport Road connects Dhaka central city with the airport and its surrounding neighbourhoods located to the north of the city, with not much residential development along the road. Subsequently, however, the city has pursued several attempts to prevent cycle-rickshaws from being used on other major roads, perhaps because of, or in relation to, the Airport Road precedent. Informed by a World Bank-funded Dhaka Urban Transport Project (1996–2005), the DCC adopted an "NMT-Free Arterial Network – Phased Implementation Plan" to withdraw cycle-rickshaws from 11 major roads of Dhaka, constituting a total of 120 km (STP 2005; Menchetti 2005; Bari and Efroymson 2005c; Rahman *et al.* 2009; Hasan 2013). The first of a series of planned interventions was implemented in 2002 when two major parts of Dhaka's inner-city road network – Panthapath (Russell Square to FDC) and part of Mirpur Road (Gabtoli to Russell Square) were made rickshaw-free. In December 2004, the DCC banned cycle-rickshaws from the rest of the Mirpur Road (Figure 8.2).

Studies examining the outcome of this policy found that such measures caused hardships to the transport disadvantaged (actually increasing their number) while contributing to increased motorization and traffic congestion (Zohir 2003; HDRC 2004; Bari and Efroymson 2005a,c). Evidence of increased hardship with only marginal improvement in traffic conditions perhaps moved the World Bank, in early 2005, to reverse its policy position on NMT use on Dhaka's major roads (Rahman *et al.* 2009; Bari and Efroymson 2005a; Menchetti 2005). The World Bank's perspective on NMT seemed to move back and forth during the period of the DUTP, particularly with regard to the relationship between NMT and the poor: for example, "despite its economic importance to the poor ... the potential of nonmotorized transport is often unmobilized or even actively suppressed" (World Bank 2002: 125). However, the damage to NMT operation and use was done. Transport planning in Dhaka continued to focus on expanding the role of MT and reducing support for NMT, a pattern reflected in the STP of 2005.

Despite the change of heart by the World Bank, the STP recommended reviving the NMT ban on major roads and has redefined the role of NMT as a feeder service to motorized transit and as a means for neighbourhood circulation only. The banning of NMT on major roads without providing separate, adequate, and continuous lanes for NMT fragments the NMT network, splits short trips, affects the viability of NMT as both a transport mode and economic activity, and further limits the personal mobility of cycle-rickshaw passengers and cyclists. Since 2004, the Dhaka Metropolitan Police (DMP), which is the responsible authority for traffic control, has also made and continues to make many sporadic and/or spontaneous attempts to restrict the movement of cycle-rickshaws at several intersections and road segments within the city (Hasan 2013).

Though the restriction on NMT was preserved in the 2005 STP, some of the newer policy regarding NMT at least more explicitly acknowledges the possible role of NMT in the transport system. However, a mismatch exists between policy intent and funding (Bari 2007). For example, the STP includes a

"Pedestrian First Policy" as part of a balanced multi-modal transport system, yet 28.3 per cent of the total STP investment has been allocated to the construction of roads and elevated expressways, while 0.44 per cent has been proposed for NMT facilities (0.22 per cent for pedestrian-oriented developments and 0.22 per cent for cycle-rickshaw and bicycle improvements).

Investment in road construction will produce large, costly infrastructure projects that serve mainly to benefit elite society and influential political groups, not only in terms of mobility, but importantly in relation to the accumulation of capital and the consolidation of political power. Overall, a clear bias toward large, visible, and expensive infrastructure and services has been observed in developing and implementing the STP, potentially satisfying the needs/desires of the socio-economic elites and the political ambitions of the government, while discounting the needs of the majority of road and sidewalk users.

The cycle-rickshaw economy

NMT is important for the generation and maintenance of economic and social systems in Dhaka. Cycle-rickshaws not only dominate the transport system of Dhaka, they are also one of the largest sources of employment for the poorest able-bodied members of society (Gallagher 1992; Zohir 2003). The cycle-rickshaw industry generates unskilled and semi-skilled employment opportunities that include: cycle-rickshaw pullers, workers in roadside repair and maintenance shops, employees working in the small-to-medium-scale industries involved in NMT manufacturing, and roadside suppliers of consumables for rickshaw operators. In Dhaka, approximately 800,000 people are directly involved with pulling cycle-rickshaws (Mahmud and Hoque 2012). While it is difficult to locate recent information, Replogle (1991) reported that nearly one-fourth of all employment in Dhaka can be attributed to cycle-rickshaw pullers and those involved in ancillary service provision including manufacturing, servicing, and repairing cycle-rickshaws, and serving and supplying operator needs. Since the early 1990s, Dhaka's economy has changed considerably. One of the challenges in gathering accurate recent data is the informal aspect of the rickshaw economy.

State-sponsored removal of cycle-rickshaws harms large segments of the working poor and impacts the well-being of people already struggling at the economic and political margins. The livelihoods of the pullers and their familial networks are fundamentally altered due to limited prospects for substitutable unskilled or semi-skilled employment. Somewhat ironically, the government partially bases its NMT policy on the grounds that the occupation of cycle-rickshaw puller is an example of an "inhumane profession" (DITS 1994). Alternative employment in farming, textile factory work, and day labour is arguably no less inhumane, particularly given the increasing number of cases in which factory workers have been killed in fires, sometimes due to being locked in their places of work by their employers (BBC News Asia 2013). In Bangladesh, there are already millions of unemployed people; it is unrealistic to assume that thousands of jobs can be created to employ people who would lose their jobs due to the banning of cycle-rickshaws. And if jobs could easily be created, it might be considered in the public interest to target these for other jobless people (Efroymson and Rahman 2005). Overall, banning cycle-rickshaws on Dhaka's major roads is harmful to the economic well-being of the most vulnerable urban dwellers.

NMT and the transport disadvantaged

In Dhaka, as elsewhere, the major groups of the transport disadvantaged are the poor, the elderly, the physically disabled, women, and children; these groups are disadvantaged with regard to access to, and usage of, the transport system and represent, collectively, a sub-population that experiences difficult mobility challenges. Bangladeshi society is largely patriarchal, with men acting as household heads, earning formidably higher incomes on average than women, and women having limited opportunity for paid work outside the home (Efroymson *et al.* 2007).[2] Like many places, such a division of labour sees many women occupied by unpaid household work – including many activities that require travel out of the home, i.e. shopping and supervision of children's travel to and from school. Efroymson *et al.* (2007) found that the problems generated by the patriarchy are particularly acute for women from middle-class households, who face restrictions on participation in paid work outside the home, yet do not earn enough income to employ others to undertake household tasks.

Women commonly use cycle-rickshaws in Dhaka, particularly for the many household-related short trips that constitute much of their patterns of daily mobility. According to the STP (2005), 40 per cent of loaded cycle-rickshaws are used by women and children, or people with goods. Women rarely if ever ply rickshaws, but instead remain mostly dependent on this mode for conducting relatively far-from-home daily activities – empowering or oppressive as these activities and travel may be. Research has found that women in particular are suffering from the cycle-rickshaw ban, and that they typically find no satisfactory alternative (Bari and Efroymson 2005b; Zohir 2003; HDRC 2004). The lack of state support for NMT, and the contemporary attack on NMT supported initially by the World Bank, impose forms of "immobility" on women, making it difficult to complete daily tasks. The related and necessary mobility required to carry out such activities is devalorized by state policies favouring motorized modes and automobility in particular. What is the alternative to the cycle-rickshaw? Some women have shifted to buses, and about 70 per cent of women bus users are new users (Zohir 2003). Public buses, however, may pose a hazard, with some suggesting that they are insecure, unreliable, congested, and unsafe for women (Shefali 2000; Zohir 2003). Female commuters face adverse travel conditions, segregated seating – despite rising patronage – and even sexual violence, particularly due to the attitudes and actions of some bus operators and male passengers (Peters 1999; Zohir 2003).

Class and income also play a significant role in influencing the transport choices people have. It is well known that low-income households have less access to private motorized transport (Giuliano 2005). According to the DUTP (1996), in Dhaka, low-income households spend 17 per cent of their monthly income on transport, while high-income households spend only 8 per cent. Transport interventions that promote the use of NMT usually contribute directly to the welfare of people who cannot afford motorized transport (World Bank 1998; Replogle 1991). Low-income households in Dhaka (monthly income < Tk 20,000 or < $250) conduct the vast majority of their household trips on foot (23–37 per cent) or use cycle-rickshaws (31–41 per cent). In contrast, only 7 per cent of all trips by the highest-income households (monthly income > Tk 100,000 or > $1,250) are walking trips, 27 per cent are by cycle-rickshaws, and 34 per cent are by private automobiles (DHUTS 2009). With regard to the combined trip-making of all low- and middle-income households, the cycle-rickshaw is the dominant mode. If cycle-rickshaws are banned on Dhaka's roads, low- and middle-income households will be disproportionately impacted, perhaps forced to use more of their time budget to walk over long distances.

Cycle-rickshaw pullers are among the poorest and most underprivileged members of Dhaka society. Typically, before coming to the city, they lived in rural areas and relied upon agriculture for subsistence. Push factors contributing to migration include loss of agricultural land due to land-erosion from rivers; frequent recurrence of natural disasters, such as floods, that undermine agricultural development and cause food crises; and rampant unemployment within the agricultural economy. Cycle-rickshaw pulling provides relatively easy access to the urban labour market and, arguably, an escape from extreme rural poverty (Begum and Sen 2005).

A considerable number of the pullers have left their families in their rural village home and migrated to Dhaka seeking employment. Families in remote villages often depend entirely upon the puller's income. Banning or curtailing the use of cycle-rickshaws in the city means a direct attack on the ability of the poorest to survive. The ban induces enormous job losses among the pullers and impoverishment of their dependants (Efroymson and Rahman 2005). Anti-rickshaw policies potentially trigger migration of cycle-rickshaw puller families, whose place in the village is economically sustained by a flow of capital from city to village. This is a (potential unintended) consequence of anti-NMT policy that clearly contrasts with government policies intended to stem the tide of rural to urban migration.

Conclusion

This chapter has examined the role of non-motorized transport within the rise of the system of automobility in the global South. We have identified a close connection between the vision of modernity and the production of automobility in some cities of the global South, and, not surprisingly, as a consequence a rising tension between the presence and place of NMT in these cities and an ongoing experiment to "modernize" political, socio-technical, and built environments to accommodate motorized passenger transport.

Across the global South there is a nascent system of automobility operating on an arguably grander scale than its Northern counterpart, with the effects of rapid motorization piling up very quickly. There is also evidence of a very rapid decline and replacement, particularly in rapidly modernizing cities such as Beijing, of the NMT legacy. At the same time, evidence could be found of increasing mode share for walking in some Indian cities, and the presence of large bike-sharing initiatives in some Chinese cities. So this exercise of attempting to characterize intersections between automobility and the global South is somewhat fraught with conflicting or seemingly contrasting examples that appear to arise from a geographically and temporally uneven process of growth and change.

Yet in many cities such as Dhaka, solving transport problems has become a chief task confronting the government, and imposing restrictions on NMT use is often viewed as part of the solution to congestion and poor level of service. However, NMT is the primary means of transportation for a large proportion of city dwellers, especially the low- and middle-income groups and women, and a major source of employment for the urban poor. In Dhaka, as in many other places, the politics of mobility becomes a politics of forgetting, immobility, and exclusion, particularly where the inter-linkages between gender, class, labour, and transport are concerned. Discouraging the use of NMT by imposing restrictions cannot solve the city's transport problems, many of which have been directly caused by the imposition of the system of automobility onto a developing-nation city.

It is useful to recognize here that in a large number of cities in the global South, including Dhaka, growth was never planned for, at least in the Western sense. Residential and non-residential buildings and streets in Dhaka, for instance, in most cases were not built to support the system of automobility. Minimum parking requirements for non-residential buildings and public spaces often remain very low, and planned street parking is largely non-existent, creating intense competition for space among travel modes whether in their mobile or immobile states. Dhaka's intensive mixed land use plays host to many short trips that can easily be completed on foot or using other forms of NMT. There is simply no direct replacement for the use of NMT in many large developing-nation cities such as Dhaka. Instead, the failure to advocate for parity between the supposed place of NMT in planning, and to secure the funding required to support it, remains a critical limitation to the development and nurturing of legacy urban mobilities, and the reduction of social inequalities, within and among cities of the developing world.

Notes

1 Development of this chapter involved a synthesis of information drawn from several sources including a literature review, the second and third authors' life experience and field observations, expert opinions, and a policy scan conducted by the authors. Secondary socio-demographic and transport data have also been drawn together from several sources.

2 In this section, and although we recognize that gender is socially constructed and that essentialism can often fall flat with regard to considering gender and mobility, we use, not unlike Hanson (2010: 8) the "messy" terms *women* and *men* as tools to guide our discussion of patriarchy and mobility, particularly of those women who occupy the so-called middle class.

9 Automobility, adaptation, and exclusion

Immigration, gender, and travel in the auto-city

Paul Hess, Helen Hao Wen Huang, and Mirej Vasic

One salient characteristic of contemporary global and mobile capitalism is the massive international migration of populations, particularly from countries of the global South into some of the most auto-dependent countries in the world as measured by per capita auto ownership and use. This chapter focuses on how immigrant households adapt to the North American auto-city, with a case study of Toronto, Canada. Many immigrants aspire to the "American dream" of homeownership. In Canada, this is reflected in fairly rapid trajectories to homeownership for many immigrants, although immigrant households may take on levels of debt that make them very vulnerable to economic shocks (Hiebert 2010; Walks 2013b). Immigrants also increasingly settle directly into suburban auto-dependent environments that are part of this dream, in part because they rely on existing social networks to find housing and are priced out of gentrifying, central neighbourhoods where there are alternatives to automotive transport. These settlement patterns thereby reinforce and reproduce the patterns of the auto-city and thus automobility, but at the same time many new immigrants can only dream of the "coercive freedom" of the car to manage living in such auto-space (Urry 2004). Rather, many cannot afford a private vehicle, and those who can undergo the licensing process, which takes time. As such, recent immigrants, regardless of their economic status, often must rely on alternative means of transportation, at least for part of their transportation needs.

The outer suburbs and much of the inner suburbs of automobile-dependent cities, including the Greater Toronto Area (GTA) in Canada, are often not well served by attractive transit options. Networks are sparse with lower service frequencies and speeds relative to central cities, while suburban transit stops require long walks for access (Lo *et al.* 2011). Additionally, spatially dispersed employment, retail, and service locations mean that distances tend to be longer than in the central city. Indeed, settlement patterns characteristic of post-war cities planned and built for the automobile, coupled with the changing geography of work flowing from the economic restructuring of cities in the developed world, present major implications for how households travel to meet basic needs and raise challenges for economic and social integration.

One outcome of this can be a "spatial mismatch" between jobs and homes, resulting in employment opportunities located far away from where new immigrants reside (see Chapter 6 by Mendez *et al.* in this volume). This problem cannot be dealt with using transit networks primarily designed for commuting downtown (Blumenberg and Shiki 2003; Blumenberg 2004b; Mees 2009). Furthermore, transit tends to run in sync with the typical nine-to-five business day, failing households with work hours that extend beyond prime operating times or who are reliant on transit for non-work travel (Cervero *et al.* 2002; Garnett 2001), a situation that disproportionately describes many new immigrants. This disparity becomes even greater among households with children, where often women must work non-standard hours (Dobbs 2005; Blumenberg 2004a; Presser and Cox 1997). In general, the mobility needs of carless suburban travellers receive little attention from transit operators (Kawabata 2003; Garrett and Taylor 1999). Within this context, recent immigrants, even those that may not have low incomes but have particular travel patterns and needs, are not well understood or accounted for (Frisken and Wallace 2008; Hemily 2004). Furthermore, the interaction between the immigration process, in which migrants adapt their cultural traditions in order to function in the labour markets and communities of their adopted country, and structured transportation networks of the auto-dependent city may provoke new kinds of social exclusions.

Exploring the household location decisions and transportation strategies of recent immigrants in such auto-dependent environments helps give an insight into the ways that the motor vehicle, as the dominant mode of travel, permeates almost all social institutions and experiences, but does so in varied and complex ways mediated by cultural practices. This chapter focuses on how immigrant households adapt to the North American auto-city, using the GTA as a case study. This chapter begins by reviewing the existing literature on immigrants, gender, and travel behaviour, and then moves to examine the mobility experiences of immigrants in the city. The specific experiences of two different immigrant groups are analysed through a survey of recent Chinese and Iranian immigrants residing in auto-dependent suburban areas of the GTA.[1] The empirical work informs an understanding of the complex ways culture and the internal dynamics of inter- and intra-household relationships are shaped by automobility, and how automobility may be related to particular kinds of social exclusion experienced by immigrants and women.

Immigrants and cultural, ethnic, and gender patterns of travel

For immigrants in highly automobile-dependent countries such as the United States (US) and Canada, the expense of owning a car, lack of credit history for auto financing, licensing procedures and barriers (as discussed by Reid-Musson, Chapter 10 in this volume), and previous lack of driving experience, among other factors, can make it difficult or unfeasible for immigrants to own and/or operate a car upon settlement (Blumenberg and Evans 2010; Smart 2010). It is not surprising then, that US and Canadian research on recent immigrants' travel patterns finds lower use of personal vehicles compared to non-immigrant populations along a number of dimensions, and that this difference lessens over time (Myers 1997; Heisz and Schellenberg 2004; Cass *et al.* 2005; Chatman and Klein 2009; Blumenberg and Song 2008; Mercado *et al.* 2012; Ma and Srinivasan 2010).

In the US, immigrants are "more spatially mismatched" with places of employment than is the non-immigrant white population (Liu and Painter 2012a). Tal and Handy (2010) report that recent immigrants generate low auto mileage compared to non-immigrants, controlling for a number of demographic variables. Likewise, Chatman and Klein (2009) report that immigrants commute less by auto, but that when they do, they travel farther than native-born populations. For non-work trips, US research suggests immigrants make shorter and fewer trips using a single-occupancy vehicle compared to non-immigrants, but again, car use is found to increase with extended residence in the country (Chatman 2013). Using 1990 and 2000 census data, Ma and Srinivasan (2010) also find cohort effects, with more recent arrivals (those who arrived between 1990 and 2000) having a higher car ownership rate than earlier cohorts. Research on the travel behaviour of immigrants to Canada is not as developed, but most findings are consistent with U.S work, with lower personal vehicle use for work travel compared to native-born populations (Heisz and Schellenberg 2004; Mercado *et al.* 2012).

In concert with the pattern for motor vehicles, immigrants use public transit and other alternative modes such as walking and cycling more than the non-immigrant population in both the US and Canada, with use of alternative modes declining as driving rates increase over time (Chatman and Klein 2009; Kim 2009; Taylor *et al.* 2009; Purvis 2003; Blumenberg and Evans 2010; Tal and Handy 2010). For example, Myers (1997) reports that after ten years of residence, immigrants markedly reduce their use of public transportation and increase car use, especially in the case of women, who start to drive alone as they live longer in the country. Canadian research again shows similar patterns (Heisz and Schellenberg 2004; Mercado *et al.* 2012). In Toronto's very suburban, auto-oriented municipalities, one-quarter of immigrants use public transit (Heisz and Schellenberg 2004). However, usage rates for those who have lived in Canada for more than 20 years are akin to those of the native-born population, even when holding age and income constant (Heisz and Schellenberg 2004). This suggests that access to a car plays a key role in assimilating immigrants into the North American auto-city. In terms of alternative modes, Blumenberg and Smart (2010) find that among recent immigrants, 9 per cent of trips are taken by foot, bicycle, and public transit, compared to 3.5 per cent among the US population as a whole, and Smart (2010) finds that immigrants are twice as likely to cycle as non-immigrant Americans, but this likelihood quickly decreases, with the odds of cycling halving with the first four years of residence.

Immigrant groups in Canada and the US also cope with lower automobile access by carpooling or ridesharing to offset the costs of auto ownership and to provide rides to individuals who require them but are unable to drive due to financial constraints or lack of skill (Chatman and Klein 2009; Liu and Painter

2012b). Like public transit use, ridesharing rates are higher among recent immigrants, compared to older and non-immigrant households, but decrease with time in residence (Cline *et al.* 2009; Myers 1997; Pisarski 2006; Purvis 2003). Immigrants have both high ridesharing rates among members of the same household (internal ridesharing) and between members of different households (external ridesharing), and ridesharing is organized for the purposes of getting to school or work, shopping, running errands, and going to church or social gatherings with friends, neighbours, family, ex-spouses, co-workers, and sometimes even strangers (Blumenberg and Smart 2010; Lovejoy and Handy 2011). Blumenberg and Smart (2010) also suggest that as time of residence in the US increases, immigrants are more likely to choose ridesharing over transit. They hypothesize that ridesharing is a response to low levels of resources that lead immigrants to make use of familial networks and use their social capital to "maximize the utility of their limited resources" (2010: 442).

Ridesharing between household members and between different households is clearly shaped by culturally bound social relations, as is all travel behaviour. For example, Blumenberg (2009) finds statistically significant differences in carpooling rates among various ethno-cultural immigrant groups, even when controlling for other socio-demographic variables such as income, with Hispanic immigrants being mostly likely to rely on external carpools, but also assimilating to cars and solo commuting faster than other groups. Such differences are seen with all modes, with, for example, Smart (2010) reporting that US immigrants from East and South-east Asia are more likely to cycle than use a car compared to other immigrant groups and non-immigrant Americans, and Heisz and Schellenberg (2004) reporting that transit commuting by immigrants in Toronto and Montreal is most likely among immigrants from the Caribbean, South-east Asia, Central and South America, and the US, and least likely among immigrants from East Asia, Europe, West Asia, and Oceania.

Gender, of course, adds another extremely important layer of complexity. It has long been established that transportation patterns are highly gendered (Fava 1980; Rosenbloom 1978; England 1991; McQuaid and Chen 2012; Best and Lanzendorf 2005; McGuckin and Murakami 1999; Turner and Niemeier 1997). Especially in the early post-war period characterized by the rapid rise of automobility, women had less access to the family car, and women residing in the auto-mobile suburbs reported higher rates of isolation, depression, and underemployment as a result (Fava 1980). Women have also been less likely to commute by carpool and transit than men (Kim 2009; Myers 1997). When coupled with low-income, however, women commute more using the public transit system than low-income men do (Mensah 1995; Limtanakool *et al.* 2006). In terms of time spent commuting, research in the US and Europe suggests that men travel longer and women travel shorter distances when they have dependants (McQuaid and Chen 2012; Best and Lanzendorf 2005; McGuckin and Murakami 1999).

Among the scant research that examines both immigrants and gender, Blumenberg (2009) finds that while native-born women are more likely than native-born men to drive alone to work, female immigrants are more likely to rely on alternative modes of transport than men. As time in the US increases, however, male immigrants drive alone more as they assimilate faster to the travel patterns of the native-born. In contrast, Smart (2010) finds that immigrant women are half as likely to ride a bicycle as American women. In Canada, Heisz and Schellenberg (2004) also report this "gendered commuting pattern," finding that women are more likely to commute to work by public transit, with gender differences larger for recent newcomers than long-term residents and non-immigrants. In fact, they find a 17 per cent difference in the number of male and female immigrant public transit commuters.

Reasons for gendered immigrant travel patterns include differences between male and female experiences as drivers when they arrive (Tal and Handy 2010), income and employment constraints that may result from the lower wages of women (Mercado *et al.* 2012; McQuaid and Chen 2012), and higher household maintenance responsibilities for women, which constrains their work travel because they must also undertake household maintenance trips including shopping and escorting children (Best and Lanzendorf 2005; Parks 2004). A number of studies also show differential access to the private car, with male mobility for employment treated as a priority over female mobility (Tal and Handy 2010; Lovejoy and Handy 2011; Hamilton and Jenkins 2000; Anggraini *et al.* 2008), often leaving women to fulfil both work and non-work travel by other means. These types of patterns are likely filtered through culturally specific gender expectations around driving, employment, and household responsibilities.

Such differential use of automobiles raises serious concerns about how suburban autoscapes drive patterns of social exclusion (for example, Litman 2003; Cass *et al.* 2005; Bergmann and Sager 2008; Currie *et al.* 2009). There are many ways to define and measure social exclusion, but the literature generally refers to people's ability to participate in the broad range of social activities and institutions of society, including employment, public and private services, and social activities. In terms of transport, the issue is in the ways that transport disadvantage and social disadvantage may intersect (Lucas 2012).

Surveying immigrant automobility

As one of the globe's foremost immigrant reception metropolises, the Greater Toronto Area (GTA), Canada's largest urban region and its pre-eminent global city, represents an excellent case study for assessing the mobility of recent immigrants and their varying access to transportation, employment, services, and amenities. According to 2006 data, the Toronto Census Metropolitan Area of over five million people hosts nearly a quarter of Canada's recent immigrant population, and foreign-born residents represent 46 per cent of the total metropolitan population, more than any other large city on the globe except for Dubai (Statistics Canada 2007). The region's major influx of immigrants has taken place within a context of increasing auto-ownership rates and rapid suburbanization of population and employment into areas poorly served by transit (Miller and Shalaby 2003). All of these factors have fundamentally altered the metropolitan dynamics of the region. Research examining the settlement patterns of recent immigrants shows that while the less affluent often find housing in lower-cost apartments in the older, inner suburban areas, wealthier newcomers increasingly settle in newer, outer-ring suburbs (Hiebert 2000; Murdie and Teixeira 2003; Walks and Bourne 2006), with both groups bypassing traditional immigrant reception areas in the older, now gentrifying, central city (Walks and Maaranen 2008).

The GTA comprises a single-tier municipality[2] (Toronto) and four regional municipalities[3] (Durham, Halton, Peel, and York), which have a number of lower-tier municipalities within them. The Toronto Transit Commission (TTC) serves downtown Toronto and its older post-war "inner" suburbs (which are all contained within the City of Toronto boundaries, formed in 1998). The TTC consists of a grid network of surface routes (buses and streetcars) that feed into a 70-km radial subway system. The remainder of the GTA is served by transit agencies associated with the regional municipalities, which provide a bus service with varying levels of frequency. A regional commuter rail system focused on downtown Toronto also serves some corridors extending to the outer suburbs. Most transit is poorly integrated across different municipalities, making much regional travel challenging and costly.

Most existing transportation datasets do not provide information that is specific to immigrant travel beyond transportation mode to work. Census data remains limited (the Canadian census does not even provide data on automobile ownership). The authors therefore developed surveys to go beyond standard data sources to explore how different groups of recent immigrants in suburban areas of the GTA considered travel when they made residential location decisions, as well as their rates of vehicle ownership, attitudes towards transit and automobiles, and patterns of car-sharing within and between households. Iranian and Chinese immigrants were selected for the survey because of researcher familiarity, language skills, and connections to the communities.[4] The survey collected data on socio-economic status, modes used for work, attitudes towards various modes, the division and scheduling of household tasks, and the household dynamics around automobile use. It is worth iterating that we do not mean to imply that either group is homogeneous in terms of culture or attitudes. Instead, the survey is intended to explore broad differences in patterns as reported in the literature and raise questions regarding immigrant mobilities in the auto-city. Some 97 surveys from Iranian participants and 95 from Chinese participants, living in the suburbs of the GTA and aged 25 or older, are analysed for this chapter.

The Iranian group surveyed for this work is clustered in an area bounded by North York (one of the former "inner" suburbs, now part of the City of Toronto), Thornhill, and Richmond Hill (within York Region), areas with the highest settlement of Iranian immigrants in the region. The Yonge and Sheppard subway lines provide higher-order transit to parts of North York; however, most parts of the jurisdiction rely on TTC bus service. Thornhill and Richmond Hill rely on York Region bus service, which connects to TTC bus and subway service in North York. The Chinese immigrants surveyed for this work live predominately in Scarborough (another former suburb now within the City of Toronto), North York, and Markham (within York Region), with some living in Mississauga (within Peel Region) and Richmond Hill. The Chinese community is relatively more spread out compared to the Iranian group. Chinese respondents settle in areas to the east of the Iranian enclaves with some partial overlap with Iranian respondents. The locations of both sets of survey respondents (see Figure 9.1)

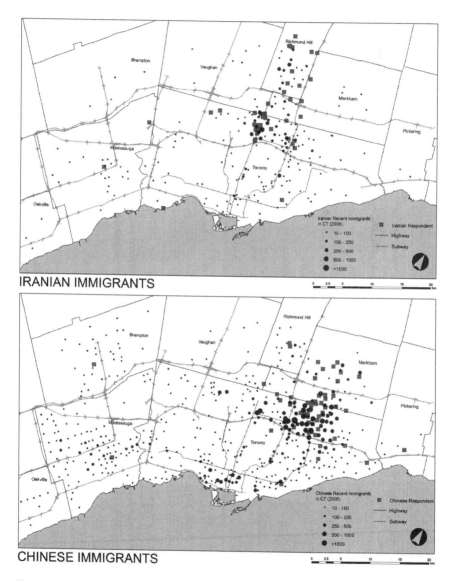

Figure 9.1 Survey respondents compared to location of residence of Iranian- and Chinese-origin populations, as reported in 2006 census of Canada (source: Created by the authors from Census of Canada, 2006).

are shown with the geographic location of Chinese- and Iranian-origin popula-tions in the region as reported in the 2006 census.

There are important differences between the samples that impact transporta-tion behaviour. About two-thirds of the Iranian sample is female, while closer to half the Chinese sample is female. The Iranian sample is also much less likely to be married (60 per cent of respondents versus 81 per cent of Chinese respond-ents), and less likely to live in detached houses (38 per cent versus 52 per cent), although 10 per cent of Chinese respondents did not answer this question. Household size for both groups is fairly large, with 42 per cent Iranian and 32 per cent of Chinese respondents reporting that they live in households with four people or more. As a group, the Chinese respondents also have slightly higher percentages of young children and lower percentages of households with elderly members.

The two sampled groups also have different income profiles, with 75 per cent of Iranian respondents reporting household incomes of $60,000 or more and a median between $80,000 and $89,000, which is above the median household income in the GTA at $77,685. In contrast, more than one-quarter of Chinese respondents reported having very low or no income (but non-response on this question was high). Similarly, more than two-thirds of Iranian respondents also reported working full-time compared to only 21 per cent of Chinese respondents, whereas a sizeable 56 per cent reported not working.

Settlement location

In order to gain a sense of what informs location choices, respondents were asked to comment on why they decided to settle where they did in the suburbs of the GTA (see Table 9.1). Cultural reasons ranked highly, especially for the Iranian group, with both knowing other people from their "ethnic community" and having relevant "ethnic shops and services" valued. Both groups also valued nearby family and friends, while Chinese respondents particularly valued "good schools". Both groups also valued "a very safe neighbourhood," but in much higher numbers in the Iranian groups – the mostly highly marked factor by this group – than the Chinese group. Interestingly, more Iranians than Chinese high-lighted affordability, even though their incomes appear to be higher.

Local work accessibility does not appear to have been a dominant consideration on residential location decisions for their either group, with less than 10 per cent answering "I have a job near where I live." Likewise, fairly low numbers of both groups listed "It is easy to drive to my job from here" as an important considera-tion. For the Chinese group, with low employment rates, the low importance of job accessibility may be understandable. Driving access is also affected by automobile ownership and access, as discussed below. Transit accessibility to work, on the other hand, was seen as fairly important, marked by nearly a third of Iranian respondents and more than one-fifth of respondents from the Chinese group.

There were also strong gender differences in attitudes towards location deci-sions and accessibility. In the Iranian group, many more women than men

Table 9.1 Reasons for location choice and attitude towards having a car

	Iranian	n = 97 (%)	Chinese	n = 95 (%)
Reason for location choice				
I have family in the area	21	21.6	18	18.9
I have friends in the area	15	15.5	17	17.9
I know many other people from my ethnic community who live in this area	37	38.1	10	10.5
There are many shops and services in this area related to my ethnic community	45	46.4	25	26.3
I have a job near where I live	7	7.2	9	9.5
I live here because it is affordable compared to other places	20	20.6	11	11.6
The schools are good here	4	4.1	26	27.4
It is a very safe neighbourhood	43	44.3	21	22.1
There is good transit service here to get to work	28	28.9	21	22.1
It is easy to drive to my job from here	12	12.4	5	5.3
Importance of having a car in Canada				
Very important	48	49.5	68	71.6
Fairly important	30	30.9	15	15.8
Somewhat	19	19.6	4	4.2
Not important	0	0.0	2	2.1
No data	–	–	6	6.3

Source: Authors' survey.

marked transit accessibility to work as being important, with the reverse pattern in the Chinese group, likely because few Chinese women in the sample have paid employment. Auto accessibility to work was seen as important by about one-quarter of men in the Iranian group, but of much lower importance for both Iranian women and either men or women in the Chinese group.

A majority of the Iranian group agreed that they would like to move to a neighbourhood with better transit, as did a third of the Chinese group. This was more marked by men in the Iranian group and slightly more by women in the Chinese group. Trading a bigger house and yard for less transit accessibility received relatively little support, with 14 per cent in the Iranian group and 18 per cent in the Chinese group, but was more supported by women in the Iranian group.

When respondents were asked whether they considered the availability of transit when moving to their current location, the Iranian sample was relatively evenly split, but more than two-thirds of the Chinese group answered "yes." Asked whether they considered the importance of owning a vehicle in their neighbourhood choice, only a third of the Iranian group, compared to over half the Chinese group, did so. Thus, as a whole, the Chinese group seemed to consider both modes simultaneously in their location decision.

That said, more than three-quarters of the Iranian group and almost half of the Chinese group reported that they would rather live in an area that is less automobile-oriented and more walkable. When choosing their location, the ways that transportation accessibility and walkability factors into decision-making are clearly complex. Many respondents from both groups did not fully consider the impact of living in a very auto-dominated area when finding housing, some may have been responding to cultural and social norms, even, as shown below, without the ability to fully partake in them due to having limited ability to drive.

Auto access, attitudes towards auto ownership, and household dynamics

The Iranian group is auto-licensed at slighter higher rate than the Chinese group, but given differences in income rates, it is surprising that this differential is not greater. Males are licensed at slightly higher rates than females for the Iranian group (69 vs 63 per cent), and, in the Chinese group, more than twice the percentage of males than females have licences (73 vs 31 per cent). Thus, many respondents, and especially Chinese females in the sample, are not able to drive even though they live in highly automobile-dependent environments.

More surprising than licensing rates are the number of households that do not own vehicles (Table 9.2). Just less than 40 per cent of both groups report being in a household with no vehicle, and another 49 per cent of Iranians and 43 per cent of Chinese respondents report being in a household with only one car. Comparing number of vehicles by household size further suggests that, while vehicle-ownership rates generally rise with household size, many households contain adults that very often do not have access to a vehicle. For example, 70 per cent

Table 9.2 Number of vehicles by household size by cultural group

Household size	Number of vehicles				
	0	*1*	*2*	*3+*	*Total*
Iranian					
1	13.3%	9.2%	1.0%	0%	23.5%
2	14.3%	6.1%	2.0%	0%	22.4%
3	1.0%	6.1%	3.1%	1.0%	11.2%
4 to 5	8.2%	25.5%	3.1%	1.0%	37.8%
6+	2.0%	1.0%	1.0%	1.0%	5.1%
Total	38.8%	48.0%	10.2%	3.1%	n = 98
Chinese					
1	8.9%	1.1%	0%	0%	10.0%
2	8.9%	7.8%	0%	0%	16.7%
3	13.3%	22.2%	3.3%	1.1%	40.0%
4 to 5	7.8%	12.2%	10.0%	0%	30.0%
6+	0%	0%	1.1%	2.2%	3.3%
Total	38.9%	43.3%	14.4%	3.3%	n = 90

Source: Authors' survey.

of four-to-five-person Iranian households report that they have only one vehicle, with most of the rest not having any. Clearly, adults in these households must rely on walking, transit, and other non-automotive modes, or must negotiate vehicle access for their transportation needs. In automotive-dependent environments, this can be very onerous, and, as is shown below, the burden mainly falls on women.

Low automobile access does not appear to be by choice, as almost 80 per cent of Iranian respondents and 86 per cent of Chinese respondents indicated that a car was fairly or very important for getting around in the Canadian city (as shown in Table 9.1). For those households that do have vehicles, virtually all of the Iranian and Chinese households in the sample acquired a vehicle in the first year of their arriving in Canada. Both groups report that they think it is "very convenient" to own a car, although car ownership is also seen as expensive, and many respondents, especially in the Chinese group, also report that they cannot afford a car.

Transit mode and travel to work

Responses related to commute mode show fairly similar patterns for the two cultural sample groups, with transit most commonly used, followed by driving (see Table 9.3). Ridesharing is much more common in the Iranian group. These differences are minor, however, in comparison to those related to gender. Among Iranians, transit use is higher for males, but females are much more likely to commute by transit in the Chinese group. Iranian males are somewhat more likely to drive to work than female Iranians (difference of 6 per cent), but the

Table 9.3 Usual work mode, use of vehicle, household responsibilities by gender (%)

	Iranian		Chinese	
	Male	*Female*	*Male*	*Female*
Usual work mode				
Drive	34.5	28.6	56.8	11.1
Get a ride with a family member	0	28.6	13.6	0
Take public transit	55.2	38.8	32.4	70.4
Bike	–	–	0	1.6
Walk	10.3	4.1	0	6.3
Most use of vehicle				
Me	45.7	19.4	79.4	24.0
My spouse	20.0	35.5	8.8	64.0
My child	–	11.8	8.0	–
Responsible for shopping				
Me	35.3	68.3	46.3	55.2
My spouse	29.4	4.8	12.2	10.3
We share shopping or go together	35.2	27.0	39.1	31.0

Source: Authors' survey.

difference in rates in the Chinese group is remarkable, with 57 per cent of males reporting driving versus only 11 per cent of Chinese females. This pattern needs to be better understood, especially as such a large proportion of the Chinese cohort reports not working at all, but this accords with licensing rates among this sample. In general, women have less access to automobiles than men, dramatically so in the Chinese group. Automobility is thus shown to be associated with gender inequalities, with the automobile reinscribed as a gendered technology (Walsh 2008; Uteng 2009; Best and Lanzendorf 2005).

This is also reflected in how cars are shared within and between households. Our survey probed further into differences in male and female access to and use of cars within households (Table 9.3). In households with male–female couples, both males and females agree that the males are much more likely to use the car. In parallel to the gender and commuting patterns, this is stronger in the Chinese group, with almost 80 per cent of males stating they have most access to a car.

Car access has, of course, a large impact on carrying out household maintenance activities that involve travel. The division of household responsibilities was explored through several questions. Men and women in spousal couples also generally agree that women are much more responsible for household maintenance activities like grocery shopping and taking children to school. The results for grocery shopping are shown in Table 9.3. While women are more responsible in both groups, interestingly, this is shared more within the Chinese group. Still, the general patterns raise questions for how women manage these activities in automobile-oriented environments without access to a vehicle. In a previous study, women reported that they mix walking with taking transit to balance travel expenses and accessing dispersed locations like shopping areas (Hess and Farrow 2012). We expect the households in this sample, which have a larger income range, to use a variety of strategies, one of which is ridesharing.

Ridesharing

To get a sense of how people who do not own a vehicle and yet live in a very auto-oriented area negotiate their daily-life travel, respondents were asked to comment on the extent to which they give and receive rides. Results show that among the Iranian group, females are more likely to be the driver in a carpool scenario not involving immediate family than are men, with more than half reporting they give rides to others "often" or "sometimes." In the Chinese group, giving rides "sometimes" is fairly common, with 44 to 45 per cent of both men and women reporting they do so (Table 9.4). Given the much higher access men have to vehicles in the Chinese group, however, this still suggests that, in relative terms, it is women that are choosing to give rides to those outside their households.

In terms of receiving rides, within both sampled groups women report receiving rides, both to places the driver "would have gone anyway" and to places the driver "would not have gone anyway," at much higher rates than men. There are varying numbers of both men and women within each cultural group who wished they were receiving rides. Not surprisingly, given the higher auto access

Table 9.4 Ridesharing (%)

	Iranian		Chinese	
	Male	Female	Male	Female
Do you give other people rides to destinations that you are also going to?				
No	78.3	44.4	54.8	52.0
Sometimes	21.7	38.9	45.2	44.0
Often	0.0	16.7	0.0	4.0
How often do you receive rides from others to locations they would travelled to anyway?				
Often	0.0	7.5	2.5	3.6
Sometimes	16.7	29.8	22.5	32.1
Rarely	33.3	31.3	37.5	35.7
Never, but I wish I could	16.7	14.9	20.0	14.3
Never, but I don't feel I need to get rides from others	33.3	16.4	17.5	14.3
How often do you receive rides from other to locations they would not have gone to anyway?				
Often	0.0	5.8	2.5	7.0
Sometimes	0.0	14.5	12.5	24.1
Rarely	21.4	29.0	35.0	24.1
Never, but I wish I could	21.4	29.0	15.0	10.3
Never, but I don't feel I need to get rides from others	57.1	21.7	35.0	31.0

Source: Authors' survey.

of men, there were many more men within each group that felt they did not need to rideshare. These differences give only a hint of the dynamics at play within and between households as members, especially women, seek mobility.

Immigrant automobilities

Immigrant communities must adjust to living in highly automobile-oriented suburbs in cities such as Toronto. Yet several important themes emerge from this analysis that are worth examining. First of all, location choice in suburban, automobile-oriented environments is clearly driven by cultural reasons as much as or even more than the accessibility and transportation requirements of the location, including the desire to encounter people from similar ethnic back-grounds and access ethnics shops, to locate close to friends and family, as well as differential perceptions of safety. Given the importance of the automobile for travel in suburban environments and the fact that many immigrant households appear to have limited access to cars, that automobile travel is not more con-sciously considered is striking. It is important to further explore whether people are making a joint choice of housing and accessibility within a household. Most people in the survey, especially Iranians, reported that they would prefer to live in a more walkable, transit-accessible environment, but this does not rank highly in their actual location choice. Whether respondents discounted more central, transit- and walking-oriented neighbourhoods because of perceived safety, housing quality, or preference of interior or exterior space (see Walks 2006), or are simply compelled to live in the suburbs due to affordability, has important implications for the relationship between cultural adaptability and integration and automobility. As the auto-city creates new logics of safety and alters traditional trade-offs between accessibility and space, these feed into cultural appropriations necessary for negotiating its contradictions.

With large, multi-generational households, many surveyed households have complicated travel needs. Cars were generally seen as being important to life in Canada, and about two-thirds of Iranian men and women in the sample have driver's licences, as do almost three-quarters of Chinese men. Less than one-third of Chinese women do, however, and for both groups many households either have no car or have more adults needing to access jobs and services than they have cars in their household. This can be seen in the high transit usage for work trips, which accounts for almost half of those travelling to work, even though transit service is limited. Thus, many adult household members do not have access to a vehicle at times when they might need one, or households do careful scheduling and engage in intra- and perhaps inter-car sharing (see below). Unsurprisingly, the desire for car ownership seems to be high, and of those households that acquire a car, many do so in the first year of residence. The fact that auto-ownership rates are low is not only related to lack of experience and capacity (licensing, etc.) but even more to the expense of ownership.

Importantly, the burden of *not* owning or having regular access to a vehicle is borne mostly by women. In the Iranian group, men and women are employed in roughly equal ratios, with men being only slightly more likely to have full-time employment. Yet, both men and women in the survey agree that it is women who are mostly responsible for household maintenance activities, that men are about twice as likely to more often use a household auto, and many more men than women report that they have almost full-time access to the household vehicle. In the Chinese group, women are less likely to work and, additionally, have much lower car access. Of those that work, only 11 per cent drive compared to 57 per cent of Chinese men in the sample, and men and women agree that a male spouse is roughly three times as likely to have most access to a household car. The survey also begins to help understand strategies used by women in coping with lower car access in these environments. Women in the Iranian group, who have better auto access than the Chinese women in the sample, report giving rides to others at much higher rates than do men. In both groups women are also more likely to report being a passenger in a carpool, compared to men.

Conclusion: driving immigrant adaptability

These initial findings largely correspond to those in the existing literature, but further emphasize the complexity of immigrant residential location and transportation choices, and intra- and inter-household behaviours around vehicle access and use, as they are shaped in complex ways by culture and gender. Indeed, the barriers and accessibility constraints faced by immigrants as they increasingly settle in automobile-oriented suburban environments gives an insight into the pervasiveness of automobility in the organization of multiple aspects of social organization, including household dynamics and gender negotiations around the logics of automobile travel. In many large, economically dynamic "global" cities, including Toronto, there has been a shift towards downtown high-rise condo development and rapid gentrification of older, low-rise neighbourhoods that are well-serviced by transit and have highly walkable environments (Rosen and Walks 2013). Focus on such ideas as "recentralization" and "peak car," as discussed elsewhere in this volume, may disguise the less visible reality that enormous growth continues to occur in auto-oriented suburban areas. Even in Toronto, which likes to boast of its dizzying production of central high-rise condominiums, more than five times as much growth is occurring on the periphery (Gordon 2013).

In places where car dependence is high, levels of public transportation may be lacking, and the physical surroundings are unfriendly to alternative transportation modes, households with limited access to cars face constrained access to jobs and services. These transportation barriers disproportionately affect new immigrants in the auto-dependent city, particularly when these individuals are moving to the suburbs in greater numbers. The inability of many to own a car upon arrival stands in their way to full accessibility for the purposes of social inclusion as well as economic opportunities. The disadvantages of not fully participating in the system of auto-based mobility act to perpetuate and exacerbate social disadvantages faced by new immigrants.

Notes

1 These two groups were chosen due to researchers' familiarity and links to the communities that facilitated recruitment. Neither should be seen as homogeneous groups in terms of socio-demographic status, nor in terms of culture or ethnicity; both populations contain important internal variation. As larger groups, however, they are from sufficiently different cultural areas that the authors believe it is useful to explore, if only in broad terms, whether there are differences in settlement preferences and mobility strategies as they face living, often for the first time, in a highly auto-dependent context.

2 A "single-tier" municipality is a municipality where there is only one level of municipal government.

3 A "regional municipality," also referred to as an "upper-tiered" municipality, is similar to a county, but generally provides more services.

4 Data was gathered through an online survey and a paper survey. The survey reached participants through immigrant settlement organizations and community groups in the Iranian and Chinese communities. In particular, recent Iranian immigrants were reached through an email list provided by the Toronto Iranian Association; recent Chinese immigrants were reached through the Centre for Information and Community Services and local Chinese websites. As the survey was voluntary, the sample was not random and should not be seen as fully representative.

10 Automobility's others

Migrant mobility, citizenship, and racialization

Emily Reid-Musson

After their first day of work in Canada, ten Peruvian men were killed as they travelled in a van on a dark February evening in 2012. The collision occurred at a rural, four-way intersection near Hampstead, Ontario. The men had spent a 12-hour workday vaccinating chickens, bookended by an hour-long commute each way to and from their worksite. By virtue of the van driver's error a truck hit the van, killing ten of the 12 men. Major farm worker transportation-related fatalities are not without precedent in the United States (US) and Canada (CBC News 2011; California Highway Patrol 2002; Work Safe BC 2007). But this collision brought to public attention the presence of approximately 20,000 workers in Ontario's large agricultural sector who work and live under *permanently temporary* "migrant" status (Binford 2013; ESDC 2013a: 36–39; Rajkumar *et al.* 2012). Over two decades, there have also been nine fatalities and numerous major injuries incurred by migrant farm workers as bicyclists in south-western Ontario. Unlike distinctly work-related fatalities, migrant farm workers typically use bikes to leave their work-live sites. Non-work transportation-related vulnerabilities fall within a grey area at the limits of what meagre workplace protections currently exist. These fatalities are frequently characterized as "accidents" resulting from individual responsibility and error.[1]

This chapter focuses on cases from the US and Canada where precarious citizenship is in part made, contested, and negotiated via automobility, at scales removed from but certainly interfacing with formal citizenship policy. In both countries, employers in low-wage sectors rely on undocumented migrant workers as well as guest worker programmes, though the scale and proportion between these populations in each jurisdiction differs. US guest worker programmes (e.g. H2-A and H2-B visas) are relatively small, dwarfed by the undocumented worker population (Passel 2006). Conditions in US and Canadian guest worker programmes vary, but both are characteristic of distinctly modern, globalized forms of state-managed labour migrant arrangements (Hahamovitch 2003). As "foreign workers" according to Canadian law, guest workers face no overt legal constraints on their entitlements to drive or to own private vehicles, or on their physical mobility. Yet, they rarely gain independent access to private cars. In contrast, undocumented migrants in the US are more likely to rely on and use private cars despite lacking federal immigration status. The purpose of highlighting migrants' divergent automobility entitlements serves to discuss how multiple, inter-scalar and inter-jurisdictional forces co-constitute migrants' citizenship and physical mobility through automobility. Responding to calls to de-essentialize automobility and mobility studies (Cresswell 2010; Henderson 2006), this chapter investigates how, through automobility, migrants' rights and mobility are both extended and curtailed.

The chapter is organized in three sections. First, critical mobilities and automobility research provides a theoretical basis to discuss how migrants' mobility and racialization are mutually produced. Second, I review US-oriented literature documenting how migrants' access to auto insurance and licensing has become a site of re-scaled immigration control. Finally I examine how migrant farm workers' local travel and mobility is produced and negotiated in relation to automobility in rural south-western Ontario. I document how workers, employers, and local non-migrant groups tacitly confront and interpret the embedded, dominant character of automobility in rural spaces as it organizes national–racial differences through physical, everyday mobility.

Automobility, im/mobility, and racialization

A central goal of mobilities research is to de-essentialize mobilities in particular historical and geographical contexts (Adey 2010; Cresswell 2010). How mobility is governed, how it is experienced and felt, what it means, and how it is represented is somewhat difficult to divide into sub-disciplinary silos like transportation geography, migration research, or political geography. Racialization is intimately affected by and reciprocally constitutive of modern human mobility, providing an ideological framework for the transatlantic slave trade and the dispossession of indigenous people, two of the most significant instances of forced migration in modern human history. More accurately understood as a process, *racialization* underscores how racial formations are necessarily coded and given meaning in relation to place and space (Roberts and Mahtani 2010; Saldanha 2009). Others have argued that "mobility regimes" or "global apartheid" regulate migrants' travel through exceptional techniques of deterrence, detention, and constraint, while the hyper-mobility of capital, commodities, and elite travellers unevenly compresses space–time (Hardt and Negri 2004: 160–166; Mountz 2011). The disposability and regulation of migrants as risky surplus populations mimic and reproduce colonial power in new ways. Feminist geographers have examined how gender ideologies rooted in notions of passivity, docility, and domesticity legitimize migrants' socio-spatial immobility and confinement in the institutionalized governance of feminized labour migration (Conlon 2011).

Automobility concretely expresses and symbolizes a far-reaching modern liberal desire for and normative ethic of free individual mobility, concretizing liberal theory in everyday life. Much like liberalism itself, automobility rests on distinctly illiberal foundations (Packer 2008; Rajan 2006). The "coercive freedom" of the car produces complicated attachments to the often very real material power that owning and driving a car can provide (Sheller 2004).

Automobility is legally circumscribed by a regulatory apparatus at various levels of the state, but shapes and is shaped by otherwise distinct spaces and institutions. It can determine livelihood, belonging, and membership within particular jurisdictions. It is an often-overlooked site where citizenship – the social and legal regulation of national belonging – is concretely articulated. While the passport acts as a technology regulating cross-border travel, the driver's licence is one of the most important legal documents facilitating physical and social mobility within national jurisdictions. It is an important de facto social right in the Marshallian sense. In turn, access to cars shapes migrants' reliance on non-car modes of travel like biking, walking, and public transit (Blumenberg and Smart 2010; Smart 2010). There has been insufficient attention paid to North American "non-drivers" in automobility studies, and little research that deliberately and critically examines how automobility, citizenship, and racialization dovetail as co-constitutive and spatialized processes (although see Henderson 2006).

The intertwined construction of race, automobility, and citizenship legitimize how mobility is governed at different scales, and how mobility (like space and place) is constitutive of racial difference. This chapter considers how geopolitical interventions into non-citizens' mobility act as contemporary racial formations from which migrants' physical, everyday movement is inseparable. People's everyday physical movement, encompassing transportation and social reproduction, is inseparable from the regulation and practice of longer-distance human migration and travel. Whether and how you commute to and from home and work depends on one's embodied ability to move, how mobility is regulated within and across national jurisdictions, the terms of bilateral labour migration agreements, and a host of other factors. The two sections that follow highlight the connections between migrant status, racialization processes, and automobility in the construction of citizenship.

Immigration control, automobility, and migrants' livelihoods in the US

"What do we want?"
"Driver's licenses!"
"When do we want them?"
"Now!"[2]

Automobility and migrant rights have become increasingly politicized as multiple levels of the US government respond to pressures from grassroots anti-migrant and pro-migrant activists to include or deter undocumented migration (Ansley 2010; Pabon Lopez 2004; Slack 2007). Undocumented migrants' access to car licensing has been curtailed in nearly all US states, while numerous local police and traffic authorities have gained immigration enforcement powers. These patterns have been unevenly witnessed across US cities and states, but can be understood as a broad, post-9/11 shift towards new sub-national forms of immigration control as they intersect with the securitization of automobility governance. As previously mentioned, the driver's licence is a technology of citizenship. Undocumented migrants' capacity to live and work in particular jurisdictions is in part shaped by the ability to drive, irrespective of formal federal legal status. Driver's licences also act as close-to-universal forms of identification, necessary to open bank accounts, travel by plane, or rent housing (Varsanyi 2006: 246).

Until the late 1990s undocumented people in the US were able to legally drive in most state jurisdictions. Over several decades, this practice was curtailed under strong pressure from grassroots immigration policy activism. In 1986, 1996, 2001, and 2006, a range of legislative measures were passed by the US federal government alongside numerous "locally generated grassroots policies" that tackled the presence of undocumented residents (Wells 2004: 1308–1310). These measures encompassed both exclusionary and inclusionary policies. These measures reversed the 100-year-old ruling that the US federal government possessed "plenary power", or sole authority, over immigration policy (the geopolitical power over the legitimate means of exclusion, expulsion, and entry from and to the US). Notably, these shifts overturned the legal obligation of subnational levels of government to ensure the welfare of all residents of their jurisdiction on the principle of non-discrimination. Through section 287(g) of the US Immigration and Naturalization Act and the post-9/11 Secure Communities programme, municipal, county, and state agencies (such as police, traffic, and by-law agencies) have been able to enter into agreements with federal immigration enforcement authorities to act on the latter's behalf (Coleman 2012). Together, these shifts have "activated a formidable battery of legal and material resources that limit the movements, rights and opportunities of immigrants in the United States" (Varsanyi 2010: 129). These "grassroots" efforts indirectly yet intentionally target undocumented people's access to services required to sustain basic livelihoods. I argue that they are also an under-recognized facet of a transformation in the governance of automobility from a problematic of public safety to one of national security (Packer 2008: 267–292).

California was one of the first American states to pass a law in the 1990s that prevented an estimated two million undocumented California drivers from obtaining or renewing driver's licences. In 2003, a bill was passed that removed these restrictions, but Arnold Schwarzenegger overturned the bill in his first act as governor. Prior to 9/11, few states other than California required legal immigration status in order to legally drive and own cars. By 2010, most US states required would-be driver's licence holders to hold a valid social security number and 42 states required drivers to demonstrate proof of their legal presence in the country (Seif 2010). For example, North Carolina has attempted to pass numerous anti-immigrant bills, yet the only one that has passed in the legislature concerned vehicle licensing (Furuseth and Smith 2010). Further, the federal government passed the REAL ID Act in 2005,[3] making immigration status mandatory for all motor vehicle licensing departments at the state level. States that issue licences to undocumented people have to indicate by design or colour that this is so on the licence itself, in order that the licence not be used as an identification facilitating the settlement or movement of the user (Seif 2010).

This has direct implications for migrants' reliance on informal and non-car forms of transportation. In analyses of the US 2001 National Household Travel Survey, new immigrants were shown to be more likely than non-immigrants ("native-born" Americans) to travel by bicycle, even in comparison to non-immigrants within the same income brackets (Smart 2010: 153). Substantive studies into why different groups of new, low-income immigrants bike more are sparse, especially when it comes to identifying and addressing non-income-related barriers to private car access and determinants of bike use (Lovejoy and Handy 2011). Alternative transportation practices such as *camionetas* and *raitero* systems[4] have emerged where immigration control efforts and automobility disentitlements create transportation service gaps for undocumented migrants (see Valenzuela *et al.* 2005). In Los Angeles, bike activists have begun pushing planners and policy makers to consider the city's "invisible riders" (Koeppel 2005), referring to (mainly Latino) immigrants who depend primarily on bicycles for daily transportation. In major cities across the US like New York City, Chicago, and Los Angeles, there are significant numbers of bicyclists who qualify as such "invisible riders": people of colour with precarious legal status and low incomes whose absence in formal bike-planning processes is seldom noticed or questioned.

The physical mobility of undocumented people in the US today is regulated within as well as across US borders. It is increasingly difficult to obtain legal driver's licences, while some local traffic and police authorities have gained new immigration enforcement powers (Coleman 2012). This is a dramatic shift that highlights how automobility is shaped through site-specific politics, social relations, and attachments rather than as a universal phenomenon. Through automobility, migrant "illegality" is a "spatialized condition" that extends "the physical borders of nation-states in the everyday life of innumerable places throughout the interiors of the migrant-receiving state" (De Genova 2002: 439). This at once diffuse and grassroots, yet extensive re-scaling of "illegal" immigration control highlights how automobility organizes US politics, identity, and social relations. Nation, citizenship, and race are enacted in everyday mobility, but the forms that difference and mobility take are not all alike. Taken-for-granted racial-national ideologies underpin the governance and representation of secure, "normal" driving subjects. The ideological meanings that give automobility social and legal weight are subject to politically charged transformations and divergences. Shifts in the governance of automobility from a problematic of public safety to one of national security have not been racially benign. Rather, they are spatial expressions of national and racial difference directed at deepening the non-belonging of one of the most disenfranchised minority groups in US society.

Migrant immobility in rural and small-town Ontario

This section explores the social and extra-legal forces that constrain migrant farm workers' mobility in south-western Ontario (on British Colombia, see Tomic and Trumper 2012). This section draws on the author's ethnographic research in several rural Ontario communities in the south-west of the province.[5] This is the centre of the third largest food cluster in North America and a significant regional economic sector in Ontario. Total revenues in the agriculture and food-processing industries combined superseded auto-manufacturing revenues in the province in 2010 (Alliance of Ontario Food Producers 2012: 2).[6] The Seasonal Agricultural Worker Program (SAWP) is one of the longest-standing Canadian migrant worker programmes (in operation since 1966). However, within the past decade agricultural employers have benefited from access to new migrant farm worker programmes beyond the SAWP, which have added greater flexibility to an already precarious, flexible migrant farm workforce (Preibisch 2010). Most, though not all, of my research participants migrated under the SAWP and were primarily men from Mexico, Trinidad and Tobago, and Jamaica.

In contrast to those discussed in the previous section, migrants working in Ontario agriculture face no overt, legally enforceable exclusions from accessing driver's licences or private car ownership, but few gain access to private vehicles and must negotiate systemic immobility and confinement in rural work-live spaces. Further, they exercise extremely limited labour mobility, remaining tied to one employer through closed work permits, and are typically housed on employer property (Basok 2002; Preibisch 2010). As "low-skill" migrants from the global South they are not provided with any formal channels to immigrate permanently to Canada. This exclusion is maintained through racist federal legislation that prevents the settlement of poor people of colour who, lacking adequate education, language skills, and capital, are effectively barred from immigrating permanently through an overtly post-racial, "multicultural" immigration system (Perry 2012; Sharma 2006; Satzewich 1991). As such, they are bound by citizenship and work conditions that produce distinctly *unfree, migrant* forms of labour power.

Within the past decade, "temporary" labour migration has reached an unprecedented rate in Canada, accelerating under employer-driven demand for workers in low-wage, low-skill jobs (in energy, agriculture, construction, services, and care-giving) through the federal government's Temporary Foreign Worker Program (TFWP). By 2008, the number of persons entering Canada as "migrant workers" (through temporary work permits) for the first time exceeded those entering as "economic immigrants" with the capacity to gain Canadian citizenship (Faraday 2012: 10). There were 338,000 foreign workers in Canada as of 1 December 2012, according to recent government statistics (Citizenship and Immigration Canada 2013: 2). Both undocumented and authorized migration to and from Canada is understood in recent research as inter-connected, institutionalized forms of "precarious citizenship" produced through Canadian policy and law (Goldring and Landolt 2013). Unlike migration to the US, a greater proportion of migrant workers arrive in Canada with legal status (e.g. as temporary foreign workers). Yet their status often subsequently lapses as there are few viable channels with which to maintain it (typically, people become undocumented when they are unable to renew their work permits or meet "points-system" criteria necessary to gain permanent resident status). The undocumented population in Canada is roughly estimated to be between 50,000 and 200,000 (Magalhaes *et al.* 2010), less than the formal migrant worker population in Canada, but not insubstantial.

Migrants on south-western Ontario farms work long hours on the farms and the majority live adjacent to their worksites.[7] Nearly all research participants were working seven days a week with Sunday afternoons and occasional full days off. A full day of work is on average ten hours, while during times of higher production (harvesting, etc.) daily work hours can increase to 14–18 per day.[8] Only one research participant had moved out of employer-provided housing, was renting an apartment, and had bought a car (a relatively rare occurrence). The majority of migrant workers depend on employer transportation, taxis, and bikes. Employers are obliged to provide weekly or bi-weekly transportation for basic needs like groceries and money wiring. It is unclear where the legal basis for employers' local transportation obligation to workers lies. While obligations regarding the provision of housing, international travel costs, and provision for travel between Canadian airports and farms are all outlined in formal employment contracts, the only local transportation-relevant obligation of employers in the Mexico–Canada SAWP contract is to be "responsible for arranging transportation to a hospital or clinic" (ESDC 2013b: 4). For the most part, migrants' working season in Canada is spent confined to work-live spaces.

In migrant-receiving communities – small towns and rural areas – one might encounter hundreds or more men of colour in public spaces during tightly controlled weekly excursions. These communities appear otherwise largely racially homogeneous. Few migrant workers have access to private cars for their independent use, while the few public transit networks in rural municipalities that do exist follow routes and hours inaccessible to workers. Temporary work permit holders in Ontario (including migrant farm workers) are permitted to use international driver's licences in the province, or apply for provincial driver's licences and acquire car insurance. Except for one research participant (who owned his own car), others who use provincial or international driver's licences in Canada do so in order to act as drivers at work, rather than for their own transportation needs. Employers assign drivers to move other workers between worksites or for weekly grocery trips. The mobility of workers who do have permission to use employer vehicles is thus shaped by the fact that they are employed as drivers: vehicles are infrequently lent to workers for their independent use. Quite literally, local, non-migrant residents see workers infrequently. Some, though not all, local, non-migrant residents actively avoid entering towns at times when migrants are known to be there.

Migrant farm labourers' lack of motorized automobility should not be interpreted solely as a question of material or financial limitations, but rather as an expression of the extra-legal spatial and social relations that exist in agricultural workspaces. Car ownership is perceived to be unfeasible within the context of rules set down by employers, employment contracts, and federally issued work permits. Indeed, it is often assumed that the choice to acquire a car simply does not exist, though there are no legally enforceable provisions to this effect. Some respondents assumed that employers would not permit them to drive. However, when pressed, no one identified a specific instance where such a regulation or rule was outlined:

> It doesn't make any sense for me to buy a car here.... How are you going to insure that car? And just use it for maybe like ... six, seven months? I think you will have problems to insure it. Somebody will have to be responsible for the insurance.... But I don't think that all the boss[es] would very happy for you to own a car ... [B]ecause we are here under their responsibility. Yeah, so, we have to be careful.

Another research respondent from Central Mexico was able to explicitly compare multiple experiences of migration and citizenship status in relation to physical mobility and daily travel, first, as a migrant farm worker in rural Ontario, and second, as an undocumented migrant farm worker in North Carolina, Virginia, and Florida. Living and working in the South-eastern US from 1997 to 1999, he could obtain state-level driver's licences despite lacking federal immigration status. Passing his driver's test with a minimum of English, he bought a car from a Spanish-speaking car dealer. He used time off work to travel.

> I felt, like, more comfortable because with the licence I could go where I wanted, even though I was illegal and always ran the risk of being caught by immigration enforcement. But I continued to drive and felt like I had more freedom to travel in the US with my driver's licence.

After returning to Mexico from the US due to family obligations, he again faced pressure to migrate for economic reasons, this time choosing an avenue that offered the security of legal, authorized labour migration. He arrived through the Canadian–Mexican SAWP to a large Ontario farm operation. Yet, despite being "illegal" in the US he benefited from greater personal freedom of mobility there than he does as a seasonal guest farm worker in rural Ontario: "[I]t's OK because I travel with a work permit, but I don't feel I have the same freedom because I have to work every day." Further compounding the financial exclusivity of automobility in North America, migrants' lack of access to private cars in rural Ontario is the result of extra-legal, social control exerted by employers.

This pattern of constrained mobility partly explains why bikes are widely used by migrant farm workers to travel between rural towns and work-live sites. Even if workers drive on behalf of their employers, bikes are one of the only *independent* means of transportation to which they have access. Workers often leave their work-live sites by bike to socialize and run errands during evenings or other free time. Describing his first worksite, one worker noted how bicycling allowed him to negotiate employer-imposed restrictions on his local mobility: "We maybe never have the freedom to move, go around as I'd like to. But at least we could ride our bicycle anywhere we wanted. He [my boss] never had any problem with that." Bike use is embedded in everyday life and social reproduction: to get to and from restaurants and bars, to shop, to get to payphones, to remit money, and to access Internet, health, ESL, and other services. This typical account (in the words of a Jamaican participant) describes how social spaces are created around payphones:

> Sometimes you could go and hang out with your friends there. The store closed early evening, at 5, [or] 6 o'clock. Some guys would be out there, hanging out there until 8 o'clock at night.... Sometimes a little earlier or a little later. And then you get back to your bunk. A lotta guys, Jamaican and Barbados guys.

While employers might provide taxi vouchers or a company van or bus (for a group of workers at a time), they are not independent means of getting around. Bikes are, at times, used by workers to access services that they do not want their employer to know about or for which employers have refused to offer a lift. Workers are not commuting nor making "necessary" trips, yet neither are their trips purely recreational. Bike travel facilitates the formation of transnational spaces in isolated, small, and otherwise racially homogeneous Ontario towns (Cravey 2003).

Rural roads are generally quieter than urban thoroughfares, yet they can be far from bike-friendly. Most roads have gravel shoulders, are unlit, with 80 km/hour speed limits. Bikes are an imperfect means of making farm work on remote sites marginally liveable. As previously mentioned, there have been nine bike fatalities in two key migrant-reliant agricultural regions in Ontario in the last two decades (incurred by migrant workers). Four of these fatalities on bicycles have occurred in the Leamington, Ontario area since 2005 (Jarvis 2013). At least seven other migrants have been seriously injured on bikes during this period. Migrants also report that garbage is thrown at them from passing vehicles, that they face verbal and racial aggression, and that they are pushed onto shoulders by deliberately aggressive drivers. Some workers have given up bike use because of fear of harassment and physical harm:

> [I]f you had a bike you could ride to town easily, but I never bothered with that.
> [Interviewer:] Why not?
> Always scared.... [B]ecause remember those guys who used to be beat up by cars around here? ... [Y]ou know, it happened, a lot. About six years ago, we have three guys hit off a bike around here. Two got killed and one hurt. It was on purpose.

This quotation refers to Charles Morris, who killed two cyclists and seriously injured another in September 2005, driving 100–120 km/hour in an 80 km/hour zone. All three victims were Jamaican migrant workers riding to use payphones in a nearby town after work. Morris was charged with three counts of dangerous driving causing injury and death, though the judge noted that the victims were riding at dusk without lights and reflectors and were wearing dark clothes, and that Morris had no prior criminal convictions, imposing a conditional (alternative) sentence to compensate for these mediating factors. Despite the legal outcome, the respondent quoted here considers the collision to have been a deliberate attack.

Bicycle use and mobility is subject to arbitrary control by employers. Like perceived prohibitions against private car ownership, these rules have no formal basis. Respondents and migrant advocates felt that this control escalated in reaction to the fatalities mentioned above. One employer with over 100 migrant employees indicated in a survey response that his/her company "actively discouraged" bike use among workers through "warnings". In another instance, all migrant employees of a particular company were compelled to sign waivers agreeing not to use bicycles while employed there. One respondent, working on a farm where two migrant farm workers were struck and killed on their bikes in 1999, was ordered not to bike by his employer and chose to follow the directive. Yet other workers whose boss had prohibited biking actively continued to do so, preferring to disobey their employer's rules in order to travel to use the Internet or attend classes, but had to hide their bicycles nearby and were afraid of being caught. Unsurprisingly, explicit prohibitions by employers against bicycling and self-inhibiting practices, based on legitimate fears of bicycling, compound migrants' immobility and isolation.

Local responses to these events have focused on bike safety education and outreach for migrant workers. These are one of the few local-scale initiatives, which municipal governments have undertaken beyond their obligations under federal and provincial law (e.g. building, fire, and public health code inspections of seasonal workers' housing). More recent efforts seek to sensitize drivers to the need to "share the road". Members of bike-safety outreach groups explained migrant bicyclists' vulnerability through a variety of rationalities:

- Migrants are less visible on their bikes because they are dark-skinned, because they bike at night without reflective gear or lights, and because they wear dark clothing.
- Workers are from "underdeveloped" rural areas where there are few formal traffic rules for motorists or cyclists.
- Migrants ride old, unkempt bikes that lead to accidents.
- There is little helmet use, thus increasing the risk of serious injury and death.
- Since workers use bikes to go to local bars and restaurants, they are more likely than other non-migrant cyclists to be inebriated when returning to living quarters in the dark.

For instance, Crisanto Jimenez Gomez, killed while cycling in Leamington in 2012, was implicitly deemed to be responsible his death: for riding erratically and late at night, for failing to wear adequate reflective gear, and for presumably being intoxicated (though no toxicology report was made available to the public). Despite these efforts, Abraham Soto-Lopez and Alejandro Rivera Martinez were struck by drunk drivers in 2009 and 2013, respectively, and were wearing reflective gear at the time of the collisions. While none of the explanations for migrants' disproportionate vulnerability as bicyclists is patently untrue (though some are), racial stereotypes about migrants shape how responses to fatalities have been formulated. Moreover, there is no bike outreach for migrants that seek to actively consider migrants' own opinions and experiences regarding their mobility needs. There have been efforts to build better bike infrastructure to address the needs of migrant greenhouse and farm workers. Publicly, these efforts are explicitly race-neutral and universal: while better cycling infrastructure would benefit farm workers who depend on bikes, its overarching benefit would be to all cyclists and drivers alike. In addition, local efforts at managing cycling are folded into broader agro-tourism cycling promotion. Migrant workers' dependence on bicycles is used as a rationale for rural cycling infrastructure expansion only as urban tourists' recreational needs have become more economically conspicuous to local decision makers.

Migrants' social and labour mobility is subject to control, yet there are no concomitant legal restrictions upon which this control is formalized. Migrants are legally entitled to drive in Canada, unlike undocumented migrants in the US. Few, however, actually claim these entitlements and gain access to private vehicles for their independent use. Consequently bicycles are one form of non-car travel that migrants can use in rural south-western Ontario to overcome their relative immobility and confinement. Migrants' use of bicycles as an independent means of travel in rural areas provides some autonomy but is an inadequate means of transcending their legally enforced social and labour immobility. Since driving is the predominant means of transportation in rural regions, migrants are doubly differentiated on rural roads and disproportionately vulnerable to physical harm within workplace and local scales.

Conclusion

The regulation and formation of migrant subjects as vulnerable and precarious bodies and workers is enacted through automobility, as a set of bordering practices removed from both the federal scale of government and the national-territorial border. The rationalities that link automobility and unequal citizenship rest on racialized ideologies about migrants as non-belonging subjects. These are operationalized in social relations of power as well as in formal policies, laws, and regulations governing driving entitlements by various levels of the state. Henderson argues for the need to de-essentialize automobility, which he prefers to view as "a site of struggle over urban space, [where] claims of a 'love affair' with automobiles veil the deeper social meanings embedded in automobility" (2006: 294). This chapter inquired into the deeper meanings of automobility as they shape migrants' mobility. Of course, undocumented migrations and state-sanctioned "guest worker" migrations follow multiple, distinct institutional and historical forms. Like immigration governance, automobility's institutional frameworks are themselves diverse. But immigration control and migrant status is spatialized in part through automobility, giving texture to migrant inclusion and exclusion, deterrence, and membership.

In recent decades, automobility has become a highly politicized policy arena in US states, counties and cities: migrants' livelihoods and therefore their ability to live and work in particular jurisdictions have been targeted in state-level legislation reversing undocumented people's access to automobility (Coleman 2012). This has reversed a long-standing pattern whereby undocumented people could participate in neighbourhoods and cities in part because they could legally drive. Conversely, while migrant farm workers with temporary work permits in Canada are legally entitled to obtain licences and drive in Canadian provinces (in this case, Ontario), they rarely do. Moreover, unlike the US context, migrants' automobility entitlements are not politicized in Canada, perhaps because migrant agricultural workers' physical immobility is normalized, even by some workers themselves, as an inherent feature of farm guest labour contracts and the "proper place" of migrant farm workers in rural space (Bauder 2005). Their unequal mobility at local scales is not formalized in any law or policy, but is the partial result of "unfree" labour relations and precarious citizenship status embedded in the SAWP employment contracts. Yet their extra-legal exclusion from access to cars and their reliance on bicycles is also legitimized through racial ideologies. These naturalize the physical and social confinement, immobility, and vulnerability that non-citizen, migrant subjects face.

For those whose reliance on private vehicles has never been threatened or questioned, automobility is so embedded in everyday life that it is nearly invisible. For a migrant farm worker who does not have a car and is confined to a work-live site for months on end, or for an undocumented worker who depends on a fake driver's licence to get from a city apartment to an exurban construction worksite, the intersections between non-citizenship status and automobility are all too apparent. Automobility can reinforce migrants' social and physical experiences of precarious citizenship. Conversely, just as the "coercive freedom" of automobility reproduces migrants' social and legal exclusion, it can also enable more inclusive, re-scaled expressions of local citizenship (Ridgley 2008). Migrants' capacity to drive can extend their sense of belonging through improved access to family, churches, jobs, social services, and so forth. This is precisely why anti-migrant nativist policy activists have targeted undocumented migrants' automobility entitlements, as they previously existed in US legislatures. This is not to claim that access to a private vehicle is a panacea that advances the exercise of liberal mobility entitlements, but rather that there is a relation between migrant automobility and migrants' capacity to negotiate unequal citizenship, just as automobility compounds their inequality. Both rural guest worker populations in Ontario and "unauthorized" migrant populations in the US negotiate automobility as it both constricts and extends local and sub-national forms of political and social citizenship.

Notes

1 For more information on the terminology in this chapter, particularly the use of *im/migrant* or *migrant* in lieu of *immigrant*, see De Genova (2002); Goldring and Landolt (2013); Sharma (2006). The term *migrant* here does not refer to transnational, "talent", "high-skill" migrants.

2 Migrant protesters' chant in Maryland: they were calling for in-state tuition and access to driver's licences for undocumented residents (Pabon Lopez 2004: 91).

3 Public Law No. 109–13: the Emergency Supplemental Appropriations Act for Defense, the Global War on Terror and Tsunami Relief (Seif 2010).

4 These are informal jitney services used by Latino migrant workers, especially undocumented agricultural and day workers.

5 In 2012 and 2013 I conducted qualitative research with workers, employers, civil society, and government in major migrant farm worker-receiving communities in south-western Ontario (primarily Norfolk and Essex Counties).

6 According to Statistics Canada, the total revenues in 2010 for agriculture, food and beverage manufacturing, and motor vehicle manufacturing was $10 billion, $39 billion, and $43 billion, respectively.

7 Housing provisions differ according to federal labour migration programme (see Preibisch 2010). My observations here are primarily based on the Norfolk County (Ontario) context but are applicable to other major migrant farm worker-receiving regions in Ontario.

8 The Ontario Employment Standards Act and the Ontario Labour Relations Act, respectively, excludes all agricultural workers (farm employees and harvesters) in the province (irrespective of citizenship status) from numerous wage provisions, and exempts all farm workers from collective bargaining rights.

Part III
Driving Politics

11 Driving the vote?

Automobility, ideology, and political partisanship

Alan Walks

Part of the "character of domination" of the system of automobility concerns the way that it self-generates its own forces of propulsion, expelling alternative productive systems and mobilities, and locking-in future (urban) development trajectories (Sheller and Urry 2000). One aspect of this involves the political dimension, by which automobility generates political demands for policies and programmes that promote and enhance the system at the expense of competitors. Of course, the political power of the auto-industrial complex is immense, and elites who rely on the profits flowing from the maintenance of the system clearly have an interest in discrediting alternative realities and normative visions by discursively constructing them as outside the realm of political possibility and "proper" public policy debates. The successful expressions of power on behalf of the auto-industrial complex in the realm of politics is one reason why Bohm *et al.* (2006) characterize automobility as a regime, rather than merely a system. As discussed in Chapter 1, the ideology promulgated by such a regime is neo-liberalism, albeit with its "actually existing" articulations varying in relation to local and national circumstances.

However, regimes of automobility would not be able to rule without the considerable and strong support provided by significant proportions of the population residing in the auto-mobile core nations, and particularly in their metropolitan regions. All of these nations are governed as democracies, with the ruling political parties elected by a plurality or majority of the vote cast in nationwide elections. Regimes of automobility are not usually autocratic or elitist, but instead often govern through widespread consent and extensive popular backing. Such regime(s) are nonetheless hegemonic, producing what Freund and Martin (1993) refer to as "auto hegemony". This is accomplished mostly not through blackmail or police repression, at least in the core developed nations (although there is the occasional resort to such tactics when pressed, often with public support), but primarily through popular agreement and largely open, transparent political agendas. As Michel Foucault argued (1977), power is a social relation that requires for its exercise some measure of consent and inclusion. The fact that political parties representing the auto-industrial complex and espousing neoliberal public policies have been able to attract the votes of so many, including those among the working class, is an important phenomenon requiring explanation.

This chapter interrogates the relationship between automobility, suburbanization, partisanship, and ideology. It seeks to understand the political contours of auto-hegemony, and in doing so, shed light on the factors leading to the embrace of neoliberal policies and political parties among the voting public in two key developed nations: Great Britain and Canada. The chapter begins by interrogating the politics of suburbia, and the mechanisms that might lead suburban voters to diverge from their counterparts in the central cities. The chapter then moves to examine and explain the timing of these shifts in reference to the restructuring of auto-mobile fossil capitalism, and the embrace of neoliberal politics. The trajectories and importance of automobility for understanding shifts in partisanship and governance are explored, first in relation to national elections occurring in Great Britain and Canada, and then among local elections in the largest cities of these two nations, London and Toronto. The chapter concludes by discussing the implications of this research for an understanding of the political importance of automobility.

The politics of the post-war suburbs

As a number of the chapters in this volume attest, the new areas built for automobility in the post-Second World War period have a number of distinct characteristics that sets them apart from the pre-war city, including lower densities, mono-functional land use, automobile dependence, and dispersal. In many developed nations this new post-war urban form is simply labelled the suburbs, although it needs to be stressed that there is a conceptual diversity of suburbanisms, and many ways of defining the suburbs (Walks 2007, 2013a). When defined as those municipalities lying within metropolitan regions but outside the central city, the suburbs are home to upward of 65 per cent of the national population in many developed countries (Sellers and Walks 2013).

The United States (US) was the first nation to witness rapid suburbanization in the post-war period. It was also the first place where a distinctly suburban politics was identified, associated with support for fiscally and socially conservative Republican politicians (Dobriner 1963; Whyte 1956; Wood 1958). Debates early on focused on the mechanisms producing this phenomenon, with perspectives rooted in either *self-selection* processes (the hypothesis that Republican voters were more likely to desire, or be able, to move to the suburbs) or *conversion* effects (i.e. something about the suburbs leads residents to switch their support to Republicans). Such debates fed discussion on whether such trends were transitory or structural, and hence whether suburbanization might mean the "Republicanization of America" (Phillips 1969). The established wisdom by the mid-1970s highlighted the increasing social diversity of the suburbs (Berger 1960; Gans 1967), and argued that this should eventually temper right-leaning tendencies (Murphy and Rehfuss 1976). Yet suburban residence as a factor predicting of Republican partisanship actually gained in strength and importance since that time (Gainsborough 2001; Sellers 1999, 2005, 2013). Certain large post-war suburban districts have evolved into the main bastions of a particular form of neoliberal Republican politics – often associated with media comment-ators Rush Limbaugh and Glen Beck, gated communities, and the "Tea Party" movement, among other things – that is deeply antagonistic to the welfare state, affirmative action, gun control, taxes, or the extension of public transit (McGirr 2001; Peck 2011).

Yet the rightward shift in suburban political leanings is not confined to the US. Such trends have received attention in Great Britain, for instance, starting with the discovery back in the 1960s of suburban "working class Tories", and trends toward "class dealignment" in British politics (Crewe *et al.* 1977; Nord-linger 1967). Although early analyses predicted the decline of city–suburban political differences as the suburbs became more socially diverse, mirroring the US scholarship (Cox 1968; Goldthorpe *et al.* 1968), it was in fact the suburbs that gave Thatcher's Tories their majorities through the 1980s (Crewe 2001; Johnston *et al.* 1988, 1993; Walks 2005a). A recent study comparing voting behaviour across 11 different nations found that voters in most types of suburbs (the only exceptions are poor suburbs containing racial or immigrant majorities) were statistically significantly more likely to vote to the right of their central cities, disproportionately preferring neoliberal or Christian right political parties, and in the case of poor non-minority suburbs, extreme far-right parties (Sellers *et al.* 2013). Accompanying the rise of large metropolitan regions has been the metropolitanization of politics, with intra-metropolitan political distinctions overlapping with, and increasingly supplanting, more established political cleav-ages based on regional territories, class, ethnicity, and religion (Sellers *et al.* 2013).

Outside the US, suburban political distinctions and their divergence from established central-city political preferences date from the late 1970s onwards. In both Great Britain and Canada, for instance, voters in central cities and their suburbs show no statistically significant differences in their party preferences before this time (Walks 2005a,b). In both nations, as in the US, it was during elections in the year 1979 that shifts in suburban support first pushed right-wing Conservative parties into electoral victories over their rivals, with trends since that time revealing ever-widening divergences (Figure 11.1). Coupled with the rising populations and levels of representation among the suburbs, such trends have meant that suburban voters increasingly have the potential to decide elect-oral outcomes, and political parties have to increasingly target their platforms and policies at suburban voters (Walks 2004a,b, 2005a,b; Wolman and Marckini 1998). This has often mean accepting select aspects of neoliberal policy plat-forms, even among parties on the left, as suburban residents are more likely to support privatization and reductions in welfare state spending (Walks 2004b, 2006, 2008). The shift of Labour in the UK to the "centre" and its embrace of neoliberalism under Tony Blair were largely dictated by the need to win over large portions of the suburban vote, or else be shut out of power (Crewe 2001; Johnston and Pattie 2000, 2011; Walks 2013d).

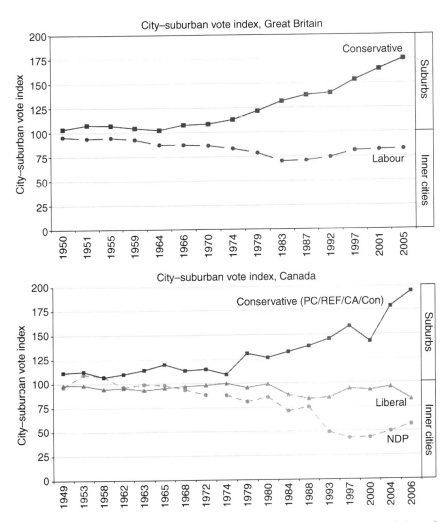

Figure 11.1 Post-war trajectories of city and suburban partisanship, Great Britain and Canada elections, 1949–2006 (source: Calculated by the author from aggregate electoral returns for each year in national/federal elections).

Notes

The suburban–city balance of the vote index is calculated as the ratio of the proportion of the total vote going to the political party in the suburbs, to the proportion of the total vote going to that party in the inner cities, multiplied by 100. See Walks (2004, 2005) for more information on the methodology for classifying electoral districts into the suburbs or inner cities. Data for Great Britain is for the 15 largest metropolitan areas. Data for Canada is for 11 key metropolitan areas (the 9 largest metropolitan areas, plus Halifax, NS and Victoria, BC). In Canada between 1988 and 2004, the main party of the right, the Progressive Conservative Party (PC), was challenged by another right-wing party, the Reform Party (Ref), which changed its name to the Canadian Alliance (CA) in 2000. These two parties merged in 2004 to form a single Conservative Party (Con). NDP = New Democratic Party, the main party of the left in Canada. The Liberal Party in Canada is considered the party of the centre. See Walks (2013b,c), for more detail on the party structures and the factors driving voting patterns in each country. The party structure in Northern Ireland is different from that in the rest of the UK, necessitating the restriction of this analysis to Great Britain.

Driving city–suburban divergence?

Suburban support for neoliberal policies and the city–suburban divergence of political attitudes raise questions regarding potential place-based political effects and the precise mechanisms driving them. According to Castells (1978), capitalist urbanization was supposed to lead to cross-class political alliances supporting the welfare state as a solution to economic crises. Not only did this largely not occur, but suburbanization has been an important factor in the election of the neoliberal governments that have selectively acted to dismantle the welfare state (Sellers *et al.* 2013). While self-selection of those with neoliberal values into particular suburban locations is clearly occurring (Walks 2006; Peck 2011), when two-thirds of the population now resides in post-war suburbs the self-selection hypothesis can no longer hold water as a general explanation for the entirety of suburban political proclivities. The suburbs have for years now constituted the "silent majority" (Lassiter 2006). Some conversion effects must also be at play. Furthermore, as Walks (2006) has shown, to the degree that self-selection is occurring, it is now mostly characterized by that of supporters of the communitarian and environmentalist left *out* of the suburbs and into the inner cities.

A number of potential "conversion" mechanisms link place and politics. The most traditional have focused on the effect of social interaction and communication flows among neighbours and other social networks (Cox 1969; Huckfeldt *et al.* 1993; Huckfeldt and Sprague 1995). Miller (1977) termed such a mechanism "conversion by conversation", arguing the majority view was more likely to win over the minority in any given locale. Yet this cannot explain why the suburbs should shift rapidly to the right over the 1970s, nor why the suburbs should just as rapidly change their support to other parties, as they did with Labour in the 1997 UK election. Furthermore, little actual evidence has been found to support anything more than a residual effect of social interaction (Curtice 1995; Walks 2006). In place of social interaction, place-based variables such as residential turnover, city size (Oliver 2001), population density and mix (Curtice and Steed 1982), municipal political fragmentation (Danielson 1976; Teaford 1997), and homeownership (Wood 1958; Pratt 1986; Saunders 1978, 1990; Fischel 2001) have all been posited as having effects on partisanship. Homeownership, often associated with post-war suburbanization, in particular has a long pedigree in political analysis as an instrument for dampening working-class radicalism, inculcating a culture of possessive individualism, and making owners weary of taxes and threats to property values (ibid.). Housing is clearly an asset that proffers unto its owners particular political and economic interests (Saunders 1990). Ronald (2008) talks of an "ideology of homeownership", culminating in George W. Bush's "ownership society" and the UK official policy shift towards "asset-based welfare". It could be argued that predatory finance took advantage of such an ideology in their targeting of prospective homeowners for specialized mortgage products in advance of the financial crisis (Aalbers 2008; Christie *et al.* 2008; Schwartz 2008; Walks 2010a).

Housing is not only an asset, but also a consumption item provided either in the private market or collectively. Analysis of the political importance of collective consumption derives from Castells' (1978) and Dunleavy's (1979) seminal work. Dunleavy argued that consumption-sector interests were rising in importance due to the role of the welfare state in subsidizing social reproduction (as per Castells 1978). However, unlike Castells, Dunleavy saw that economic interests in consumption could lead to ideological divergence with respect to the role of the welfare state and levels of service provision. In particular, lifestyles in a number of communities had become aligned with a "privatized mode of consumption" (particularly concerning private housing and private transportation) and they could be expected to support private market solutions (or be more easily manipulated into it), in contra-distinction to the "public/collective mode of consumption" prevalent in the central cities where one might expect greater support for welfare state expansion. A related perspective sees political alignments as rooted in the phenomenology of the urban experience, with socio-spatial practices either masking or transgressing ideological assumptions (Eagles 1992; Burbank 1995; Cresswell 1996; Goonewardena 2005; Walks 2006). Such practices and perceptions pertain not only to consumption interests, but also to local social composition, regional production interests, and the efficacy of public polices (see also Savage 1987; Cutler 2007; Cox and Jonas 1993).

Such perspectives are relevant to the political importance of mobility. Auto-mobility and the experience of driving help reproduce an ethos of individualism, self-reliance, and competition, while alternative mobilities (particularly public transit) are claimed to enhance feelings of tolerance, cooperation, and mutual respect, due to the sense that everyone is in the same boat (Rajan 2006; Sewell 2009). Divergent mobilities bring with them divergent citizen subjectivities as well as divergent personal interests. Yet despite a 40-year history of political contention around transportation infrastructure planning, and popular media discourses built around the political proclivities of auto-mobile publics (from SUV drivers in the US to the "two cars attitude" of suburban voters in the UK; see Clapson 2003), as of yet little scholarly analysis has examined transportation or mobility as independent mechanisms driving partisanship or political ideology. Of course, there are a number of different possible suburbanisms, and suburban diversity rather than homogeneity is the norm (Walks 2013a). Many commentators unjustly stereotype the suburbs, and suburban residents (Nicolaides 2006; Davidson 2013). Regardless, post-war suburbanization could not have occurred as it did in most developed nations without the automobile, and without the steady flow of affordable energy that powered them as well as the other machines that have become modern household necessities (from lawnmowers, to hot-water heaters, to washing machines) (Huber 2013). What, then, are the relationships between automobility, ideology, and partisanship, and how does this relate to the rise of neoliberalism? The rest of this chapter examines this question.

Driving ideology: from liquid automobility to neoliberal suburbanism

As discussed in the first chapter in this volume, the values promoted by contemporary automobility, including freedom, automony, individualism, self-reliance, self-responsibility, and unfettered mobility, are among the core ideals of liberalism (Rajan 2006; Paterson 2007). These have been internalized as natural rights within the pro-car neoliberal "automentality" that has arisen, particularly in the US but also in the UK, Canada, and elsewhere, to challenge any limitations on driving (Rajan 1996; Paterson 2007; Seiler 2008). Any perceived attack on the automobile is framed as an attack on personal and political freedom, and the core values of a "free" society, as argued in popular works by Bruce-Briggs (1977), Lomasky (1997), Johnston (1997), and Dunn (1998) (see also De Place 2011). In many US cities, internalization of a natural right to auto-mobility has provided cultural justification for the secession of the white middle class into auto-dependent suburbs, and political cover for limiting the extension of transit out of the central city (Henderson 2006).

It was during the 1970s that political positions began to crystallize around topics of relevance to automobility. The OPEC oil crisis, the emergence of deindustrialization, and the publications of the Club of Rome's *Limits to Growth* (1972) and Worldwatch Institute's *Running on Empty* (Brown *et al.* 1979) all worked to stimulate normative debates concerning the sustainability of contemporary urban development patterns and potential policy interventions for reducing automobile and energy use. It was largely in response such debates that the pro-car backlash originally emerged, claiming that moves to increase fuel efficiency and limit car use constituted a "war on the automobile" (Bruce-Briggs 1977; Rajan 1996). However, the pro-car intellectual lobby had more far-reaching affects then merely defending automobility, because their arguments logically extended to the justification of the whole way of suburban life that was increasingly being criticized as unsustainable and unjust, and for which planners and policy makers were now seeking reform and control. The rising popularity and acceptance of neoliberalism as a political philosophy during the 1970s cannot be separated from the entrenchment of auto-mobile suburbanisms right at the time that both deindustrialization and media discourses began calling into question the future viability of fossil-fuelled capitalism. This politicized the largely taken-for-granted assumption that automobility, suburbanization, and Keynesian economic management were synonymous with progress and development, and led to new political positions either justifying or attacking certain of these assumptions. The new left chose to defend environmental regulation and aspects of the Keynesian welfare state, while opposing key attributes of the "system" (of automobility), with this opposition often articulated personally in the spatial relocation out of the suburbs and into the inner cities (see Ley 1996; Walks 2006). The new neoliberal right often elected to defend automobility, the suburban way of life, and to attack Keynesianism. Peck (2011) calls the kind of politics begat by the new right in the US "neoliberal suburbanism", in that it derived out of the reflexive, dialectical relationship between the evolution of post-war politically fragmented suburbanism and new-right neoliberal politics.

According to Huber (2013), the social and spatial relations undergirding a petroleum-based economy have not only provided the energy subsidies facilitating decentralized auto-mobile suburbanisms and ways of life, but have also been key to the emergence of the politics of denial and common-sense entitlement among middle- and working-class suburban residents in the US. This is related to what Huber (2013) calls the ethos of neoliberal "entrepreneurial life". The ubiquitous but largely hidden and taken-for-granted work provided by oil's energy subsidy (via cheap gasoline and its distribution, not to mention the state highway building and mortgage-insurance programmes funded by taxes on the auto-industrial complex) made it seem that suburban residents alone were the creators of their own destinies through their own hard work, and thus entitled to manage their affairs without interference by the state or other actors. The private mode of consumption and production in suburban life, entailing a tight interweaving of private mobility, private property, driving to private workplaces, etc., coupled with the increasingly entrepreneurial form of household functions – making car and mortgage payments, balancing household budgets, investing household savings, etc. – reinforced the feeling that individuals and households were "freely" in control of building their "life as a business negotiated through the commodity form" (Huber 2013: 94). This often coalesced into an "ideology of hostile privatism" (McKenzie 1994), perhaps most articulated in the rise of private local governments and gated communities, and opposed to not just "big government" and tax-funded welfare programmes but any form of politicized or institutionalized privilege (Huber 2013). The dispersed settlement patterns begat by automobility, and the entrepreneurial-minded yet diffuse and marginal forms of economic power held by suburban households, encouraged "the idealization of an apolitical economy based on the decentralized forces of the price mechanism" (Huber 2013: 122).

Yet, because of their dependence on cheap energy, suburban entrepreneurial lives were vulnerable to the twin oil shocks of 1973/74 and 1979. Coupled with the emergence of deindustrialization and rapid inflation through the 1970s, many suburban residents now became acutely aware of the price and availability of gasoline in relation to their wages, leading to the politicization of energy policy (Huber 2013). It is thus no coincidence that 1979 was the year many "conservative" political parties and politicians promising cheaper energy and advocating roll-back neoliberalism were elected in Anglo nations (Reagan in the US, Thatcher's Tories in the UK, and Joe Clark's Tories in Canada). The message underlying the approaches of each of these parties was that consumers did not need to change the way they live; instead the problem was an out-of-control, overbearing tax-happy state willing to risk unemployment and economic prosperity in the name of misplaced social objectives and environmental regulation. In the US, Reagan's key campaign slogan was "Our problem is not a shortage of oil, it is a surplus of government" (cited in Huber 2013: 118). In Canada, Clark's

Tories were elected on a platform of cutting taxes and privatizing Canada's national oil company, Petro Canada, while promising *not* to raise the gas tax or seek a national energy programme (the latter proposed by the then-Liberal government). In more recent times, the realization that affordable energy is necessary to the continuation of the "American way of life" had been central to suburban support for US ex-president G.W. Bush's "ownership society" and the politics of "yes, blood for oil" in relation to the war in Iraq (Huber 2013).

The political ecology of automobility in Great Britain and Canada

There has been a tight symbiotic relationship between the simultaneous promotion of homeownership, resource development, and automobility among neoliberal conservative political parties, not only in the US but also in nations such as the UK and Canada. Soon after taking power, Margaret Thatcher engaged in the privatization of council housing estates through right-to-buy legislation, instituted mortgage-interest tax deductions, and began the process of privatizing public utilities, public transit, and eventually inter-city rail (the latter implemented under her successor, John Major). Council-house sales were successful in attracting a number of working-class voters who saw in them a chance to become homeowners (Saunders 1990). In regards to transportation, Thatcher was successful in discursively constructing public transit as a heavily subsidized yet poorly managed public service needing private-market discipline, while ably masking the vastly more significant public subsidies going to public roads and motorways (Dunleavy and Husbands 1985). Above all, it was the opening up of North Sea oil to drilling that provided Thatcher's Conservative government with sufficient revenues to cut income taxes and to privatize a number of public utilities through public offerings, which Thatcher hoped would stimulate a neoliberal-utopian "share-owning democracy" (Ronald 2008). The links to Huber's (2013) concept of "entrepreneurial life" are clear. The much-loathed "poll tax", which changed the property tax into a per-head tax with obvious negative implications for tenants and positive implications for wealthy homeowners, is often seen as the policy that led to Thatcher's downfall (Bagguley 1995).

The electoral defeat of the (then John Major-led) Conservative government in 1997, brought on by a significant suburban shift to Labour, occurred only a few years after the peaking of North Sea reserves (Unger 2013). This has not, however, meant that suburban voters have modified their policy preferences, or that earlier Labour policies were then vindicated by the party's subsequent electoral success. On the contrary, "new" Labour under Tony Blair avoided revising many of the previous Conservative government's policies (although they did reverse the mortgage-interest tax deductions implemented under Thatcher). Indeed, New Labour proceeded to scrap their own former policy of supporting public ownership, in favour of maintaining the privatization of utilities, council housing, public transit, and inter-city rail systems enacted under the Conservatives, while further deregulating mortgage finance under the rubric of promoting homeownership and a new relationship to the welfare state, encapsulated in the concept of "asset-based welfare" (Finlayson 2009; Doling and Ronald 2010). The 2010 Conservative victory was partially predicated on successfully winning back the suburban vote (Johnston and Pattie 2011).

In Canada, the Tories led by Joe Clark were first elected to a minority government in 1979 on a promise to allow homeowners to claim mortgage-interest deductions, in addition to privatizing Petro Canada (Pratt 1986). Yet despite clear commitments to both policies, they ironically lost the subsequent 1980 election when they broke their other promise *not* to raise gas taxes. The Conservative government came back in 1984 under a different prime minister (Brian Mulroney) who then saw out the privatization, and began the process of rolling back the welfare state. Since the mid-1990s, leaders of the Conservatives and their related parties (involving numerous name changes and mergers) have all come from Alberta, Canada's main oil-producing province. It is thus not surprising that Conservative governments elected since 2006 have been particularly aggressive in attacking environmental protections related to resource (oil and gas) extraction, and in silencing any vocal critics and even their own scientists, as well as justifying covert surveillance of peaceful protests against Tar Sands development in terms of the "threat to national security" such protests represent (Leahy 2013).

However, in Canada it has mostly been the provincial governments that have enacted the neoliberal roll-back of the welfare state, since most collective-consumption items (health, education, housing, ground transportation) fall under provincial jurisdiction. This began with the election of a Conservative government in Alberta under Ralph Klein in 1993, and continued as subsequent provincial governments slowly fell to the right. One the most aggressive in this regard has been the Ontario Conservative government that ruled between 1995 and 2003 under the premiership of Mike Harris. This government pursued a major restructuring of the welfare state including the downloading of responsibility for social housing to municipalities, significant cuts to income taxes, welfare benefits, and public schools, and forced amalgamations of municipalities, which they successfully portrayed as a "common sense revolution" (Keil 2002). Notably, this was the first government in the province in which a majority of its support and seats came from the post-1970 ("outer") suburbs surrounding Toronto (Walks 2004b).

The importance of automobility to these political transformations has not received sufficient attention. In fact, it is variables related to automobility that have superseded homeownership in articulating support for Conservative parties in both countries, despite their clear policy bias towards homeowners and the rapid growth of homeownership – increasing from 55 per cent in 1975 to 70 per cent of households by the late 2000s in the UK, and from 60 to 69 per cent between 1971 and 2006 in Canada (see Ronald 2008; Hou 2010). Table 11.1 presents backward logistic regression models estimating the proportion of the vote in electoral constituencies going to the Conservative parties in both the 2005 UK national election and the 2006 Canadian federal election, while controlling for a host of socio-demographic variables as well as region of residence. Both models reveal historic associations with class voting, as income and occupation variables reveal particularly strong effects, while the presence of visible minorities is associated with lower Conservative support in both countries. In each country there are strong regional effects (with support coming from the south of England in the case of Great Britain, and Alberta in Canada), as well as strong effects flowing from concentrations of primary occupations (with rural areas being more likely to vote Conservative in both nations).

Among the drivers of suburban political effects, it is variables related to automobile use or ownership that take precedence over most others. Among electoral constituencies in metropolitan areas, there is a strong correlation between the level of homeownership and the proportion that commute to work via automobile in Canada ($r=0.87$), and the proportion of households with two or more automobiles in Great Britain ($r=0.86$). Once the variable for automobile commuting is included, however, homeownership completely falls out of the Canadian model, while in Great Britain, the coefficient for homeownership remains but the sign turns *negative*. That it, those who own their own homes are *less* likely to vote for the Conservatives than are private renters (although they are more likely than council-housing tenants). In both cases, these results demonstrate that the variable for automobile ownership is the one actually driving Conservative support in the suburbs. Furthermore, when these variables for automobile ownership and use are added to the backward regression models, the previously significant zonal effects of suburban residence (Walks 2005a,b, 2013d,e) also proceed to fall out of both models.

The geography of automobility has a significant effect on suburban support for national parties of the right. In fact, if one only knew the proportion of households in each British electoral constituency that owned two or more cars and vans, one could predict with 56 per cent accuracy the spatial distribution of the Conservative vote in 2005 across Britain's metropolitan electoral constituencies (the bivariate correlation is $r=0.75$, r square$=0.56$). This is more than two-thirds of the total (81.9 per cent) variation in the Conservative Party vote that can be explained by all variables acting together (Table 11.1). The level of homeownership in a bivariate correlation, by comparison, explains only 32 per cent of the variation in British Conservative support among metropolitan constituencies. Among Canada's key metropolitan areas, meanwhile, the proportion driving to work accounts for 29.3 per cent of the variability in Conservative support, second only to the local unemployment rate (34 per cent, which, however, falls out of the regression model). Commute mode accounts for approximately two-fifths of the total amount of spatial variation in Canadian federal Conservative support that can be explained using all the data at hand (Table 11.1).

Through the mid-2000s, debates in both nations increasingly shifted towards a politics of consumption, rather than production, in particular around neoliberal constructions of the best ways of providing collective social services and raising living standards. In the UK, this became framed in terms of a "third way" service delivery, often involving targeted place-based policies and public–private partnerships (Giddens 1998; Bastow and Martin 2003). In Canada, political debates

Table 11.1 OLS regression estimates of electoral support for the Conservative Party in national elections, Great Britain and Canada

Great Britain – 2005 national election	B	Beta	Canada – 2006 federal election	B	Beta
% Population Change 1991–2001	—	—	% Population Change 2001–2006	—	—
% Age < 16 Years Old	0.945 ***	0.134	% Age < 18 Years Old	—	—
% Age > 65 Years Old	1.031 ***	0.248	% Age > 65 Years Old	0.451 **	0.274
% Couple Family Households	—	—	% Families w/Children at Home	—	—
% Lone-Parent Family Households	—	—	% Lone-Parent Family Households	-0.299 ***	-0.351
% Visible Minorities (Non-White)	-0.121 **	-0.101	% Visible Minorities (Non-White)	—	—
% Hindu	—	—	% Chinese	—	—
% Jewish	—	—	% South Asian	—	—
% Muslim	—	—	% Black	—	—
% Sikh	—	—	% Recent Immigrants	0.324 ***	0.228
% Managerial Occupations	1.076 ***	0.272	% Managerial Occupations	—	—
% Professional Occupations	-1.068 ***	-0.285	% Professional Occupations	—	—
% Manufacturing Occupations	—	—	% Manufacturing Occupations	—	—
% Services Occupations	-1.804 ***	-0.147	% Services Occupations	—	—
% Routine Occupations	-1.287 ***	-0.263	% Arts, Literary, Sport Occupations	—	—
% Elementary Occupations	0.457	0.092	% Primary Occupations	0.949 ***	0.222
% Unemployed	-1.564 *	-0.104	% Unemployed	—	—
% No Educational Qualifications	-0.177	-0.091	% Less than High School Educ.	—	—
% with University Degree	—	—	% with University Degree	—	—
% Home-Owners	-0.311 **	-0.244	% Home-Owners	—	—

% Rent from Private Landlord	–	–
% Rent from Public Authority/Council	-0.493 ***	-0.330
Suburban Residence (3-zone)	–	–
% Households w/Two+ Cars/Vans	**0.351 *****	**0.259**
% Households with No Cars/Vans	–	–
North West England	-3.538 ***	-0.079
North East England	-3.046 *	-0.049
Wales	-12.111 ***	-0.211
Scotland	-9.308 ***	-0.187
Constant	45.079 ***	
R Square	0.819	

% of Income from Investments	0.712 ***	0.146
% Low Income	-0.191 *	-0.113
Suburban Residence (3-zone)	–	–
% Who Drive to Work	**0.236 ****	**0.197**
Atlantic Provinces	-7.793 ***	-0.153
Quebec	-14.671 ***	-0.393
Manitoba & Saskatchewan	–	–
Alberta	25.217 ***	0.466
British Columbia	–	–
Constant	-11.149	
R Square	0.724	

Source: Calculated by the author from aggregate election results at the level of constituencies (UK 2005 national election, Canada federal 2006 election); 2001 Census of the United Kingdom; and 2006 Census of Canada

Notes

Units of analysis are national/federal electoral constituencies. Coefficients are the results of backward OLS regression, in which variables that do not enhance model fit are removed from the model in stepwise fashion (indicated by "–") until the most parsimonious model is attained. Data for specific visible minority status (Black, South Asian, Chinese) were not available for Scotland and Wales, and so religious identity was instead included in the Great Britain analysis. Note that the Conservative parties in the UK and Canada are completely independent of each other, although the Canadian party can partially trace its lineage to its eighteenth-century UK counterpart in the days before Canada became an independent nation. Both the UK and Canada use virtually identical first-past-the-post electoral-district-based electoral systems.

Sig. = *** p < 0.001 **p < 0.01 *p < 0.05

centred on whether public health care might benefit from private investment and delivery, whether private schools should receive public subsidies (a defining feature of the 2007 Ontario provincial election), and whether riders of public transit might receive tax credits. The Conservative government in Canada, led by Stephen Harper, won a majority in 2011 after having delivered a series of selective tax cuts for stimulating middle-class family consumption of private-sector services.

Tellingly, the politics of unionization and strikes, regional economic development, and industrial policy so prevalent in the 1960s and 1970s all but disappeared from political discourse, only resurging slightly *after* the financial crisis with the politicization of shale gas exploration ("fracking") and national pipeline extensions (e.g. Leahy 2013). As Huber (2013: 22) notes, the success of neoliberal hegemony rests with the "the quarantining of politics and agency" to the realm of social reproduction. Much of the new focus on a politics of consumption since the 1980s can be linked to financialization: easy access to credit on behalf of households works to mask and depoliticize many of the social relations around production and place. Yet, as discussed in Chapter 4, the growth of household credit itself can be traced to the historical successes of the auto-industrial complex now looking for new profitable fields of investment in a period of waning automobility.

The politics of automobility in Greater London and Toronto

Of course, while local context and ways of life can influence national electoral contests and party platforms, it is at the local level that political ecologies of automobility are most salient. The largest cities in the UK and Canada – London and Toronto – provide important case studies for analysing how the politics of automobility are being played out within the city. Both are the financial centres of their respective nations, and are lauded for a high quality of life. Both cities have experienced the most extensive and rapid gentrification in their countries, and both cities have very distinct patterns of urban mobility between the older "inner" parts of the city built up before the war, and post-war suburban areas. Both cities have also experienced a spatial restructuring of the boundaries of local governance, with the (re-)instatement of the Greater London Authority (GLA) in 2000 (the Greater London Council had been disbanded by Thatcher's Conservative government in 1986), and the amalgamation of six former lower-tier municipalities with one upper-tier regional municipality into a new single-tier City of Toronto in 1998 by the Mike Harris Conservative Ontario government (see Boudreau 2000; Boudreau *et al.* 2009; Bashevkin 2006). By the time of this spatial restructuring of governance in both cities, the majority of the population now resided in the post-war suburban rings: the outer London boroughs in the case of London, and the former municipalities of Etobicoke, North York, and Scarborough, in the case of Toronto. Tellingly, after the spatial consolidation of governance in both cities, local political debate has been dominated by a politics of mobility.

A controversial figure from his battles with Thatcher when Leader of the Greater London Council from 1981 to 1986, "red" Ken Livingstone nonetheless won election as an independent candidate for Mayor of the GLA in 2000 (the executive of the "New" Labour Party refused to allow him to run as the Labour candidate in that election, but eventually conceded for subsequent elections). The defining policies of the first Livingstone-led GLA included a freezing of public transit fares, a congestion charge to limit the number of automobiles entering central London, and the implementation of the electronic "Oyster" smartcard ticketing and transfer system for accessing the London transit network. Steve Norris, the Conservative Party candidate for the first two GLA elections, continued his opposition to the congestion charge, vowing to scrap it during the 2004 election campaign and promising an amnesty for non-payers if elected. While Livingstone won the 2004 election, the election victory came with the loss of a number of the outer London boroughs to the Conservatives, where the congestion charge was particularly unpopular (Edwards 2008; Hosken 2008). During his second term, Livingstone announced his intention to raise the congestion charge from the initial £5 to £10, and to implement a CO_2 emissions-based charge that would cost upward of £25 for the most-polluting vehicles, while also extending the western boundary of the congestion-charge zone. The Conservative challenger, Boris Johnson, won the 2008 election on a promise to scrap the emissions-based scheme, to limit the congestion charge, and to remove the western extension (ibid.). Johnson won mainly by taking additional votes away from Livingstone in the outer London boroughs (Figure 11.2).

Meanwhile, the first elections of the new City of Toronto, starting in late 1997, revealed a deep-rooted political split between the post-war suburbs on the one hand, and the older sections of the former "inner city" municipalities of Toronto, East York, and York built up before the Second World War on the other (Boudreau 2000). The first and third Mayors of the amalgamated City of Toronto (Mel Lastman and Rob Ford, respectively) were able to win election on the support of voters living in the post-war suburbs where a majority of the population lives. The second Mayor (David Miller) largely derived his support from the inner city, but was able to successfully extend his appeal into poorer suburban areas to ensure his electoral victory (Boudreau *et al.* 2009). It is the 2010 election that is particularly interesting for the overt politicization of issues directly related to automobility. This is the election in which a seeming long-shot, Rob Ford, rose to become the third Mayor of the amalgamated City.

Ford's campaign largely consisted of two repeated slogan-promises, that he would "stop the war on the car" and "end the gravy train". The latter played on the perception (largely disproven post-election) that previous administrations had been misspending taxpayers' hard-earned dollars, as well as on discursive constructions of new bike lanes and the "Transit City" plan begun by the Miller regime (that would have extended light-rail transit into un-served areas of the suburbs) as a "war on the car". Ford even vowed to get rid of Toronto's well-loved "streetcars" (above-ground light-rail cars/trollies), on the grounds that they held up traffic. Ford cancelled the Transit City plan on his very first day as

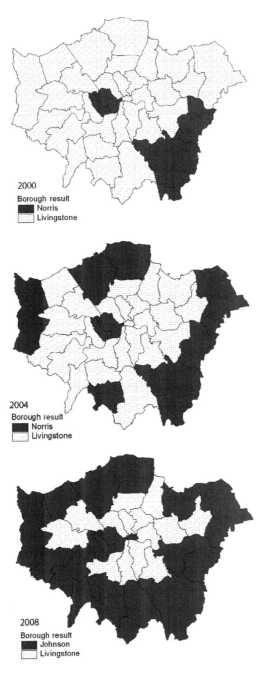

2000
Borough result
■ Norris
▫ Livingstone

2004
Borough result
■ Norris
▫ Livingstone

2008
Borough result
■ Johnson
▫ Livingstone

Figure 11.2 Mayoral election results for the Greater London Authority, 2000–2008 (source: Created by the author from official electoral returns).

Note
Results are mapped by borough.

Mayor, even though the provincial government and not the City were covering virtually all of the costs. He then proceeded to order the removal of selected bike lanes. Ford has said that it is the fault of cyclists themselves if they get killed on the road because "roads are built for buses, cars and trucks" (Mahoney 2010).

A map of the 2010 Toronto election results reveals a clear dichotomy between the pre-war parts of the city – where densities are higher, and transit use, bicycling, and walking are common – and the post-war suburban parts of the city, which are largely automobile dependent (even in those locales where densities are not low) (Figure 11.3). Ford, a member of the Conservative Party, was able to win virtually all the polling divisions in the post-war suburbs, and hence their wards, while areas built up before the Second World War reveal pluralities for his main (Liberal Party) rival. Ford's popularity among many residents of the City's post-war suburbs (which he himself branded "Ford Nation") persisted through his first term, despite numerous indiscretions involving rude public behaviour, evidence of drug use, and various "friends" with criminal records (Doolittle and Donovan 2013; Donovan *et al.* 2013). Ford has been said to govern and rule using "uncompetence" – the proud and wilful promotion of ignorance and misunderstanding of municipal policies, legislation, and respectful behaviour – often on display during the local radio show that he hosted with his brother (Tossell 2012). Whether such uncompetence was wilful or not, it could not hold any traction with a large section of the voting public unless it also aligned with their interests and subjective experiences. The post-war suburbs in

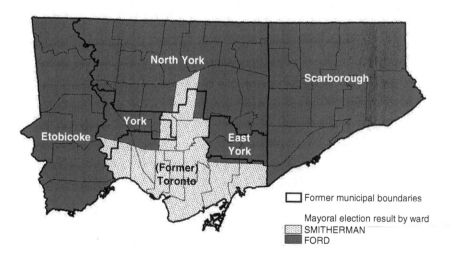

Figure 11.3 City of Toronto 2010 mayoral election results by ward (source: Created by the author from official electoral returns of the 2010 Toronto election, by ward).

Notes
Shown also are the names and boundaries of the former municipalities within the current City of Toronto as they existed before amalgamation in 1998.

Toronto have been deindustrializing, making unemployment a growing issue, while traffic congestion has become a persistent problem and public transit is in a clear state of decay. Ideologically, Ford's approach reflects the neoliberal "automentality" described in the introductory chapter. Ford sought to eliminate or reduce a number of city social programmes, cut property taxes and wages for city workers, and privatize social housing and garbage collection. Filion (2011) refers to Ford's politics as "Toronto's Tea Party". However, it is Ford's approach to transit policy that foremost defined his regime. In the face of clear need for new transit infrastructure, he argued for either underground subways or nothing, because the street should be left for the free flow of cars (ibid.). The celebrity friend he selected to introduce him at his mayoral inauguration did so with a diatribe against "all the pinkos out there who ride bicycles" (Grant 2010: A1).

The emergence of politico-ideological divides related to transportation policy is reflected in ecological analysis of the variable correlations in each respective mayoral election (Table 11.2). While class politics has always been a factor in mayoral elections in Greater London, since 2000 mobility is shown to have an important effect on support for the Conservative (and other) candidates, alongside immigration status, age, and family status. Instead of the number of automobiles per household (as in the rest of the UK) being the important factor, in Greater London the main political divergence is expressed as a dichotomy between those with and those without automobiles. On its own, knowing the proportion of households that owned at least one car explained 30 per cent of the variation in support for Johnson in 2008 across London boroughs (out of a total of 89.5 per cent explainable using all variables together). The strength of this variable in the multivariate model, furthermore, more than doubled between the 2004 and 2008 elections, increasing at a more rapid rate than any other variable and reflecting the increasing politicization of transportation policy and its differential articulation in the space of the city over that time.

Automobility is an even more important factor driving support for Rob Ford, the mayoral candidate hailing from the Conservative Party in Toronto (in what are officially non-partisan elections). The proportion driving to work is the most important variable accounting for support for Ford in 2010 among all those available for testing in the model (Table 11.2). Just knowing this proportion, on its own, allows one to predict with an incredible 76.3 per cent accuracy the percentage of each ward's electorate voting for Ford. Ford's promise to "stop the war on the car" clearly helped garner the votes of residents living in auto-dependent areas, many of whom have been suffering from both financialization (rising indebtedness) and deindustrialization (unemployment and underemployment among immigrants and manufacturing workers), propelling him into the mayorship. Of course, a significant proportion of those residing in each city do not actually vote. Turnout in the 2008 London election was only 45.3 per cent, while in the 2010 Toronto election it was 58.4 per cent. However, in neither election do variables associated with automobility reveal any relationship with turnout. Automobility's effects are instead related to ideology and partisanship.

However, despite a polarized political landscape, mobility is a complex human phenomenon and tensions rooted in the politics of mobility have many outlets and potential resolutsions. In London, the incoming Conservative Mayor not only chose to honour the bicycle-share programme initiated by his Labour predecessor and maintain a (reduced) congestion charge, he has initiated a pan-London bicycle network that would connect the high-density world of inner London with the more auto-mobile outer London. Here, the preference for autonomous mobility is being reinterpreted and re-channelled into alternative auto-mobilities that connect the city. In Toronto, meanwhile, Ford had to acquiesce to a (slight) tax increase to facilitate federal government contributions to subway train extensions into the suburbs, with the debate then turning to questions around which are the best routes and how best to serve different social groups (Gee 2013; Moore *et al.* 2013). Despite these complexities and hybridities, the primary focus notably remains on free and autonomous movement. London's bike network will not significantly affect the flow of automobiles, while the proposed subway train lines in Toronto would either flow underground or be raised above ground so as not to restrict car traffic.

Conclusion

Automobility is not merely a techno-economic system with effects on the social behaviour, nor a monolithic regime vesting a ruling elite with the power to implement policies that maintain the system. Instead, automobility has political effects in the realm of everyday life, with powers and interests diffuse and fragmented among various political and non-political actors, including residents and voters displaying varying levels of participation. The evolution of the auto-city in the post-war era has produced distinctly auto-mobile suburbanisms, dependent on maintenance and expansion of the auto-industrial complex for livelihoods and on the automobile for mobility. The spatial dovetailing of politico-economic interests in the realm of employment with those in the realm of mobility produces new ideological interests and assumptions that then get reinscribed into the cultural meanings and political trajectories of the city.

It is in cities that combine gentrifying, globalizing, and financializing cores with auto-mobile suburbs where such divergences rooted in automobility can be most expected to emerge, and to be articulated in competing political visions. It is in such cities that mobility is most contested, and that sufficient power and interest are vested in competing political visions. However, the latter depends greatly on local context, and the resolution of tensions resulting from the politics of mobility are contingent on local historical processes, patterns of settlement and urban forms. While the more automobile-dependent metropolises such as Toronto, and others in North America, increasingly flirt with auto-mental neoliberal politicians and politics – declaring any attempt to promote alternative mobilities as a "war on cars" (De Place 2011), other cities with greater choice and diverse legacies are able to constrain and re-channel such expressions, even in the face of populist neoliberal governments, and to experiment with hybrid automobilities.

Table 11.2 Regression estimates of electoral support for Conservative Party mayoral candidates: Greater London (UK) and Toronto (Canada) mayoral elections

Greater London Authority Mayoral Elections					City of Toronto Mayoral Election – 2010		
London Candidate	Steve Norris		Boris Johnson		Toronto candidate	Rob Ford	
	2004		2008			2010	
	B	Beta	B	Beta		B	Beta
% Population Change	—	—	—	—	% Population Change 2001–2006	—	—
% Age < 16 Years Old	2.384 **	0.617	6.210 ***	1.165	% Age < 18 Years Old	—	—
% Age > 65 Years Old	-1.090 **	-0.965	-1.953 ***	-1.351	% Age > 65 Years Old	0.896 *	0.170
% Married (aged 16+)	-1.995 ***	-0.622	-2.816 ***	-0.636	% Married	-0.679 *	-0.338
% Lone-Parent Households	0.534 **	0.648	0.623 *	0.548	% Families w/Children at Home	—	—
% Foreign-Born	—	—	—	—	% Foreign Born	0.256	0.223
% Chinese	—	—	—	—	% Chinese	—	—
% South Asian	—	—	—	—	% South Asian	—	—
% Black	—	—	—	—	% Black	—	—
% Muslim	—	—	—	—	% Housing built before 1945	—	—
% Managerial Occupations	—	—	—	—	% Managerial Occupations	—	—
% Manufacturing Occupations	—	—	—	—	% Manufacturing Occupations	0.679 *	0.253

% Services Occupations	–	–	–	–
% Routine Occupations	-1.025 *	-0.576	-1.563 *	-0.637
% Self-Employed & Small Empr.	–	–	-2.898	-0.343
% Unemployed	–	–	4.231	0.479
% No Educational Qualifications	–	–	–	–
% with University Degree	-0.812 **	-1.114	-1.188 *	-1.180
% Home-Owners	–	–	–	–
% Single-Detached Housing	–	–	–	–
% Rent from Public Authority	–	–	–	–
% Households w/Two+ Cars	–	–	–	–
% Households with No Cars	**-0.456 **	**-0.670**	**-1.275 **	**-1.356**
Constant	109.0		184.6	
R Square	0.895		0.916	
n (number of boroughs)	32		32	

% Services Occupations	–	–
% Arts, Literary, Sport Occupations	-0.988 *	-0.249
% Primary Occupations	–	–
% Unemployed	–	–
% Less than High School Educ.	-0.331 *	-0.171
% with University Degree	–	–
% Home-Owners	–	–
% Single-Detached Housing	–	–
% Ave. Household Income (by $10k)	-0.672	-0.135
% Low Income	–	–
% Who Drive to Work	**0.579 **	**0.477**
Constant	26.346	
R Square	0.930	
n (number of wards)	44	

Source: Calculated by the author from official electoral returns from each municipality, as well as the 2001 UK census, and the 2006 Census of Canada.

Notes

Electoral returns were calculated at the level of London boroughs for Greater London, and wards in the City of Toronto. Coefficients are the results of backward OLS regression.

Sig. = *** p < 0.001 **p < 0.01 *p < 0.05.

Ideological divergences rooted in contested automobilities are not only present among local politics; they are articulated in new ways as they scale up to the national level. Since the 1980s, national political debates increasingly became framed around neoliberal perspectives on private versus public/collective service delivery in the realm of consumption, while avoiding questions regarding deindustrialization, employment restructuring, labour, economic development, and energy policy thrown up by the problems of automobility. The ubiquity of automobility in the auto-city, coupled with the increasingly easy credit terms offered to households, has worked to naturalize neoliberal discourses of private service delivery that see households as masters of their own domain and governments as guarantors of consumers' rights, while masking the social relations that facilitate the production and social reproduction of automobile suburbanisms. Yet, the contradictions of automobility have continued to build. The waning of the auto-industrial complex and the crises of deindustrialization, and now financialization, labour precarity, and congestion transgress naturalized assumptions, forcing competing political narratives into the open and compelling political parties to improvise ideological programmes that they hope will articulate with the emerging frustrations experienced among their primary constituents. This has increasingly meant a divergence between the politics practised in the post-war suburbs, and those in the pre-war city. Through this process, the growth of the auto-mobile suburbs has produced the political constituency supporting the expansion of neoliberal policies and automentalities, self-generating the political forces of its own maintenance and propulsion. Yet as the contradictions of automobility mount, they become increasingly politicized and open to questioning, rupture, and transformation.

12 Freeway removed

The politics of automobility in San Francisco

Jason Henderson

As discussed elsewhere in this volume, the time for rethinking automobility is urgent. Sustainable greenhouse-gas levels will not be achieved in a globally equitable way if the United States, as forecasted, increases its motor vehicles from the present 250 million automobiles to 325 million vehicles in 2050. If China had the same per capita car ownership rate as the United States, there would be more than one billion cars in China today – which is equal to the worldwide rate (see Chapter 2 by Martin in this volume). And the reality is that for the foreseeable future an electric car is a carbon-burning car. If developed nations expect China, India, and other developing nations to realistically address global warming, they will need not only to provide leadership, but also decrease their appetites for excessive, on-demand, high-speed automobility. North Americans in particular must undertake a considerable restructuring of how they organize cities, and that must include rethinking automobility.

Reflecting this urgency, this chapter considers what can be learned from San Francisco, California, which in many ways is at the cutting edge in challenging automobility in the United States. San Francisco has one of the nation's highest rates of car-free households (29 percent), some of the highest rates of transit ridership (17 percent) and bicycling (4–6 percent), and is known for its walkability (SFMTA 2011). It is also a city where challenges to automobility have included the removal of freeways and reform of residential off-street parking policies, both of which ostensibly limit automobility.

However, juxtaposed against these metrics and policies is a very high density of automobiles – approaching 9,000 per square mile – possibly one of the highest concentrations of cars found anywhere in the world (SFMTA 2010). While during the twentieth century San Francisco preserved its walkable Victorian and early modern neighborhoods it also injected automobiles, parking, freeways, and wider streets into the urban fabric. This juxtaposition makes it more challenging to reallocate streets and is accompanied by a political backlash against rethinking urban space to reduce driving. The fact that so many people still own cars in San Francisco reflects that automobility is not simply contingent on the built environment. Political backlash to policies limiting automobility reflects the fact that automobility is also ideological. It is therefore useful to examine, as I do here, ideology as it relates to automobility.

Ideology is a system of ideas and representations that dominate the minds of individuals and of groups participating in politics, and includes ideas about what the scope of government should be vis-à-vis automobiles and urban space, how decisions should be made, and what values should be pursued with respect the automobile. How we get around does have ideological undertones, and can be what the geographer David Harvey (1973: 18) described as an "unaware expression of the underlying ideas and beliefs which attach to a specific social situation." Seizing on this I want to sketch how automobility is conceptualized vis-à-vis three competing ideological groupings emerging from contemporary US political discourse: progressives, pragmatic neoliberals, and conservatives.[1] Briefly, with respect to automobility, progressives believe government should actively discourage and limit automobility for social and environmental reasons, pragmatic neoliberals hold that automobility should be shaped by the market, and the conservative politics of automobility in the United States posits that government should unquestioningly and proactively accommodate automobility. Considering these ideologies by way of two case studies of San Francisco – freeways and parking – can inspire, inform, but also caution the politics of possibilities for contesting automobility in cities.

Freeway removal

On October 17, 1989, the magnitude 6.9 Loma Prieta Earthquake shook the San Francisco Bay Area. The loss of life and the destruction were devastating, including 42 people killed when an elevated freeway collapsed in working-class West Oakland, across the Bay from San Francisco. Yet to many people in San Francisco the disaster had a silver lining. It was an opportunity to consider removing some of the vestigial freeways that were built in the 1950s and 1960s. After an extremely contentious political debate that lasted a decade, in 1999 San Francisco voters decided to remove a portion of the earthquake-damaged Central Freeway, located in Hayes Valley, in the center of the city just west of City Hall (see Figure 12.1).

The lessons of the Central Freeway debate come at an opportune moment. Much of North America's aging urban freeway system is approaching 50, 60, and even 70 years of service. The American Society of Civil Engineers (2009) warns that one-third of the nation's major roads and freeways are unsound and that 26 percent of bridges are structurally deficient. The nation's urban freeway infrastructure, especially bridges and elevated freeways, must be overhauled, and this will bring disruptions to traffic flows while also being very costly. Given this gloomy picture, some people are asking, why not just tear down some of the old freeways and rethink urban transportation (Congress for the New Urbanism 2012)?

The politics of possibilities for removing freeways is gaining traction. Milwaukee replaced a freeway stub with a boulevard. In New Orleans there is a movement to remove an aging elevated freeway that blighted a once-thriving African American neighborhood. In Seattle there is a debate about removing the

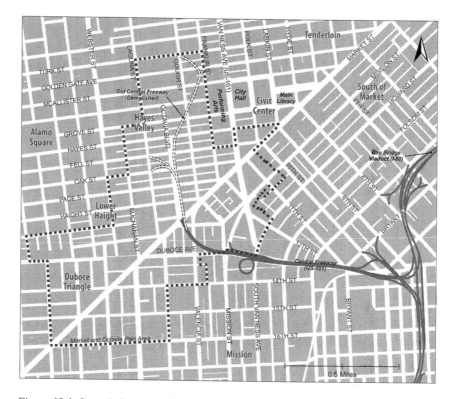

Figure 12.1 Central freeway alignment and Market and Octavia plan area (source: Michael Webster, San Francisco Planning Department).

crumbling Alaska Way Viaduct on the city's waterfront. Similarly, the Gardiner Expressway in Toronto has received attention for potential removal. In Washington, DC, the Bronx, New York, Syracuse, Louisville, and Providence, removing, tunneling, or realigning of freeways has been proposed. Globally a highway segment was removed in Seoul, South Korea and freeway battles have spread in places like Chile (the Coordinadora No a la Costanera Norte) and in India (flyovers in Mumbai). So it is perhaps a good time to ask: What can the experience of removing the Central Freeway tell us about potential outcomes in other cities, and the broader rethinking of automobility?

Progressive freeway removal

In 1959 the Central Freeway was completed through Hayes Valley but the northern segment was canceled during San Francisco's "freeway revolts" from 1959 to 1966 (Jones 1990; Issel 1999; Johnson 2009; Henderson 2013). This meant that the Central Freeway terminated in Hayes Valley and by the late 1960s more than 150,000 vehicles per day were traveling through the area (San Francisco

Planning Department 1971). Already an economically and politically marginal neighborhood, the intense traffic led to the further decline of Hayes Valley, and by the 1970s the underside of the Central Freeway was notorious for prostitution and drug dealing. Shortly after the 1989 earthquake, the far northern vestigial of the Central Freeway had to be quickly demolished because it was unstable. Coupled with the demolition of a segment of waterfront freeway also damaged in the 1989 earthquake, many people saw the benefit of removing the freeway (with this short segment gone), and the idea of further removal spread.

Meanwhile, a new stratum of well-educated, politically progressive middle-class residents who were seeking inexpensive rental housing or were willing to invest some sweat equity in renovating properties was transforming the Hayes Valley neighborhood. Accompanying this transformation were small businesses that lined local streets with arts and craft, clothing, and other shops. As well, a newly invigorated, citywide progressive mobility discourse, led in part by a vocal bicycle advocacy movement, was gaining traction, and by the mid-to-late 1990s had achieved political parity with other progressive causes.

Progressive mobility discourses invoke concerns about the environment and social justice, and seek to use government to limit the impacts and extension of automobility in the city. This is achieved through reallocating street space to public transit, bicycles, and pedestrians, as well as by requiring fees and taxes that support public transit and affordable housing that is proximate to good mobility. This progressive mobility discourse informed the arguments of citizens and activists who were seeking freeway removal.

Hayes Valley residents and progressive activists initially lobbied for complete or near-complete removal of the remaining sections of the Central Freeway. They showed their removal proposal to various community groups and organizations around the city and received a largely positive response, suggesting that this option should at least get a fair hearing. But city traffic engineers and the city's consultants literally laughed at the activists. They insisted the damaged freeway remain partially open and that a new, modern, seismically advanced freeway eventually replace what lingered. Freeway removal was not a fait accompli.

Conservative backlash

In 1994 the Northridge earthquake in Los Angeles damaged part of that city's freeway system. There was little public debate about rebuilding freeways, and the damaged Santa Monica Freeway was quickly rebuilt. In San Francisco proponents of rebuilding the Central Freeway were awed by the speed of post-earthquake rebuilding in Los Angeles. Against that backdrop, a group of pro-freeway neighborhood and business interests organized in 1996 to counter organizing on behalf of progressive political factions. This emerging rebuild coalition included a conservative politics of automobility based in some of the city's single-family residential neighborhoods and shared by some business groups.

San Francisco is far from a bastion of the conservative ideologies associated with American politics, especially regarding social issues such as gay rights or religion. A mere 9 percent of registered voters are Republican, often, but not always, a proxy identifying the presence of conservative politics (Marinucci 2012). Thus the term *conservative* must be used here with reservation and contingency. Yet there is a pronounced conservative element in San Francisco's politics of automobility, particularly in the hill districts and west side, where automobility is central to many people's everyday life due to prevailing low-density urban form.

While progressive groups in general believe government should discourage automobility, the conservative politics of mobility posits that government should proactively accommodate uninhibited movement mainly by car, even when that requires generous subsidy or undermines broader, collective environmental and social goals. Dovetailing with Matthew Huber's (2009) theorization on the cultural politics of US gasoline prices, the conservative politics of automobility is entangled with imagery of work, home, lifestyle, mobility, and freedom that invokes nationalist cultural claims that automobility (and cheap gasoline) is crucial to maintaining the American Way of Life. The conservative politics conjures rigid cultural arguments such as "automobility is American" or "Californian," considers automobile usage a natural right, and insists that government requires abundant space for automobility throughout the city to accommodate people's love affair with automobiles and their absolute need for them.

Conservative discourses of automobility – what Walks in the introductory chapter terms "automentality" – are steeped in the normative vision that people need and have a right to cars, rooted in the essentialization and assumed inevitability of automobility. Implicit in this discourse are certain conservative conceptualizations of family and responsibility. The conservative politics of mobility addresses the day-to-day moralities involved in coordinating family life and social networks in an automobilized society. Using the family car is associated with caring for and loving one's family and friends (Sheller 2004).

The highlighting of one's personal responsibility toward one's family results in the internalization of the necessity of driving. This can translate into lack of interest in collectively solving large-scale problems such as congestion, pollution, and the inequality that stems from automobility. Instead, it is supposedly responsible to move oneself or the family through the city by car – that is, to secede – and fill one's daily needs atomistically. Meanwhile, automobility enables one to circumvent, if not secede from, the perceived evils of the city. This mode of thinking results in demand for the freeway, which enables high-speed bypassing of threats or inconvenience. In the case of the Central Freeway debate, conservative discourses on automobility therefore stress that government should rebuild the freeway. Yet before discussing the conflict at the ballot that emerged from the progressive–conservative divide over the freeway, the role of neoliberalism should be considered.

Pragmatic neoliberals and freeway removal

Part of the politics surrounding the removal of the Central Freeway concerned competing neoliberal visions around using market forces to shape transportation policy. Specifically, built into the final legislation removing the freeway was a financing scheme stipulating that the land parcels beneath the former freeway be sold at market rates and that the proceeds used to help finance construction of a boulevard to replace the freeway. Real estate firms would purchase signature properties adjacent to the new boulevard in the increasingly gentrifying Hayes Valley. This is quite different from other potential uses for the space as advocated by different progressive factions, such as turning over all of the parcels to social housing or other public goods, a path that was not taken.

The financing scheme reflected an emerging neoliberal turn in cities and signaled a retreat from Keynesian policies for public finance. Neoliberalism accompanied a substantial retrenchment in federal funding of urban programs during the 1980s, forcing localities to take on more of the burden for providing local needs like transportation. In California this retrenchment was accentuated by the passage of Proposition 13 in 1978, which froze local property taxes to 1975 values, limited the rate for reassessing property values, and required a two-thirds supermajority vote in a plebiscite for any local special tax increases, such as for transportation (Walker 2010). The combination of federal and state defunding of urban infrastructure meant that cities like San Francisco had scarce resources to take on the removal or replacement of a freeway.

As the Central Freeway was debated in the mid-1990s, the neoliberal financing arrangement gained traction among some progressive factions, signaling a rapprochement with the city's neoliberal land development class. Many progressive advocates recognized that money, not just conservative opposition, stood in the way of freeway removal. Activists in Hayes Valley and in the broader progressive mobility movement came to cautiously accept the notion that adopting the neoliberal approach of land sale could be the best way to achieve their vision of removal. However, while progressives were overwhelmingly in favor of freeway removal and conservatives adamantly opposed to it, the neoliberal land development class had a disjointed, ambivalent view of the freeway, reflecting a broader ambivalence toward automobility.

On the one hand, the Central Freeway provided high-speed, high-volume automobile access to specific locations important to the neoliberal class, such as the Civic Center performing arts venues, which were adorned by the philanthropic wealth of capitalist elites. The Civic Center was a major regional destination, and the elite did not take the bus to the symphony or the opera. On the other hand, freeway removal, coupled with developmental opportunities in the former path of the freeway, meant there was much profit to be made on urban infill and densification in a centrally located part of the city that was relatively well served by public transit. The consideration of the Central Freeway required pragmatic neoliberals to weigh these competing outcomes, and the decision to rebuild or remove the freeway was not as clear-cut to them as it was to either progressives or conservatives. But neoliberal ambiguity offered openings to the progressive camp and a political consensus was emerging between the progressive opposition and neoliberal land developers that the freeway would be torn down north of Market Street but, as a compromise, rebuilt to the south. Traffic still flows to, or near to, the key regional destinations, but significant development potential was also made available.

Many in the progressive camp knew that if they were to achieve freeway removal they had to have political allies beyond their normal bailiwick of environmentalists and sustainable transportation advocates. Therefore, progressives sought to promote freeway removal not simply as a tactic against automobility but also as an economic development tool that would benefit private developers. In sum, this was a tacit settlement between pragmatic neoliberals and progressives that shunted aside the conservative vision of a full freeway rebuild. The rapprochement would be tested in three rounds of balloting over the Central Freeway.

Dueling ballots

In 1997 conservative activists, frustrated with years of indecision and fearing that the city would eventually side with removal, put the freeway question on the ballot.

Activists placed 1,000 pro-freeway signs on major roads throughout the city. The signs proclaimed "Open Central Freeway," and displayed a telephone number that, when called, was answered by a service that asked for the caller's name and phone number. That spring and summer, they collected this information and organized a petition drive to put the question of rebuilding the freeway on the ballot. They gathered over 28,000 signatures and qualified the rebuild for the ballot that November.

This inaugurated three years of dueling ballots pitting conservative rebuild proponents against progressive removal advocates, with pragmatic neoliberals generally siding with progressives. The rebuild campaigns were financed by a collection of political organizations including merchants' groups, the Coalition for San Francisco Neighborhoods, a conservative umbrella neighborhood organization, the Republican Party, building and construction trades unions, the association of realtors, and an array of local politicians who advocated conservative positions about automobility.

The removal campaigns were led by neighborhood activists in Hayes Valley, progressive mobility organizations, key environmental organizations such as the Sierra Club, the San Francisco Democratic Party, progressives on the Board of Supervisors (the equivalent of a city council) and in the State Assembly, the city's gay rights organizations, architecture and historic preservation organizations, and housing advocates such as the San Francisco Tenants' Union. Joining this progressive bloc were some pro-business neoliberal organizations, including San Francisco Planning and Urban Research (SPUR), a prominent developer-oriented think tank, and the San Francisco Chamber of Commerce. Neoliberals and progressive factions established this loose ad hoc coalition in spite of their deep differences on other transportation issues (such as how to finance public transit). During the four ballot elections, the progressive–neoliberal rapprochement was epitomized in a campaign poster that read, "Chamber of Commerce and Sierra Club Agree: No on H (rebuilding the freeway), Costly, Unsafe, Gridlock."

On the first ballot in November 1997, albeit with low voter turnout (28 percent), 53 percent of voters favored rebuilding the Central Freeway. Progressives were at first generally disorganized, and there was no major attraction on the ballot such as a high-profile mayor's race or national candidacy to draw more progressive voters. In view of the low voter turnout and a close election, a core group pushed forward and a second proposition – calling for removal – qualified for the ballot in 1998. The second ballot won in November 1998 by 10,000 votes, receiving 54 percent of citywide votes in a much higher voter turnout (55 percent) than the previous year. Progressives were better organized, tightened their message, and focused on progressive precincts where voter turnout in November 1997 had been low. Conversely, the west-side pro-freeway activists were less organized and less politically active leading up to the second freeway ballot, perhaps believing that the people had spoken and no one would take the second initiative seriously.

As removal seemed imminent, the pro-freeway faction reconnoitered in 1999 and circulated petitions for yet another ballot initiative to rebuild the entire freeway. Exhausted from signature gathering, the progressive–neoliberal camp convinced four allies on the Board of Supervisors to place a counter-initiative on the ballot instead. That initiative expanded the political tent for freeway removal even further, and there was now a fourth ballot! In 1999 a mayoral campaign also tapped into the wider progressive movement and the tone of the campaigns included a concern about how to keep San Francisco from morphing into a Silicon Valley bedroom community. There was an active anti-gentrification movement, particularly in the Mission District, and tenants' rights and affordable housing advocates helped with the removal campaign.

In November 1999, the freeway removal ballot won with 54 percent of the vote citywide, and the rebuild ballot failed, receiving 47 percent favorable votes. The electoral debate over removing the Central Freeway, at least north of Market Street, was over. But the struggle over automobility was far from over.

Post-freeway parking reform

With the Central Freeway debate finally settled, new housing proposals accelerated in San Francisco, while the dot.com boom transformed parts of the city into hip, urbane alternatives to the low-density, homogeneous office parks in Silicon Valley. These neighborhoods were proximate to the city's southbound freeways, and since most software and Internet jobs in sprawling Silicon Valley were accessible only by car, there was pressure for new housing with ample off-street parking.

With gentrification a top local political issue, in Hayes Valley the stakes over off-street parking policy were high. The neighborhood had 22 newly vacated parcels (amounting to seven acres) where the freeway once ran overhead, and further afield there were many more underutilized parcels, including surface parking lots, gasoline service stations, automobile dealerships, and a six-acre former college campus. Cumulatively this land amounted to the possibility of 6,000 new housing units and 10,000 new residents. Not wishing to engage in a parcel-by-parcel skirmish over parking, city planners, progressive mobility activists, and developers considered an area-wide, consensus-based land-use plan that would define the parameters of development, including heights, bulk, density, and parking.

Thus was born the Market and Octavia Better Neighborhoods Plan (MOBNP), which combined progressive visions of limiting automobility with neoliberal ambitions for urban redevelopment (the name of the plan reflects the intersection of Market Street with Octavia Boulevard, an intersection once crossed by the elevated Central Freeway; see Figure 12.1). Now officially part of the city zoning code, the MOBNP was groundbreaking because of its blunt challenge to automobility. The plan slows the movements of cars heading to and from the regional freeway grid, by eliminating some (but not all) sections of the wide sets of one-way couplets that fed toward the freeway in Hayes Valley. It reduced the vehicular carrying capacity of streets by reintroducing pedestrian crosswalks that were eliminated in the 1950s (as part of smoothing the flow of the one-way couplets). And it calls for the city to eventually study removal of the rest of the Central Freeway. The plan also calls for transforming alleys, which are now largely used for parking, into pedestrianized green spaces and public plazas. The tenor of the MOBNP decidedly promotes the possibilities of car-free living and it is with regard to parking that the MOBNP pushes the envelope furthest.

Parking reform is a central component of how urban mobility is being rethought. In *The High Cost of Free Parking*, Donald Shoup (2005) outlines how parking policy reflects a vicious cycle that results in more driving. The derivation of conventional minimum parking requirements in the United States begins with the assumption that mobility means driving everywhere, for everything, all of the time. It is assumed that parking must be provided for every type of land use and function. As a result zoning laws everywhere in the United States, even in Houston, which is known for its lax zoning laws, have rigid parking requirements, often in excess of what is realistically necessary. This oversupply of parking effectively brings the price of parking for an individual motorist to zero, thus contributing to the lower price of driving and to more driving.

Under the typical guidelines for residential off-street parking found in San Francisco, housing would be required to have one off-street parking space for each new unit. These parking "minimums," as codified in San Francisco's zoning code, mean that 6,000 new parking spaces would be constructed on the MOBNP area under the conventional city zoning code. If the new housing in the MOBNP were built according to mainstream American parking standards (with ratios typically greater than 1:1), more than 10,000 parking spaces would probably have to be constructed. The MOBNP dispenses with conventional parking standards and eliminates parking minimums, while a range of "maximums" reduce the allowable ratios for parking to lower than 1:1. In the areas of the MOBNP closest to the downtown, the permitted parking maximum is one space for every four residential units (0.25:1). In the mixed-use neighborhood commercial corridors along Market Street and other transit-served streets, the permitted parking maximum is one space per two residential units (0.5:1), and in the remainder of the MOBNP the maximum is three parking spaces for every four residential units (0.75:1) (SFPD 2008: 8). Additionally, the MOBNP bans curb cuts for driveways on streets with transit service or identified as neighborhood-serving commercial corridors.

The MOBNP reflects some of the core values of progressive mobility, including the notion that government should regulate and limit automobility, in this case through limiting parking. Progressive organizations promoted a vision of the city that privileged housing and public space over abundant parking. Yet, as in the case of freeway removal, it was not easy to eliminate parking minimums and establish tighter caps. For one, the ambiguity among pragmatic neoliberals toward parking, like that over the removal of the Central Freeway, lent an air of confusion that prolonged the debate.

Neoliberal parking

The relationship between parking and neoliberalism can be contradictory. On the one hand, it might seem self-evident that neoliberals would want to deregulate parking by eliminating minimums and allow developers the flexibility to build as they see fit. In a dense, transit-rich city, there are substantial reasons to redevelop without being burdened with parking provisions. This enables more housing provision per land unit and hypothetically more profit, in that parking by itself does not bring a high rate of return. Indeed, many in the development community accept the reduction of parking minimums because it implies less government. But neoliberal developers are less than enthusiastic about the more stringent maximums promoted by progressives, which cap the amount of parking that can be built. Many disdain any government-imposed caps – which they define as undue government intervention and regulation – on allowable parking.

When the final adoption of the MOBNP went before the San Francisco Planning Commission in 2007, the neoliberal mayor – Gavin Newsom – had his appointees to the commission, which constituted the majority on that body, vote to increase the parking maximums in the original plan, giving developers more parking allowances than progressive activists had wished for. Progressive activists then had to call on the Board of Supervisors, where they held a majority, to successfully reinstate the tighter parking caps.

Parking undergirds a larger conflict over housing and gentrification. Developers and realtors in San Francisco recognize that there is a re-urbanizing class stratum interested in the consumption of the city and its livability. San Francisco attracts and incubates a new bourgeoisie and a petit bourgeoisie (or "creative class"), including executives and management, self-employed consultants, engineers, and especially tech workers in biotech, software, and Internet social networking firms (Knox 1991; Florida 2005). From a real estate angle, the most pronounced of these patterns of consumption in older cities like San Francisco has been re-urbanization in the form of gentrification and historic preservation. Arts and music, bars, restaurants, a "café culture," museums, and other traditionally urban amenities are considered to be central to the lifestyle of the creative class.

Significantly, this stratum, while centered on a lifestyle choice to live in a compact, walkable city, includes a car for commuting, recreation, and shopping as well as intensive air travel and frequent high-consumption holidays like driving to Lake Tahoe in the Sierras for skiing or hiking. Neoliberal developers recognize this class's penchant for "the good life," and seek to minimize the negative aesthetics and more extreme externalities of automobility but not to meaningfully alter its primacy in everyday life. The irony is that as this stratum consumes the city as a spectacle or lifestyle choice, the very *tout ensemble* of the city is withered away one garage and one displacement of an affordable housing unit (by inserting parking beneath Victorian flats with secondary units) at a time.

After the MOBNP was finally adopted in 2008, developers pushed back and the debate over parking ratios did not end. Built into all San Francisco zoning codes is the option for a conditional use permit to increase the amount of permitted parking in cases where a developer can supply evidence of the need for additional parking. In winter 2008–2009, in an early test of the parking policies of the MOBNP, progressives lost an appeal of a conditional use permit for excess parking in the rapidly gentrifying North Mission. When the Planning Commission, in a 4–3 vote, allowed the excess parking, progressive transportation activists and environmentalists unsuccessfully appealed the decision before the Board of Supervisors. The appellants needed eight votes but mustered only seven, reflecting the political alignment of the board at the time, which had a progressive majority but not a supermajority (eight out of 11 votes is needed).

Even though progressives lost this particular battle, their willingness to challenge conditional uses for excess parking was potent. In all subsequent large development projects since 2010, the city planning staff has recommended against requests for excess parking and a new performing arts venue called San Francisco Jazz was permitted with zero parking. By 2011 developers stopped asking for conditional uses for excess parking within the MOBNP area and in early 2013 a project with 69 units was approved with zero parking – the first truly car-free market-rate housing development in the post-freeway removal era. With over 1,800 units under construction or approved within the MOBNP area by mid-2013 – all with less than 1 : 1 parking – reduced parking has not stifled development in this part of San Francisco. Yet in much of the rest of the city, beyond downtown and some inner neighborhoods, parking policy remains decidedly conservative.

Conservative parking

Conservatives have played a substantial role in shaping the course of parking reform in the city. In the early 2000s, as various progressive and neoliberal groups promoted parking reform in the MOBNP, albeit for very different reasons, a conservative backlash rose to defeat a parallel citywide parking reform effort. Briefly, in 2003–2004 the city planning department drafted a new *Housing Element*, which is required by the State of California as part of the city's general plan. *Housing Elements* must show where cities plan to enable the construction of low- and moderate-income housing, in this case based on forecasts of future employment and population trends as calculated by regional modeling.

The plan was to rezone corridors that contained high-capacity transit, allowing modestly higher heights and density, and, significantly, reducing parking requirements. The plan also called for legalizing secondary units, which are often small apartments on the ground floor of a single-family unit of housing. To lessen competition for on-street parking, progressives proposed that secondary units be legalized within two blocks of transit lines and that on-street parking permits be limited or forbidden to tenants of secondary units. The secondary unit proposal suggested that 150 affordable housing units could be added throughout the city annually, but that this required relaxing off-street parking ratios.

The entire *Housing Element* was contested by conservative neighborhood activists, including veterans of the Central Freeway debates, who corralled activists to speak out at Planning Commission hearings and who eventually met with the neoliberal but politically ambitious mayor to demand that the *Housing Element* be stripped of the parking reform and density proposals. Heeding the conservative backlash (and seeking higher, statewide office), the mayor and the majority on the Planning Commission (which he appointed) diluted the *Housing Element* and erased language aimed at reforming parking and encouraging densification in transit corridors. The *Housing Element* was adopted in 2004 without parking or density reform. Progressives, rather than appeal to the Board of Supervisors, reconnoitered to focus on the area-specific plans for downtown and the MOBNP, for which there was now less conservative opposition.

The strategy of focusing on specific geographic areas was tenuous, however, because the conservative parking backlash continued, and conservatives next rallied around a ballot measure in 2007 that mandated parking for all new development in San Francisco regardless of where. The ballot measure, popularly known as the "Parking for Neighborhoods Initiative," would have amended the San Francisco planning code to require more parking and nullify the recent and anticipated gains made by progressives (SFPD 2007). The proposed initiative would have abolished maximums, replacing them with required minimums. This meant that developers had to provide parking, whereas under progressive reform parking was an option but not a requirement. The initiative would have frozen all remaining citywide parking to the pre-reform standards, thereby pre-empting years of efforts by progressives, planners, and neighborhood activists to reduce the impacts of parking in the inner-ring neighborhoods.

Progressives obviously cringed at the initiative, but some neoliberals sought to thwart it as well. Led by SPUR, pro-development neoliberals recognized that the exchange value of the city, and thus their profit, was at stake with this measure. The proposed increases in required parking would have actually dampened office and retail development in the downtown because the street capacity there would not have been able to handle the increase in cars, thus creating a diseconomy. The diseconomy would ultimately shunt more office, residential, and retail outward, aggravating sprawl and decreasing the exchange value of the downtown.

Recognizing this, and despite their nuanced differences regarding parking caps, a coalition of progressives and neoliberals organized to defeat the ballot initiative. The coalition was successful when, in November 2007, it was defeated by 67 percent of voters (voter turnout was a low 36 percent). The end result, however, was and still is simply a holding of the line on parking reform. As of mid-2013 no real expansion of off-street parking reform has been attempted in San Francisco beyond the northeast quadrant of the city.

Mobility out of balance

When progressive mobility activists initially suggested removing the Central Freeway many local officials laughed at them. They were told that removing the freeway was impossible. Yet progressive organizations would not accept this narrowly defined set of possibilities. They organized and built alliances, engaged in a long political struggle, and remained persistent. Through a compromise that included bones for neoliberal interests in redevelopment and partial rebuilding, the freeway through Hayes Valley was finally removed. Similarly, the reform of parking is one of the most radically important elements of contesting the spaces of automobility, and was achieved through a progressive–neoliberal rapprochement, albeit geographically limited to only certain areas.

Although progressive groups have promoted freeway removal and parking reform as part of a broader agenda of reducing car dependency, the necessary substitute investment in public transportation, bicycle infrastructure, and pedestrian improvements has lagged. For example, along the dense north–south Van Ness Corridor that traverses the MOBNP, 46 percent of households in the corridor are car-free but transit service is deplorable and bicycling is difficult. Buses operate at 5.2 miles per hour, crawling along in mixed traffic, and the few bike lanes that exist in the area are fragmented at intersections to allow cars to make turns.

To address this imbalance, in the late 1990s the city began considering bus rapid transit (BRT), running in exclusive bus lanes and with higher-quality bus service and reliability, as well as bikeways, including on streets crossing the MOBNP area. BRT and bicycling are keys to the success of the MOBNP, which acknowledges that with 6,000 new housing units on the horizon, reduced parking, and freeway removal, viable transit and bicycle capacity is a necessity. In 2002, 75 percent of San Francisco voters approved a funding stream for BRT and a citywide bike plan, and subsequently progressive mobility advocates repeatedly lobbied the city for BRT on Van Ness (and other major streets) over the next decade, and also lobbied for bicycle lanes throughout the area.

Yet due to foot-dragging and fear of conservative backlashes to reallocating car space for transit and bicycles, the city embarked on a series of excruciatingly detailed studies of over 140 street intersections to measure precise impacts that BRT on Van Ness Avenue would have on delaying automobiles, producing a document that was thousands of pages long and took years to complete. Similarly, bicycle plans languished with years of precise study of potential future automobile delay at each intersection on streets where bikeways would be constructed. Meanwhile, fiscal crisis during the "great recession" resulted in severe cuts to public funding for transit at every level of government, and planning for BRT slowed – despite city approval of multiple new housing developments in the MOBNP area.

While 1,800 housing units were under way or permitted within or near the MOBNP area by mid-2013, the most optimistic debut for BRT was 2018. A second BRT line that would provide much-needed east–west capacity (mimicking the flows of automobiles on the Central Freeway) is also behind schedule. Piecemeal bikeways are years from being connected into a comprehensive network. In 2013 a loud, vocal group of merchants, invoking conservative discourses about automobility, objected to fully separated "cycle-tracks" on Polk Street, parallel to the Van Ness BRT, and the city retreated from implementing a much-needed north–south bikeway.

Backfilling the dearth of new transit capacity, a decidedly neoliberal mobility solution has emerged, at least for commuting. Premium-service, privately operated bus operations have proliferated among large high-tech software and biotech companies in the Bay Area, enabling well-to-do commuters to avoid interaction with the unreliably slow city transit system or other public transit systems, while enjoying more direct service than that of conventional public transit. In 1980, 9 percent of commuters in San Francisco left the city every day to go to work. In 2010, outbound commuters approached 25 percent, suggesting that parts of San Francisco were functioning as a bedroom community for suburban employment centers like Silicon Valley (SFCTA 2011). In noteworthy ways the provision of private transit is an immediate reaction to poor transit reliability and lack of connections. But the poor public transit service results in part from land-use decisions on behalf of these very corporations, which are characterized by dispersed, automobile-oriented campuses that are disconnected from adjacent communities and lack robust transit.

Many employees in the technology sector shun living in suburbia and prefer the city. Although most reverse commuters drive, increasingly thousands are using luxury buses provided by third-party contractors for Google, Yahoo, Facebook, Apple, Genentech, eBay, and an array of Silicon Valley firms. Every weekday employees are shuttled between San Francisco and suburban corporate campuses, and urban-based workers are provided an easy commute that also allows them to work on the bus. Many of these $2 million, 45- to 50-passenger buses have wireless Internet access, iPod and laptop plug-ins and docking stations, televisions, restrooms, leather seats, and tabletops for working. The shuttles have rules of etiquette, such as limiting cellphone conversations to work-related calls and speaking in a low voice. Each passenger is guaranteed a seat, and most buses fill with commuters who spend their time working, surfing the Web, or watching television.

Over the long term, the privatized commuting arrangement may accentuate a regressive transit policy. Fundamentally, it creates an erstwhile pro-transit constituency among tech workers and wealthy professionals who favor environmental awareness and, more broadly, a less auto-dependent lifestyle. Yet, this class is seceding from the public, and in due time may embrace policies that starve public transit as they resist taxes and fees that fund public transit.

All indications are that premium private transit is expanding. As Bay Area transit agencies strain under declining revenues, deferred maintenance, and deep federal and state cuts, one Bay Area transit contractor has rapidly expanded to over 200 vehicles and more than 100 drivers, increased daily ridership to more than 6,000 people a day, and grew 30 percent a year between 2005 and 2010. Moreover, in 2011 and 2012 real estate listings for some neighborhoods began to mention their proximity to private shuttle bus routes, and anecdotally some observers have suggested that such proximity increases monthly rents by up to $400, while the *Wall Street Journal* reports a 20 percent premium on new condominium sales in that area (Said 2011; Keates and Fowler 2012). All of this foreshadows a potential transit future in which a premium system serves the wealthy in first-class coaches – and in premium livable neighborhoods – and a dilapidated, economy-class system serves the lower classes that are gentrified out of the core.

The lesson from San Francisco is laid bare. The removal of freeways and reduction of parking may be consistent with visions of alternative, progressive mobilities, but is complicated by the desires of neoliberals to profit from attractive new development opportunities and the broader gentrification and displacement that are occurring in San Francisco. This conundrum is overlaid with underfunded public transit, apprehension by city officials to adequately reallocate street space, and an emerging ad hoc private transit system for wealthier commuters. A tacit and localized progressive–neoliberal détente over land use enables dense new housing development on former freeway parcels, but the neighborhood became inaccessible to many working-class people while the politics of transit funding exposes rifts and remains a challenge for progressive efforts to contest automobility.

The central location and removal of a segment of freeway contributes to the MOBNP area's desirability as a place to live car-free, but ironically its proximity to the new rebuilt freeway segment just south of Market Street means there is tremendous pressure to build new housing that accommodates people commuting by car or private corporate commuter bus to Silicon Valley and other suburban job centers. New luxury infill housing is often marketed by realtors for its walkability, bikeability, and easy access to the freeway, but it is also part of a transformation of some San Francisco neighborhoods into exclusive bedroom communities. Without a balance of public investment in public transit and social housing, the contestation of automobility, while a laudable goal and met with some success in San Francisco, remains a thorny and complex endeavor.

Note

1 Elaboration of this framework can be found in Henderson (2013). What I am calling "conservatives" here are those (in the United States) who have internalized the "auto-mentality" discussed by Walks in Chapter 1. To distinguish those bearing this ideology, and the pro-business neoliberals in San Francisco, the pre-fix "pragmatic" has been added to the latter in this chapter.

13 Political cycles

Promoting velo-mobility in the auto-mobile city

Alan Walks, Matti Siemiatycki, and Matt Smith

One of the burning questions for those studying the political, economic, social, and cultural contours of automobility concerns how it might be opposed, reformed, and renegotiated. Given that the structure of the city has significant influence on travel decisions and choices, it is important to consider how the auto-city might be reformed to make it more sustainable and resilient. While there is no single solution, given that the wider political-economic system of which automobility is a part is articulated unevenly and differently in each place, it is clear that non-motorized forms of transport will have to gain in acceptance if automobility is to be truly reformed. The bicycle is both a political symbol of opposition to automobility, and a pragmatic mode of transport for a less auto-dependent urban world, and its expression in the rise of velo-mobility constitutes an important component of a socially and environmentally sustainable future.

The retrofitting for velo-mobility will require social, cultural, and physical changes to the geography of cities. Perhaps the most basic reform, upon which new transport practices, planning policies, legal cultures, and employment relations must eventually be built, involves the provision of cycling infrastructure. A network of connected bike lanes are needed to allow those who might commute to their jobs the ability to do so, improving the accessibility of origins and destinations and allowing for an alternative to automobility. Safe and legal bike parking at multiple places of origin and destination are necessarily for storage. As Pucher and Buehler (2008) note, cycling needs to be made "irresistible" in comparison with motorized transport before those who are not otherwise risky or limited in their transport options will decide in sufficient numbers that cycling provides a viable alternative. Many of the technical and financial aspects of such changes are quite minor, often involving very small investments in roadway signage, lane markings, changes to kerb lines, and bike racks at key locations. Buehler and Pucher (2011, 2012) have conducted research on the kinds of infrastructure improvements and planning policies that make cycling safer, more convenient, and viable for commuting across different cities. However, it is not the economic but the political costs that have limited their implementation. Indeed, implementing even the most minor and inexpensive changes has often proven very difficult politically.

The solutions to these problems and the decisions about how to approach them are ultimately political. The purpose of this chapter is to examine the politics of planning for velo-mobility in the auto-mobile city. After historically situated the bicycle as a mode and a symbol, we then focus on three key case studies outlining the diverse kinds of political struggles that characterize debates around cycling infrastructure, and examine how political barriers preventing the implementation of new cycling infrastructure were overcome. The first of these concerns classic battles over cycling in Amsterdam, Netherlands, which have since made that city an international beacon for enlightened velo-mobile policy making. The second case study involves the implementation of bike lanes on a major bridge in Vancouver, Canada, a city with a "green" reputation yet where even minor reforms have been often met with rabid reactions from local politicians. The final case study concerns one of the most contested and scrutinized cities in the developed world, New York City, where a number of battles have been fought simultaneously, and with some unexpected recent success. Each of these three cases involves the interplay of complex local settings, political cultures, and social movements.

Political cycle: the rise of velo-mobility as a mode of opposition

The bicycle has played a number of roles related to the development of, and opposition to, an auto-mobile modernism. In fact, the bicycle is seen to have set the way for the rise of automobility (Furness 2010). As early as the turn of the twentieth century, cyclists and bicycle clubs worked alongside the emerging automobile associations to push for road construction and improvements, including widening, paving, and straightening (Hamer 1987; Norcliffe 2001), literally paving the way to the kind of automobile suburbs that would later allow the middle class in the larger cities to flee the city (McShane 1994). As with the automobile, the bicycle affords the rider the individual freedom of mobility under the promise of technological progress, and unlike a horse does not need to be fed or housed. The bicycle was one of the first durable and inexpensive mass-produced items marketed in the United States, and the organization of bicycle manufacturing acted as a model for the emerging automobile industry later on (Petty 1995, cited in Furness 2010: 18). Before the advent of the motorcar, the chain-and-gear "safety" bicycle, which is the progenitor of today's typical bicycle (replacing the difficult-to-ride high-wheeler in the 1880s), was seen as a modern liberator, both of restrictive patriarchal and gendered norms related to women's mobility, association, comportment, and dress, and of other typical constraints imposed by rural and city life (Marks 1990; Norcliffe 2001; Strange and Brown 2002).

After the price of the safety bicycle became affordable to the masses, the bicycle came to be celebrated for its revolutionary potential. Over the turn of the century and into the 1930s, cycling was promoted by socialists as a way of improving the quality of life among the working classes and rural peasants, and the bicycle became an essential icon of the socialist agenda in many European countries, though less so in North America (Jones 1988; Pye 2003). However, this faded during and after the Second World War as automobile ownership became affordable and was promoted as a substitute for the bicycle by the state for reasons of economic development and international power (not only by the capitalist states, but socialist, or in the case of Hitler's Germany and Mussolini's Italy, fascist states as well). Over the early post-war period, the bicycle was increasingly relegated to a child's toy and associated with a phase of childhood in the most auto-mobile nations.

In contra-distinction to the days of the high-wheelers in the 1870s when bicycles were an elite luxury sport in ascension, by the 1960s the bicycle had become a marginal mode of transport among adults, used mostly by the poor. In a number of nations, cycling activity and production entered a period of rapid decline, which was only interrupted by a temporary burst of activity during the early 1970s that peaked during the oil crisis (at time when automobile production rates also started to decline; see Figure 13.1). Furthermore, cycling was made invisible in most post-war transport planning processes, even in the Netherlands, which is considered the pre-eminent "cycling nation" (Aldred 2012: 86). It was mainly due to the state encouragement of cycling in China that total global bicycle production continued to rise through the 1970s and 1980s, but with the reforming of China's economy and its insertion into global capitalist production

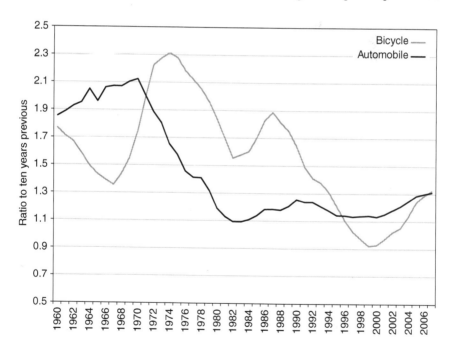

Figure 13.1 Change in global bicycle and automobile production, 1960–2007 (source: Calculated by the authors from data published by the Earth Policy Institute (Roney 2008)).

Notes
Shown is the smoothed five-year moving average ratio of total production in each year to total production ten years previously. Bicycle data includes "e-bikes".

networks in the late 1980s, the automobile has increasingly taken precedence in China as well. While a number of European countries and cities were able to reverse the decline in bicycle use (see Pucher and Buehler 2006, 2008), until very recently the trend among the most auto-mobile nations, and even cycling nations such as China, has been a continued decline and marginalization of cycling (Table 13.1). In most North American cities, among the most auto-dependent on the globe, cycling continues to reveal very low mode shares.

It was largely during the counter-cultural ruptures occurring in cities of the developed world in the second half of the 1960s that the contemporary political significance of the bicycle was forged. Many of the struggles of the time were waged on issues related to the dominance of the hydro-carbon economy in the context of a rapid and disorienting urbanization, including not only urban renewal and the imposition of an auto-mobile modernity on the city (highways, bridges, etc.), but imperial wars fought over oil and other resources, conflict over nuclear weaponry, and growing signs of environmental degradation as revealed in Rachel Carson's *Silent Spring* (1962). The autocratic and paternalistic tendencies of the state in their attempts to quell self-expression and dissent only encouraged a search for potent oppositional symbols, and among those appropriated was the bicycle. Much of the new-left critique was pointed at "the system", typically defined in terms of the military-industrial-educational-consumerist complex. Of course in retrospect what was being targeted for critique was the broader system of automobility, and all this entails politically, socially, economically, and culturally. It was only once cycling was no longer a common part of everyday life, but instead appearing "out of place" in the new modern auto-mobile society, that cycling truly became politicized (Aldred 2012: 87).

The bicycle began to feature prominently in counter-cultural and environmental forms of protest, and in alternative conceptions of social reform. The bicycle became socially constructed as the clean, free, and green alternative to a decidedly "ungreen", polluting, and stultifying automobility (Horton 2006). Various forms of "biketivism" and "vélorutionaries" have since the late 1960s sprung up, influenced by myriad different political agendas and philosophies – including anarchism, Marxism, left-liberalism, and eco-socialism (Furness 2010). As an object of political opposition, the bicycle has come to be seen as a "signifier of utopian futures" and an object of play and subversion, in productive tension with the political importance of its utilitarian mode of everyday mass transport (Aldred 2012: 87). Horton (2006: 45–46) outlines a number of reasons surrounding why and how the bicycle embodies a number of key values dear to environmental and social activists alike: the bicycle rejects the compulsion to ever-increasing speed, instead substituting a slowness that contests the "rhythms" of automobility, and one that can be discursively constructed as "appropriate technology". As a vehicle, it provides "flexible, individual, and 'private' travel", yet one that simultaneously enables the cyclist to demonstrate "responsibility to the planet" and thus environmental bona fides. The bicycle is (rightly or wrongly) perceived as equitable and democratic, affordable and cheap, while being not "especially classed or gendered", and thus "open to all". By requiring

Table 13.1 Bicycle mode shares across the globe, selected metropolitan areas

Metropolitan area/city	Country	Bike (%)	Year
Amsterdam (City)	Netherlands	38.0	2008
Copenhagen	Denmark	29.0	2004
Muenster	Germany	27.0	2006
Freiburg	Germany	22.0	2006
The Hague	Netherlands	22.0	2004
Utrecht	Netherlands	21.0	2004
Alborg	Denmark	17.0	2004
Rotterdam	Netherlands	16.0	2004
Munich	Germany	13.0	2006
Cambridge (Shire)	UK	10.0	2000
Berlin	Germany	10.0	2006
Frankfurt	Germany	10.0	2006
Boulder, CO	USA	6.9	2000
Stuttgart	Germany	6.0	2006
Victoria, BC	Canada	5.6	2006
Inner London councils	UK	4.0	2000
Portland	USA	3.5	2005
Stockholm	Sweden	3.4	2004
Canberra	Australia	3.1	2005
Perth	Australia	3.0	2005
Nottingham (Shire)	UK	3.0	2000
Paris	France	2.5	2007
Minneapolis	USA	2.4	2005
Ottawa	Canada	2.1	2006
Melbourne	Australia	2.0	2005
Brisbane	Australia	2.0	2005
Greater Manchester	UK	2.0	2000
Outer London councils	UK	2.0	2000
Vancouver	Canada	1.7	2006
Washington, DC	USA	1.7	2005
Montreal	Canada	1.6	2006
Denver	USA	1.4	2005
San Francisco	USA	1.4	2000
Quebec, QC	Canada	1.4	2006
Calgary	Canada	1.3	2006
Adelaide	Australia	1.2	2005
Edmonton	Canada	1.1	2006
Toronto	Canada	1.0	2006
Chicago	USA	0.7	2005
Sydney	Australia	0.6	2005
Boston	USA	0.6	2000
Los Angeles	USA	0.6	2005
New York	USA	0.5	2005
Philadelphia	USA	0.3	2005
Houston	USA	0.3	2000
Pittsburgh	USA	0.1	2000

continued

Table 13.1 Continued

Metropolitan area/city	Country	Bike (%)	Year
Beijing	China	37.0	2000
Beijing	China	16.0	2009
Hangzhou	China	43.0	2000
Hangzhou	China	37.7	2009
Guangzhou	China	10.9	2003
Shenzhen	China	4.0	2010

Source: Borjesson and Eliasson (2012), Buehler and Pucher (2010), Census of Canada (2006), CUSTReC (2007), Nadal (2007), Pucher and Buehler (2006, 2008), Trivector Traffic (2006).

body power and exertion, and in being open to the elements, the bicycle is seen as a wellspring of health and fitness, and benignly associated with nature. And although a form of private transportation, the lack of enclosure means that one can communicate and must often interact with others while moving, leading to a more public, human-scale, peopled mobility (Horton 2006: 45–46).

Furthermore, Horton (2006) notes that in contrast to other materialities, there is virtually no problematization or criticism, no "negative history", of the bicycle. "A zero-emission vehicle which does not contribute to congestion and rarely maims other creatures, the bicycle seems to have no adverse consequences" (ibid.: 46). Hence, cycling can be performed as a mark of distinction with a "green morally exemplary identity", and by simply putting in the effort to cycle the rider can be seen, and personally feel, to be constructing the alternate world to which s/he aspires (ibid.).

Yet, as Furness (2010: 215–217) discusses, the bicycle industry is not much different from other global manufacturing industries in having many large multinational corporations who pursue shareholder profit by chasing after low wages, poor labour conditions, and minimal regulations. Any sustainable transportation future will have to deal with these issues. Moreover, there are multiple underlying sources of political and community tensions related to bicycle usage in cities. A main source of tension undoubtedly is contestation over scarce road space that may be reallocated from private motor vehicles to dedicated bicycle paths, and the winners and losers associated with such decisions. Additionally, the attitude that cyclists may hold of a "green morally exemplary identity" has created tensions with non-cyclists. Cyclists disobeying traffic laws (i.e. running red lights or failing to obey stop signs; biking the wrong way on one-way streets) with impunity and, in some cases, being confrontational with motorists has contributed to conflicts not only with motorists and truck drivers but also with pedestrians, another active and "green" mode of non-motorized transport. The demographics currently attracted to urban cycling are another potential source of community tension: while cycling has the potential to have broad appeal, it tends to be a travel mode that is attractive to and is used more prominently by affluent white men (Steinbach *et al.* 2011).

Clearly, the bicycle is well equipped to satisfy at least some of the requirements of an environmentally friendly, human-scale alternative auto-mobility, alongside programmes to expand transit usage and pedestrian activity. For this to happen the city itself first needs the basic infrastructure in place to make cycling and other non-motorized modes of travel a viable alternative for a sufficient number of people. The major looming political question facing opponents of automobility is under what conditions, and through what political coalitions and social movements, can support be gained for significant investments in cycling facilities? The three case studies that follow provide examples of three different cities in which, through different tactics, arose key cycling innovations and infrastructure.

Amsterdam: from white bicycles to public bike schemes

Amsterdam is known as one of the world's foremost cycling cities, and the Netherlands (along with Denmark) one of the world's premier cycling nations. As of the mid-2000s, more than one-quarter of trips in the Netherlands (and 18 per cent of those in Denmark) were being taken via the bicycle (Pucher and Buehler 2008). In the City of Amsterdam, 38 per cent of all trips are made by bicycle (Buehler and Pucher 2010). Carstensen and Ebert (2012) argue that this pattern was largely set during the inter-war years, when the national cycling unions that formed around the turn of the century found common cause with car drivers and automobile associations in pushing for both paved roads and cycle path networks. Unlike the less successful approaches taken in other European countries, in which the bicycle advocacy organizations promoted bicycles for racing – a specialized, temporary activity usually sequestered from everyday life – in the Netherlands the Dutch Cyclists' Union (DCU) promoted cycling for touring and tourism purposes. They thus sought to build an extended cycling infrastructure, integrated with but often separated from that for motorists and pedestrians. By the 1920s the DCU had been able to use their strategic alliances with the automobile associations to successfully advocate for a national-scale separated bicycle path network (ibid.). This was partly made possible by a "bicycle tax", which originally went to financing the general deficit as well as new road-building projects in the 1920s, but later was used by cyclists and their organization to exert political leverage, claim agency in traffic planning, and "create a tradition of traffic engineering devoted to cycling paths and regulation" (Carstensen and Ebert 2012). By the Second World War, cycling was sufficiently common and embedded within everyday life in Amsterdam that the Nazi theft of bikes for the steel and rubber has been characterized as one of "the worst one can do to a Dutchman" and memorialized in the anti-German gibe "give me back my bike" (cited in Mapes 2009: 66).

However, as in the rest of Europe, the early post-war period witnessed a rapid decline in the proportion of cyclists in Amsterdam, and most areas bombed during the war were rebuilt with the automobile in mind. While perhaps not quite as dramatic a drop as in Sweden (from 70 per cent cyclists in 1940 to only

1 per cent in 1970) (Emanuel 2010, cited in Carstensen and Ebert 2012), the decline in bicycle use in the Netherlands was, by the early 1960s, quite advanced. As with Copenhagen and other European cities, many old streets in the centre of the city became highly congested with automobiles, and many plazas were being used as parking lots. Plans were even put forward in the 1950s to fill in some of Amsterdam's well-known canals to use as roadways (Mapes 2009: 66). While governed in a generally non-confrontational way, both the national political apparatus and that of the city were essentially in the hands of the pro-growth, pro-automobile, centre-right coalition of political parties led by the conservative People's Party for Freedom and Democracy. Deference to authority and Pillarization – the separation of society into separate social "pillars" – was still the norm. Labour revolts, and protests against the new urban renewal policies that removed old urban fabric and cycling paths, were sometimes put down brutally by the police (Kempton 2007).

Political battles to safeguard the role of the bicycle in Dutch society dovetailed with the rise of counter-cultural social movements, which aligned with protests against the war in Vietnam, a collapse in support for Pillarization, and a rejection of nuclear energy (Jamison *et al.* 1990). A key event in this history, and one that would galvanize future movements in places like Copenhagen, and eventually Paris, London, and New York, was the White Bicycle plan, one of the many "white" plans promoted by Provo, a small, pacifist, pro-environmental anarchist collective formed in 1965 (see Kempton 2007). Under the original plan, the municipality of Amsterdam was encouraged to purchase upward of 20,000 white bicycles each year and make them available free of charge (without locks) to the public for trips within the city, to be left wherever they were last used for the next potential rider. Although rejected by the City, Provo went ahead with their plan, acquiring and painting white 50 bicycles and distributing them across the city. When the city police confiscated the bikes under an obscure and usually unenforced by-law requiring bicycles to be locked when in public, Provo took the few bicycles that were returned to them, added locks, painted the lock combination codes on their frames, and redistributed them across the city. After a month, the number of white bikes still in distribution was small, and the plan largely fell apart (except in the Hoge Veluwe National Park where free white bicycles are still in use for park patrons). Provo also fought against the rising police violence and harassment of the time: another of their "white" plans (similarly rejected by the City) advocated that the police disarm, dismount their squad cars for white bicycles, and hand out free fried chicken and contraceptives (ibid.).

While both Provo and the White Bicycle plan eventually faded out, their guerilla-style actions and the sympathy garnered by the excessive policy response significantly influenced political debates in the City of Amsterdam regarding the role and functioning of democracy, and were instrumental in making the bicycle one of the symbols of the local new-left opposition (Kempton 2007). Young couples got married under white bicycles, rock-and-roll songs were written about them (Tomorrow's single 'My White Bicycle', for instance), and images of abandoned or confiscated white bicycles became icons for a modernity and democracy gone awry. In this way, the white bicycle movement became part of the cultural struggles framing the rise of various reform-minded left-leaning governments through the late 1960s and 1970s. The white bicycle movement was part of the backdrop of the 1967 election of a Labour Party mayor in Amsterdam who took it upon himself to reform the police, and who was instrumental in building the metro system and instituting pilot projects of "woonerf" multi-modal laneways that significantly slowed down car traffic within the city. An important facilitator of political shifts in the Netherlands at the national level is the proportional representation electoral system, which bolsters both electoral choice and turnout because no vote is wasted, and encourages smaller and younger parties fighting for specific policies to enter coalitions with the larger established parties. During the 1970s eco-socialist organizations and political parties, as well as environmental organizations influenced by Schumacher's "small is beautiful" thesis, such as Kleine Aarde (Small Earth), fought for human-scale environmental policies (Jamison *et al.* 1990). Later in the 1970s, when faced with recession Dutch governments pursuing austerity found it cheaper and politically easier to subsidize bicycle infrastructure as an alternative to new expensive roads, and car-free Sundays were implemented as a response to the OPEC oil embargo in 1973 (Mapes 2009; Carstensen and Ebert 2012).

Importantly, Amsterdam's white bicycle experiment provided the first model for a public shared-bicycle scheme, in which bicycles are borrowed or rented as part of a public service administered or regulated by a local collective or authority, and can be accessed by anyone at marginal rates for trips within the city (De Maio 2003; Beroud and Anaya 2012). Variations on this provided the models for "second-generation" bike-sharing schemes, including those in Copenhagen, Denmark, and the "yellow bicycle" schemes in La Rochelle, France (1974) and Portland, Oregon (1994). From the early 1990s a number of European cities adopted bike-sharing programmes. However, it was not until the early 2000s that electronic technologies made it much simpler to operate self-service public bicycle schemes, leading to a veritable explosion of this kind of infrastructure (Beroud and Anaya 2012). According to the United Nations (Midgley 2011) upwards of 375 public bike-sharing programmes are currently found in 33 different countries, through which over 230,000 bicycles circulate. Among the more successful of the recent crop of schemes are the Vélo'v programme begun in Lyon in 2005, and Paris's Vélib in 2007, the largest bike-sharing system outside China (the largest is found in Hangzhou). The European schemes provided the archetype for the first of the current "third" generation of bike-sharing schemes in North America, including Montreal's Bixi in 2009 (extended to Toronto in 2011), Boston's Hubway system in 2010, and New York's CityBike scheme in 2013, now the largest in North America. The white bicycle plan was also the inspiration for the membership-driven community bike-share programme in Toronto (2001–2006) and Washington, DC's Capital Bikeshare programme. While still in the experimental stage, and understandably fraught with persistent problems (including theft, vandalism, and undersized networks), such public bicycle-sharing schemes hold promise as one method to get more bicycles and bike riders onto city streets. Stating this, in many cases such bike schemes are being pursued for shareholder profit by private corporations, and there is a danger that the promise of democratic transport, as embodied in the original white bicycle plan, could be transfigured and appropriate by private capital. This is a history that is still being written.

Vancouver: political bridging

By North American standards, Vancouver is a leading city for cyclists, with the third highest share of work trips made by bicycle among the continents' large cities, the largest proportion of female bicycle commuters, and the lowest rate of cycling accident fatalities (Pucher *et al*. 2011). However, the provision of quality cycling facilities in Vancouver has been decidedly mixed. The City's public transit service has added safe bike locker facilities at some transit hubs, and installed bicycle racks onto buses and permitted bicycles onto trains in the region. The City has taken measures to improve the on-road safety of cyclists by painting bicycle boxes ahead of motorized traffic at key intersections, and has promoted calming traffic on local roads frequently used by cyclists through speed bumps and other street alterations. Nevertheless, Vancouver has among the fewest on-street cycle-only lanes and off-road bike paths of any major North American city (Pucher *et al*. 2011). The city has also been a late adopter of the global movement towards bicycle-sharing programs. Indeed, outside of the dense downtown and surrounding inner city, much of metropolitan Vancouver follows the pattern of the auto-city, with vast spread-out land uses and fast-moving arterial roads. Moreover, water bodies bisect the region and the downtown core is on a peninsula with limited bridge access, heightening public awareness about the scarcity of road space and a sense of competition for that space between different models of travel.

The struggle to make the city and its surrounding region more bicycle friendly has spanned three decades, born out of competing visions about who has legitimate claims on public road space, and more broadly which political interests in the city should be prioritized. Vancouver has long been a central hub in the global environmental, urban livability and sustainability movements, with an active and engaged citizenry and significant cluster of environmental and social justice non-governmental organizations. Greenpeace was founded in Vancouver in the late 1960s; the city hosted the first United Nations World Urban Forum in the 1970s; and scholars working in Vancouver developed key concepts in sustainable development such as the "ecological footprint" (Wackernagel and Rees 1996). This culture of environmental consciousness has been mobilized by policy makers and non-governmental organizations to successfully expand sustainable transportation options, developing city policies and programmes to prioritize walking, cycling, and public transit ahead of automobile travel. Citizen revolts prevented large freeways from being built into the downtown, and limited the size and extent of Vancouver's freeway network (Mees 2009).

However, as in other large Canadian cities, Vancouver has been a site of vigorous political debate and contestation over the direction of public policy and planning towards sustainable transportation, and in particular any projects that even appear to worsen conditions for motor-vehicle drivers. Business associations, trade organizations for the freight industry, and right-leaning political parties have consistently campaigned to implement policies that maintain and further auto-oriented mobility in the region. In fact, a plurality of residents living in Canada's pre-eminent metropolitan areas, including in Vancouver, are supporters of the federal Conservative Party and their policies of privatization, low taxes, anti-environmentalism, and deregulation of the energy industry (Table 13.2). Partisanship is increasingly structured in terms of transport mode, with car drivers far more likely to support the Conservatives, whereas the majority of cyclists support parties of the left – the NDP, Canada's social democratic party, and the Green Party. Indeed, cyclists are one of the few significant bases of support for the Green Party in Canada, a pattern even more true in Vancouver than elsewhere (Table 13.2).

As in other cities, it is proposals to implement on-street bicycle lanes requiring the reallocation of general traffic road space that have been especially contentious in Vancouver. An illustrative case is the lengthy process to implement a bicycle-only lane on the Burrard Street Bridge, the oldest of the three fixed links

Table 13.2 Canadian federal partisanship by transportation commute mode (%)

Eight key CMAs	Green	NDP	Liberal	Conservative
Bike	23.7	36.8	18.4	13.2
Public transit	7.0	32.2	33.3	22.8
Car/truck as driver	3.6	14.9	33.0	45.1
All other modes*	8.2	26.4	32.7	25.2
Total	5.1	19.7	32.8	38.4
Vancouver CMA (only)				
Bike	27.3	45.5	18.2	9.1
Public transit	8.3	46.4	26.2	19.0
Car/truck as driver	3.5	20.4	35.7	40.1
All other modes*	0	26.7	33.3	40.0
Total	4.8	26.4	33.3	35.3

Source: Calculated by Walks from a 2007 RDD CATI representative telephone survey commissioned by the author of 2,542 individual respondents of working age in eight regionally representative Canadian metropolitan areas (CMAs): Toronto, Montreal, Vancouver, Ottawa, Calgary, Winnipeg, Quebec City, and Halifax. 439 of these respondents live in the Vancouver CMA.

Notes
Political parties are arranged from left to right in terms of their place on the political spectrum. Partisanship here is discerned via three questions: (1) "Which party did you vote for in the last election?", (2) if did not vote, "Which party would you have voted for?", and (3) if no party yet identified, "Which party do you feel closest to?". Partisanship is only reported for the four main political parties in English Canada, and thus totals do not always sum to 100.
* Includes, car/truck as passenger, motorcycle, taxi, walk, boat/ferry, unknown combination of modes, and "works from home".

connecting the downtown core with the southern part of the city and surrounding suburban region. The six-lane bridge was not initially designed to provide dedicated space for bicycles, and this created tensions between motorists, pedestrians, and cyclists. In the 1980s, cyclists were ordered by the city to share the narrow bridge sidewalk with pedestrians. However, this created safety concerns that were publicized in the media, including cyclists being seriously injured after falling off the sidewalk into traffic.

In the mid-1990s, a proposal by the right-of-centre pro-business council (controlled by the Non-Partisan Association, the main right-of-centre municipal party in Vancouver) was to add new cycle-only lanes onto the outside of the bridge deck. This plan had the advantage of avoiding the need to remove existing road space, but was very costly and opposed by conservationists concerned about the heritage value of the bridge. Instead, in 1996 the council was persuaded to proceed with a six-month trial to close one lane of traffic going into downtown and replace it with a bike-only lane, raising objections by business organizations, drivers, and other right-leaning groups. In the initial days of the trial, egged on by influential and inconvencienced west-side drivers, local media reported relentlessly on the traffic chaos in the morning commute caused by the driving-lane closure, branding the bike lane a disaster. Facing immense public pressure, the city council cancelled the Burrard Street Bridge trial after less than a week, and returned the lane to general-purpose traffic, forcing cyclists back onto the sidewalk.

Despite the failure of the trial, debates regarding solutions to pedestrian and cyclist safety on the Burrard Street Bridge persisted over the next decade, as both public awareness and citywide increases in cycling as a travel mode slowly inched the issue closer to the mainstream of Vancouverite opinion. Political deadlock persisted, however, largely due to deeply engrained and aggressive opposition among drivers living on the west side of the city to the removal of traffic lanes on the bridge or any solution that might impinge on road space. The NPA depended on votes from this part of the city, and the new NPA mayor in 2005, Sam Sullivan, would not support a new bike-lane project. Former NPA city councillor Peter Ladner, who had originally supported the initial trial, reported that the negative reaction from the NPA base was at a visceral "gut" level, forcing him to backtrack (personal communication, 2010).

In the 2008 Vancouver civic election, a newly formed "centrist" political party called Vision Vancouver proposed a new trial to reallocate one lane of traffic on the Burrard Street Bridge to a cycle only lane as part of their election campaign platform. The proposal was not seen as a major vote winner, and the Vision mayoral candidate, Greg Robertson, took a political gamble in targeting his message to labour, environmental, and activist groups, as well as residents living in the east and southern parts of the city where incomes are lower and cycling is more common. When Vision won the mayorship and a large majority on council, Robertson immediately pushed to implement a new one-year trial to reallocate one lane of traffic heading out of the downtown core to bicycles. This design was planned to minimize the initial impact on road traffic. Learning from past experience, the local politicians and city staff also undertook a media campaign to raise awareness about the trial and options for motorists to reduce congestion. On the first morning of the trial in 2010, the media was amassed at the bridge entrance waiting to report on the traffic chaos expected to ensue, based on the gridlock that accompanied the previous trial 12 years earlier. The positive reporting of the fact that it did not materialize was a major public-relations breakthrough for proponents of the bike lanes. The number of cyclists using the bridge went up, while travel times over the bridge by car did not increase markedly, cementing justification for making the new bike lane permanent. This was an important political victory.

In the aftermath of the opening of the Burrard Street Bridge bike lane, Vision Vancouver proceeded quickly to expand the amount of on-street bicycle-only lanes across the city by reallocating existing road space, using a similarly cautious approach. Although there is increasing agreement in Vancouver about the benefits of such cycling infrastructure, and although a body of local experience shows that predictions of traffic tie-ups can be avoided through careful planning, each new proposal is still met with fierce community opposition. In answering why there still exists considerable anxiety among car drivers to each new proposal, Gordon Price, former six-term NPA Vancouver city councillor, rhetorically explains the contours of the ongoing politics of mobility in the city:

> Is it because those who are car-dependent fear someone might force them out of their cars, make them feel guilty for their habits, or above all, inconvenience them for the sake of those who have traditionally been of lower status or who, more annoyingly, flaunt their fitness and the traffic laws with impunity and who don't, in the minds of vehicle owners, pay their way?
>
> (Price 2013)

The implementation of key interventions to improve cycling in cities such as Vancouver remains troubled by an enduring paradox related to the rising politicization of transport policy, and its expression within the local political party structure. Yet, the Vancouver case demonstrates the viability of a steady, incremental approach to the extension of cycling infrastructure within this context, one that gains support with each new cyclist added to the city's streets.

New York City: out of the bosom of conflict

New York has always been a particularly contested city, internalizing and reflecting many of the tensions that beset the United States as a whole. This includes the fact that the core of the City itself is dense and ideally suited to walking and bicycling, more like a European city than the typical North American pattern, while the post-war suburbs built throughout the larger metropolitan region are among the most automobile dependent in the country. New York City also has the most extensive public transit network in the United States, and a larger share of all commuting trips take place via transit in New York than other large US cities. New York City is well known among planners for the battles waged between the notorious Robert Moses, one-time chairman of the Triborough Bridge Authority and Planning Commissioner responsible for building extensive freeways and bridges through the City from the 1930s through the 1960s (in the process destroying many older neighbourhoods), and anti-freeway activists including Jane Jacobs (see Caro 1974). These battles have consistently been fought over how to define an appropriate urban modernism, and what happens in New York is often taken for what is suitable in other contexts.

Like all other US metropolitan areas, New York is a city (and a region) of automobility. Manhattan's taxicabs provide one of New York's iconic global images, while Moses's massive bridge and freeway network became the early global model for how to retrofit the eighteenth-century city for the automobile in the post-war era. It is within this context that the contested history of bikeways in New York must be understood. As in Vancouver, New York also tried earlier to instal bike lanes. After visiting China and witnessing the mass use of bicycles there for daily life, then-mayor Ed Koch had the City's planner install protected bike lanes on three of New York's prime north–south avenues (Fifth, Seventh, and Broadway) in the summer of 1980. However, within only days the City began hearing loud complaints from automobile drivers, and even pedestrians, who argued that the lanes reduced safety and increased congestion. Despite the fact the bike lanes had only been in place a short time, the State Governor complained to US President Jimmy Carter about the waste of money they represented, and Koch was compelled to remove them after only a couple of months (Mapes 2009: 186–187).

However, after a period of significant political struggle, New York is now once again building bike lanes. Furness (2007, 2010) demonstrates that it was necessary that simultaneous but disparate political movements, often operating in tension with one another, persisted in pursuing pro-cycling reforms through different channels. The two main organizations in the recent story are the City's main bicycle lobby group, Transportation Alternatives, a well-established pro-environmental group advocating incremental policy negotiations, and the more recent and radical direct-action organization, Time's Up!, promoting confrontation, conflict, and public disobedience. The antics of the latter made the efforts of the former appear mainstream, and allowed for a policy middle-ground to be staked out (the so-called "flank effect"). Furthermore, as is the case with many other political struggles, in New York it was only after violent confrontation and an excessive response by the state police apparatus that sufficient empathy and political will was found among the majority of voters for a compromise that would include bicycles.

As in other developed-nation cities, where the 1960s counter-culture arose in response to the imposition of cross-cutting freeways and other aspects of monofunctional automobility, the bicycle was already well established as a political mode of transport in New York by the 1970s. Drawing on the Provo white bicycle plan, novelist Norman Mailer's unsuccessful mayoral campaign in 1969 advocated distributing free bikes around the city and banning private cars (Mapes 2009: 170). During the early 1970s, in an early model of what has since become critical mass rides, "bike-ins" and "bike-brigade" protests were often held in the city, holding up traffic in the name of taming auto-mobile capitalism (Blickstein 2010). However, it was only with the violent and unprecedented police response to the 2004 Critical Mass ride in Manhattan that the issue became one that garnered empathy from the larger public and compelled a search for a new policy approach.

Critical Mass (CM) began in San Francisco in the early 1990s, involving large semi-organized bicycle tours whose main purpose is to bring cyclists together to claim the streets and express a collective solidarity (Carlson 2002). In doing so, CM rides block traffic in what is sometimes called a "commute clot", showing their strength in numbers and asserting cyclists' rights to public streets and safe riding conditions (ibid.; Furness 2007, 2010). CM is at once "a protest, a form of street theatre, a method of commuting, a party, and a social space" (Blickstein and Hanson 2001: 6). While CM rides are said to have developed a "notorious reputation" among public officials and police forces for the seemingly lawless and antagonistic anti-car sentiments, CM rides until 2004 were typically led or escorted calmly by the New York City Police Department (NYPD), which until that time had been "more or less friendly" to CM (Blickstein 2010: 893).

However, starting with a key 2004 ride that coincided with the Republican National Convention (under a then-Republican mayor), the police chose to use excessive force and intimidate the cyclists. During the convention, 264 cyclists were arrested on the day of the ride, and many of them violently "brutalized", on the grounds that they violated a rarely used law that forbid the "blocking of traffic" (despite police aid for CM for this very practice in previous CM rides, exposing the irony of CM's main slogan: "We are not blocking traffic, we ARE traffic") (see Blickstein 2010; Furness 2010). Furthermore, the police illegally confiscated bicycles from riders, including riders that were not part of the CM rides and bicycles that were legally locked up (ibid.). Police harassment and intimidation of cyclists continued for a number of subsequent years on the grounds that CM was "a danger to the public safety" (Furness 2010: 2), with the City and NYPD illegally demanding CM first get parade permits, seeking a permanent injunction against activists for Time's Up! advertising and gathering in central public squares, and successfully charging CM riders with disorderly conduct (Blickstein 2010). Between mid-2004 and mid-2006, New York City is claimed to have spent US$1.32 million arresting and prosecuting cyclists and seizing bicycles (Furness 2010: 3).

However, perhaps demonstrating the promise of democratic pluralism (and the fact that people can change their stripes), the mayor of New York City, Michael Bloomberg, moved to revise the approach of the City towards cycling in 2007. Responding to the growing sympathy for harassed cyclists, as well as a city commissioner's 2006 report proposing an extensive bikeway network in response to rising cycling fatalities and injuries, Bloomberg appointed a new bike-riding deputy mayor, a new pro-cyling transportation commissioner, and even rescinded his Republican Party membership (becoming an independent)

(Mapes 2009). The new transportation commissioner moved quickly on a pilot programme of wide separated bike lanes on key avenues (originally on Ninth Avenue, followed by Eighth, the remake of Times Square on Broadway, and then in the outer boroughs including Bedford Avenue in Brooklyn) as well as bike racks and bike "boxes" (ibid.). In May of 2013, New York City opened the largest bike-sharing programme in North America, its CityBike programme, modelled on Montreal's Bixi scheme.

While there are undoubtedly multiple factors underlying the dramatic shift in policy that has unfolded in New York (under the same mayor, no less), Furness (2007) argues that the interplay of radical and mainstream organizational strategies was ultimately necessary for the successes that were won, in part because the more radical events and protests promoted by Times Up! and the CM ride could be compared against the more moderate and incremental requests for bike lanes from Transportation Alternatives. In contrast, "where they don't have Critical Mass, they think bike lanes are radical" (respondent cited in Furness 2007: 312). The exact impact of such effects is not easy to measure, not least of all because, as Furness notes, moderate cycling activists "must actively make use of the flank effect, lest they become lumped together with the 'radical wing'" (ibid.). Yet, as had transpired in 1997 in San Francisco when the police chose to confront 5,000 cyclists out on a CM ride, the conflict and confrontation brought media attention and forced a public debate, in which the right of cyclists to safe public road space and basic infrastructure (lanes) eventually became accepted as "common sense" by the discerning public (Furness 2007). Evidence from other countries, including Brazil, suggests that CM rides (locally referred to as "bicicletada" in Brazil) are having a similar effect in spurring conflict, debate, and ultimately changing policy responses in cities such as Pelotas and Porto Allegre (see Jones and de Azevedo 2013).

Conclusion

The bicycle has been adopted as one political symbol of opposition to the dominant system of automobility. Although perhaps too uncritically accepted as free of taint, and increasingly tied to peculiarly middle-class aspirations for the city, the bicycle nonetheless is held out as the sustainable transportation alternative to a auto-mobile hydro-carbon economy with its ties to imperial wars and intra-urban segregation. To an extent, this is a result of the structure of domination of the current system and regime of automobility, which has an interest in marginalizing alternative auto-mobilities that threaten the profits and power of the auto-industrial complex, and which in turn has inadvertently remade the bicycle into a symbolic vehicle of the powerless and the anti-car opposition. If velo-mobility is ever to become sufficiently common to have any influence or potentially even supplant contemporarily automobilities, the materiality of the city itself will require considerable reform and rebuilding, and the political culture around cycling will require its own revolution. This is not to argue for new restrictions on car use, nor to propose heavy-handed forms of state-imposed social engineering on an increasingly stressed auto-mobile public. Instead, mobility must be democratized and the ability to choose alternative mobilities built right into the physical and social fabric of the city.

How best to realize the necessary political and cultural changes? The findings of our case studies show that the push to improve urban cycling at least since the 1960s has more often been tied to left-leaning politics of various forms and to certain political counter-cultures (although not universally or historically – the League of American Wheelmen during the late nineteenth century was often racist and sexist; see US Federal Highway Administration 2012). The role of activists undertaking direct transgressive actions such as Critical Mass rides have been essential in raising public awareness about the place of cyclists in the city and contributing to a rise in the political capital of the cycling movement. The allocation of road space in cities is inherently political, and early conflicts over high-profile bike lane trials in Vancouver and New York City and the student-led anarchist movements to promote cycling in Amsterdam created the political space for debates concerning existing governance structures and for more modest approaches to ultimately take hold later on. Most cities now have a cycling strategy aimed at expanding the number of local riders, many are beginning to allocate new road space to bicycle infrastructure, and in many cities there are now bike-sharing schemes that allow those who do not own bikes to cycle in the city and combine cycling with transit trips.

As the beginnings of a planning and political paradigm that accepts the legitimate place of bicycles on city streets takes hold, tensions have emerged about the future role and identify of cycling activism, and the place for antagonistic or conciliatory strategies. For cycling activists that remain true to leftist, environmentalist, Marxist, or anarchist histories of their movements, the bicycle continues to be a signifier for an alternative lifestyle. It is not only a challenge to the dominance of the automobile on city streets, but also the perceived underlying causes and products of automobility, including urban sprawl, globalized consumerism, industrial food systems, and inequitable wage labour, a fact that has been picked up by proponents of automobility and used to politicize urban policies around mobility (as discussed in Chapter 11 of this volume). At the same time, the expansion of urban cycling is now being promoted by a pragmatic cadre of planners and activists, for whom the bicycle is not a core aspect of their identity but seen as simply a more convenient, low-cost, and environmentally sustainable and healthy mode of transportation. In turn, a swelling coalition of professional planners, urban professionals, bike commuters, activists, and politicians has sought to professionalize cycling advocacy and work within the existing political, planning, and legal structures to effect change. A key objective is to develop and implement strategies to overcome political barriers or community opposition to cycling proposals. To this end strategies are being undertaken by local governments, sustainable transportation non-profit organizations, and small businesses to increase cycling among currently under-represented demographic groups including women, seniors, visible minorities, and low-income people, which can create a more diversified political constituency. It is not likely that the current state of urban cycling could have been achieved without the counter-culture ideologies and transgressive techniques of those early activists who fought to make an increasingly invisible mode of transportation visible again to motorists and political leaders. For this, cyclists can thank the history of radical political organizations, protests, and events that have occupied, even if temporarily, key sites and routes in the public sphere, challenged dominant perspectives of automobility, and placed such issues near the centre of public debates.

14 Taking the highway

Expressways and political protest

Matt Talsma

Public space theorists have long noted the importance of outdoor gathering places for a properly functioning democracy (Habermas 1989; Mitchell 2003). The political activity of taking to the streets to publicly air concerns and grievances has a long and productive history. In the late nineteenth and early twentieth centuries, it was not uncommon for masses of striking workers to parade through the streets of Canadian cities – actions that resulted in significant advances for the labour movement (Heron and Penfold 2006). Yet while these types of public political actions still enjoy wide public support elsewhere (see, for example, Chatterton 2006; McNally 2006), in North America acceptable political activity is increasingly limited to the ballot box. Keller notes that in Western societies "the idea of the public forum has dramatically faded in the twenty-first century, and non-celebratory public gatherings have declined" (2009: 13).

The recent Occupy Wall Street movement drew attention to the decline of public political discourse, using a dramatic reinvigoration of the practice of "occupation". Occupy participants set up semi-permanent camps and successfully, albeit temporarily, revived the notion that outdoor spaces of cities can be used for political dialogue, as well as living spaces. However, largely negative media coverage managed to convince an apathetic public that occupiers were either social degenerates or dangerous radicals, and the encampments were eventually shut down by authorities, in many cases through violent forced removal (see Khon 2013). However, a number of the questions raised by the occupy movement linger. How can contemporary social movements get their grievances heard? If public dialogue is to be accomplished through occupation of physical space, which space should this be? Where is the contemporary agora?

Expressions of political apathy can be partially attributed to the anti-collective subjectivities produced through neoliberalism. Neoliberal subjectivity is fashioned through an individualist and entrepreneurial market rationality that emphasizes personal responsibility over collective effort and organization (Brown 2005; Lemke 2001). Automobility is intertwined with these neoliberal subjectivities. As noted in the introductory chapter, those qualities the car provides – freedom and autonomy – are precisely those ideals so valued in liberal society (Lomasky 1997; Rajan 2006). Because the lifestyle mode of automobility has succeeded in delivering those basic liberal ideals through enabling unrestricted movement, the car has become thoroughly integrated into a discourse of natural rights, such that liberalism is both achieved and reproduced through auto-mobility (Rajan 2006). And as discussed by Walks in the first and eleventh chapters in this volume, the everyday practice of car-driving shares particularly important relationships with neoliberalism, both in the way that it governmental-izes and in the way that it atomizes subjects.

The system of automobility, and the neoliberal subjectivities that accompany it, have changed the political landscape as drastically as the car has altered the physical landscape. Yet, the political costs and implications of automobility, as well as the sites of contestation within the auto-city, remain under-theorized. This chapter explores the political implications of a culture of car dependency through an analysis of the reaction to two separate but similar events, where public political demonstration clashed directly and dramatically with the right to mobility.

Lessons from the road

Throughout the developed world, and in many places in the developing world, automobilized landscapes and lifestyles are such everyday features that they have become thoroughly normalized. For instance, as Martin demonstrates (Chapter 2, this volume), the number of fatalities caused by automobile colli-sions worldwide often exceeds that attributed to many forms of disease. Despite this, the car is rarely scrutinized for its deadly capacity. This death toll is merely regarded as an unfortunate but necessary condition in what Andrey (2000) calls the "automobile imperative".

This normalization of automobility means that many of the conditions it pro-duces are under-recognized. For example the motorist's relationship with the automobile and the road forms a specific set of "technologies of the self" – tech-niques of self-discipline internalized by the subject (Foucault 1980) – that govern a citizen's public interactions and activities. The disciplinary action of the system of automobility is experienced from an early age. Children, targeted by in-school and extracurricular safety campaigns, are taught to be wary of public space (Bonham 2006), and, like the stranger, the road is something to fear (see Chapter 5 by Buliung *et al.*, this volume). The private space of the backyard becomes one of the few places for safe play, as the spaces outside our front doors have been transformed into a dangerous realm of machinic movement (even a game of road hockey, arguably Canada's quintessential pastime, is pro-hibited in many urban jurisdictions due to liability concerns; see Alcoba 2010). In this way, the youngest generation is forced to learn about, guard against, and take responsibility for their own vulnerability to a hostile environment produced by another generation's mobility habits.

The disciplinary nature of automobility takes on a more formal quality when youth begin their instruction in driver's training. Learning the rules of the road may well be the young citizen's most important lesson, as Virilio notes: "'Good conduct' is no longer *morals* taught in school, but driver's education" (1986: 90, emphasis in original). It is here that the student learns a whole array of self-regulatory techniques, and internalizes the disciplinary regime of automobility: moving at a certain speed, staying between painted lines, obeying signs and traffic signals, and routinely checking mirrors, dashboard, and blind spots.

Obtaining a driver's licence serves as a formal induction into the power structures of civil society (as discussed in Chapter 10 by Reid-Musson in this volume). Foucault notes that power operates through subtle mechanisms: "methods of observation, techniques of registration, procedures for investigation and research, apparatuses of control" (1980: 102). Each of these mechanisms is localizable within the governance regimes of automobility. And if a prominent feature of any efficient police state is the requirement that citizens at all times carry, and produce on demand, identification papers, the regulatory apparatus of an automobilized society achieves as much in a more innocuous manner. While formal rules of road conduct govern the driver through threat of punishment, non-formal codes of conduct operate on the level of civility. A good driver is a good citizen. A breach of the non-formal code occurs when a motorist behaves dangerously – say, by tailgating or speeding – provoking an emotional response (the honk of a horn, the shake of a fist, the verbal assault). The attitudes and emotional exchanges between motorists, and between drivers and non-drivers, often serve as actions of micro-discipline – drivers policing other drivers through shaming and reprimand. These practices maintain and reinforce formal codes.

In a similar way, social norms around automobility also govern action through the construction of the motor vehicle as a responsible transport mode and as a social marker of status and success. For example, in many suburban communities the practice of the "school-run" (moving children to and from school by car) is constructed as good parenting; the responsible parent is one who ensures the safety of their children through chauffeuring. Not only does this practice reinforce the notion that walking is too dangerous, but it also works as a social pressure coercing parents to adopt the practice, lest they be perceived as negligent (Bonham 2006; see also Buliung *et al.*, this volume).

As noted in Chapter 13 in this volume, cycling too is often constructed as a dangerous and irresponsible transport mode. In urban areas cyclists are perceived as reckless and inconsiderate – a stereotype epitomized in the figure of the bike messenger (Fincham 2006). Outside densely populated urban areas, cycling is often associated with lower status and income. A similar view is applied to those who take public transport in the auto-city. The social construction of non-automobilized transport modes as irresponsible, dangerous, or a symbol of personal failure operates to coerce subjects to adopt a lifestyle of automobility, and thus entangles them within the accompanying regimes of discipline and consumption. Discourses and social norms function both to reproduce automobilized landscapes as dangerous, while simultaneously encouraging the subject to adopt automobility as a risk-mitigation strategy.

Individualism or isolation?

As noted in the introductory chapter, the freedom and individual autonomy provided by the car also produces an isolation effect, not just because many car trips
are made alone, but also because inside the car each motorist is cut off from the
external environment and from all others. The isolation of individualized and
privatized transport can have a depoliticizing effect. The chance meetings and
happy accidents that Jane Jacobs (1961) regards as essential for a properly functioning, socially, and politically engaged community are virtually eliminated by
the practice of driving. This scenario mirrors the conditions created by neoliberalism and the competitive capitalist economy. Alone in the car, the motorist
is an atom, one among many, and each is looking out for themselves, battling all
others in the daily fight against traffic.

Similarly, urban environments designed primarily to accommodate the motorist remove the potential for spontaneous sidewalk conversation and neighbourly
interaction of more traditional streets and cities. Berman (1988) argues that the
modern technology of city planning has designed urban landscapes with the
explicit intention of removing the volatility and explosive potential of the street.
The nineteenth-century boulevard was a place where the "fundamental social
and psychic contradictions of modern life converged and perpetually threatened
to erupt" (1988: 167–168). Modern planning has attempted to "kill the street", to
remove and segregate previously mixed uses, to separate rich and poor, to desig-
nate, zone, and divide spaces for work, living, commerce, leisure, and transport.
In this way the city becomes a factory, with zones and regions given unique and
specialized functions. The rationalized spatial order of the modern planned city
functions to move commodities and people with all the efficiency of the Taylorist workplace. Roads and highways operate as grand conveyor belts delivering
and distributing parts and labour in a timely manner to each specialized zone.
This separation of people and activity is both economically efficient and counter-
revolutionary: the contradictions that might otherwise produce political interest
and collective mobilization disappear through the spatial separation of
antagonisms.

Taking the highway: protest on Toronto's Gardiner expressway

This chapter examines two extraordinary instances involving non-motorized
infringement on *auto-space* taking place outside the realm of institutionalized
politics. The events with which this chapter is concerned were illegal, temporary
occupations of that most sacred and emblematic auto-space: the freeway. Both
events took place in the span of a year, on the Gardiner Expressway – Toronto's
elevated, lakefront highway, and the primary route through and to the southern
end of the city core.

Critical Mass IS/ blocks traffic

Critical Mass is a monthly bicycle ride that takes place on the last Friday of every month. Started in San Francisco in 1992, the practice has since spread to hundreds of cities around the world (Furness 2007). The Toronto Critical Mass ride has been ongoing since 1997 (Weese 2008: 5), and in the summer months can attract over 500 cyclists.

Critical Mass is a leaderless collective of cycling advocates whose monthly group ride functions as a performative critique challenging the hegemony of the car in the modern city. As legitimate road users, cyclists are engaged in a reassertion of their rights to the road through the reappropriation of road space en masse. As a demonstration, Critical Mass is unique in that the message is embedded in the practice. The Critical Mass ride takes over all of the lanes sharing the same direction, and will often disregard traffic signals in order to keep the group safely together. While this practice likely aggravates motorists already frustrated by the Friday evening commute, mass participants justify this action with the oft-repeated slogan "We are not blocking traffic, we are traffic!"

As a leaderless ride, the route the mass takes through the city is spontaneously decided and never the same. On Friday, 30 May 2008, the group, approximately one hour into the ride, found itself at the base of the Lower Jarvis Street on-ramp leading to the westbound Gardiner Expressway, which it subsequently ascended. Martin Reis, a Critical Mass participant, describes the decision as sort of a logical next step; in his words the consensus was: "Here we are, let's take the Gardiner" (Topping 2008: para. 2).

The group ascended the ramp at approximately 7 p.m., merged with rush-hour traffic, and took over all three westbound lanes, which are the only expressway lanes funnelling traffic out of the downtown core to the western and north-western suburbs. They then cycled several kilometres along the elevated section of the expressway, passing over and through the heart of downtown Toronto. Participant Derrick Chadbourne described the event as "a bit scary at first, but after we took over all the lanes it turned out to be a really enjoyable ride" (Weese 2008: 5). Police officers, tipped off by an off-duty security guard, eventually caught up with the group and corralled them down the Dunn Street off-ramp to rejoin the city proper, roughly six kilometres and four exits from where they had entered the expressway. Traffic was halted behind the demonstration for the duration of the event, which lasted about 15 minutes. Three participants were arrested (Weese 2008: 5).

Protesting state refusal to act on war crimes

Not fully a year later, Torontonians would again witness a radical and unexpected taking of the Gardiner, this time on a far greater scale and within a context much more serious. The Tamils of Sri Lanka are a minority ethnic group that had been in negotiations with the majority Sinhalese for political representation since independence from the British in 1948. Facing persistent political marginalization and discrimination, the Tamil United Liberation Front (TULF), in 1977, declared its intentions for sovereignty of the Tamil homeland (a region in Sri Lanka's north) and the right to self-determination.

Ignoring this, the Sri Lankan government instead escalated its strategy from political oppression to military violence. In response, the Tamils adopted tactics of armed resistance led by the Liberation Tigers of Tamil Eelam (an organization the Conservative government of Canada designated a terrorist entity in 2006). The armed conflict, although punctuated by short periods of ceasefire, continued up through 2009 (see McConnell 2008).

Canada is home to the largest Sri Lankan diaspora in the world, and the population of Tamil Canadians in the Greater Toronto Area exceeds 200,000 (Reinhart 2009: A12). In early 2009, the conflict in Sri Lanka was reaching a climax as the Sri Lankan military continued to press north. In an effort to draw attention to the intensification of the conflict and growing humanitarian crisis, Toronto Tamils began staging a series of demonstrations throughout the city. Over a period of several months, Tamils demonstrated at Queen's Park (the provincial legislature), on University Avenue outside the American consulate, and even executed a well-organized human chain that encircled the downtown core (Mathieu and Yutangco 2009; Ferenc 2009).

These demonstrations continued, with little coverage in the mainstream press and with no official response from the Canadian government. As the conflict continued to escalate, demonstrators become more desperate. On Sunday, 10 May 2009, Toronto Tamils awoke to reports of spikes in civilian casualties as the Sri Lankan government intensified its assault on Tamil resistance fighters, allegedly through indiscriminate shelling of an agreed-upon civilian "safe-zone". Frustrated by this news, as well as the impotence of months of peaceful demonstration, the protesters, who were already positioned outside Queen's Park at the time, took their demonstration through the streets of downtown Toronto. Marching south along Spadina Avenue, the Tamil demonstrators found themselves at the base of the Spadina on-ramp, and subsequently moved up on to the Gardiner Expressway (Figure 14.1). Over 5,000 demonstrators remained on the expressway for most of that Sunday evening, completely shutting down the route in both directions until federal representatives agreed to raise Tamil concerns in Canada's House of Commons (Van Rijn 2009).

These two cases mark a dramatic clash between the politics of automobility – a presumed right to unfettered movement throughout the city – and the democratic right of free speech and representation through public demonstration. The out-of-the-ordinary nature of these two events make them analytically useful because they provide access to opinions that might otherwise be less available. Using a media and discourse analysis of the reporting and commentary surrounding the two cases, I examine what these events tell us about the contemporary political landscape within the automobilized city. I also explore what the response to these events might reveal about the political subjectivities produced through automobility.

Data was collected from each of Toronto's mainstream daily newspapers – both print and web editions – as well as a number of Toronto blogs that focus on

Figure 14.1 A Tamil protest spontaneously moves up the entrance ramp and onto the Gardiner Expressway in downtown Toronto, May 2009 (source: Alan Lai: alan@math.toronto.edu; http://flickr.com/axio).

urban issues. I examined news articles, blog posts, editorials, letters to the editor, and reader comments posted on news and blog websites. The initial print news reporting on the events were essentially descriptive, but included comments from local authorities. In the days following the events, and more particularly the Tamil protest, the print newspapers examined ran articles dealing variously with the illegality of the protest, coverage of ongoing Tamil demonstration, columnists describing Torontonians as frustrated and wearied by the disruption, and editorial and opinion pieces responding to the protest. All together, the reporting and commentary both online and in print was largely negative, critical, and unsympathetic (see Table 14.1).

After reviewing the available data, a number of major themes were selected as the most relevant to the discussion of automobility and neoliberal governmentality. These include: the subject of rights (to the road, to protest); the question of legality and the accompanying concern over public safety; individualism and anti-collective sentiment found in the reaction; and finally uses of public space in the automobilized city.

Right to assembly or right to mobility?

The protests took place on a space explicitly built for the movement of motor vehicles; it is clear who has the legal right in this situation. The protesters

Table 14.1 Local newsmedia slant towards the Highway Protests, Toronto 2008/2009

	Neutral	Negative	Sympathetic	Total
Newspaper articles				
Toronto Sun	5	10	3	18
Globe and Mail	5	2	0	7
Toronto Star	10	7	3	20
Total	20	19	6	45
User Comments				
Torontoist	14	44	42	100
Blog TO	13	38	22	73
The Toronto Star (comments)	–	70	30	100
CP24	–	75	25	100
Total	27 (7%)	227 (61%)	119 (32%)	373

Sources: *Toronto Sun* 31 May and 1 June 2008, 11–15 May 2009; *Toronto Star* 31 May and 1 June 2008, 10–12 May 2009; *Globe and Mail*, 10–12 May 2009.
User comments sources: Tim (2008), Topping (2008, 2009), Marlow *et al.* (2009), Gazze (2009).

demonstrating on the Gardiner were not using the highway for commuting. They didn't need to be there. They weren't using that space in the way that it was intended to be used. Their presence on the roadway, then, constitutes a direct attack on the right of movement, a right that motorists have come to view as absolute (Sheller and Urry 2000). As one commenter put it plainly, "why does their right to protest trump my right to move freely around the city?" (Smith 2009: para. 25). In an invocation of classic negative liberalism, a *Toronto Sun* letter-writer argues: "Tamil protesters seem to think they are exercising a democratic right. They need to realize their rights end where the rights of others begin" (Lenarcic 2009: 21). The implication here is that demonstrators don't have the right to assembly once that right interferes with the mobility rights of others. The popular feeling that the protesters had revoked right of movement can also be seen in the language of "hijacking" and "hostage-taking" used by Ontario Provincial Police (OPP) commissioner Julian Fantino (Clarkson 2009: 5), and reiterated elsewhere by many others.

It is interesting, however, to consider the extent of the public outcry in comparison to the number of those who were actually affected by the traffic disruption. While speculative, we can assume that most of those outraged by the event were likely not present on the Gardiner or greatly impacted by the demonstration. The bitter tone of the reaction to these events indicates the value placed on the *possibility* of free movement. The flexibility of the automobile provides the *infinite potential* to pick up and go at any time, with the important condition that roads are clear and traffic is moving. A blocked road obviously removes this potential and frustrates the guarantee of automobility.

More evidence for the presumed right of guaranteed mobility is found in "what if" arguments, which were common in the reaction to both events. For example, what if an emergency vehicle needed to get through? What if someone had to go to the hospital? In a letter to the *Globe and Mail*, an expecting mother (supportive of the Tamil cause up until the Gardiner protest) expresses deep concern over the possibility of a preterm labour while stuck in traffic (Singhal 2009). "What if" scenarios indicate the extent to which the possibility of free movement at all times is perceived as a guaranteed right, and argue that the removal of the right to absolute mobility is potentially deadly. Here, not only are demonstrators attacking the right to free movement, they are also attacking their right to quick access to vital resources and services.

Breach of codes

The protest events were serious breaches of both formal and non-formal codes of road use. Non-formal codes of courtesy and consideration for motorists were completely ignored. If getting cut-off by another driver evokes an emotional response, an intentional highway blockade must magnify this many times over. Participants of both demonstrations were referred to as "crazy", "irresponsible", and "uncivil". Indeed the many comments from online users along the lines of "we live in civilization" and "this is not the Canadian way" indicate the degree to which these events were regarded as outside accepted behavioural norms.

More seriously, the events were perpetrated with blatant disregard for formal laws codified in the Ontario Highway Traffic Act. The illegality of both actions was understandably a primary concern for many commentators. The question then becomes: are there any circumstances under which dramatic acts of public disruption and civil disobedience are justifiable or ethically defensible? For the majority of those concerned the answer is a resounding "no". Although many expressed sympathy for the demonstrators ("I am a cyclist myself"; "I support the right to protest"), the illegal nature of the actions took precedence. A commonly shared sentiment was that "no matter what the cause, no matter how just the cause, the protesters broke the law ... there is a consequence to breaking the law, for society to function it must be enforced" (Davide 2009: para. 33).

Adherence to the rules of the road is simple common sense. Good citizen-drivers can take comfort knowing that dangerous drivers will be disciplined. Many felt that regulations outlawing the presence of cyclists and pedestrians on expressways should have been more fully enforced, and that the police were too accommodating to the protesters. An editorial cartoon published in the *Toronto Sun* days after the Tamil protest illustrates this sentiment well. The cartoon prescribes a three-step protocol for dealing with traffic-disrupting demonstrations, beginning with the use of a megaphone, the next step suggests the use of tear gas and rubber bullets, and, failing this, billy clubs and handcuffs (Donato 2009). Picking up on this thread, *Sun* columnist Peter Worthington (2009: 22) asks why Toronto police did not adopt more aggressive tactics, and notes that Canadian politicians are "notoriously wimpy and cowardly" when dealing with domestic dissent.

With regards to cycling in particular, there appears to be frustration on the part of drivers that cyclists are not bound to the same rules of the road as those in their cars (although this accusation tends to ignore the fact that motorists, too, are regularly in breach of the law). Furness (2007) suggests that motorists may be frustrated that cyclists are enjoying the type of freedom that marketing professionals had promised the car driver. With the exception of long-distance travel, the bicycle offers more freedom and autonomy, and higher effective speeds, than the automobile (see Chapter 7 by Walks and Tranter, this volume). In addition, by avoiding licensing requirements the cyclist evades the formal governmental technology of the state. A persistent call from some members of the public to require the licensing and registration of cyclists (see Kirsic 2010) is an attempt to rein in and incorporate the cyclist into the governance regime that motorists are subjected to.

Calling on the police to adopt more repressive strategies when dealing with protesters, including suggestions of incarceration, signals the frustration and dissatisfaction felt when others do not exercise proper self-discipline. In a curious way we can find connections between neoliberal self-governance and popular support for active police repression. Neoliberal discourse creates the perception that the rights and freedoms promised by liberal democracy have been sufficiently achieved (with the freedom of mobility being one such manifestation). Any group unsatisfied with the status quo that express this dissatisfaction by taking action outside accepted norms and codes commits a breach in required conduct of self-governance. The self-governing majority (content with their own situation but angered by disruption) authorizes police violence against those who have not been properly civilized. Here we observe the division created between those who have adopted societal norms and those who have rejected them – "between the civilized members of society and those lacking the capacities to exercise their citizenship responsibly" (Rose 1996: 45), with the former members in support of the repression of the latter.

Public safety

Of course, this disciplinary intervention by the police might be necessary if public demonstrations endanger public safety, and both events were characterized as such by public officials and the media. Sergeant Redden of the Toronto police traffic services explained that the Critical Mass ride "was a very dangerous stunt" and that it "could have been worse in that there could have been a fatality" (Weese 2008: 5). Elsewhere he elaborates, "it goes beyond common sense in saying it's not a good idea" (Freeman 2008: para. 8). In response to the Tamil protest, Mayor David Miller stressed that, although "the Tamil people have a right to demonstrate, and there are some very serious things happening in Sri Lanka … but you can't endanger public safety" (Gazze 2009: para. 15). Later, reiterating Police Chief Blair, Miller remarked that "the protestors were endangering themselves, they're endangering the public, and they're endangering the Toronto Police" (Weese and Artuso 2009: 5). Provincial Premiere Dalton McGuinty brusquely announced: "you can't block highways, you endanger others and yourselves" (Bonogoure 2009: A10). Then-federal opposition leader

Michael Ignatieff had similar words, claiming that the demonstration "put the safety of the protesters and innocent by-standers – including women and small children – at risk" (Baute and Wallace 2009: A1).

It is difficult, however, to know for sure how endangering the actual events were, and accounts vary. A first-hand account of the Critical Mass ride claims that "the act was done in a very safe manner. We took the right lane and slowly merged left as the opportunity safely presented itself" (Johnson 2008: para. 32). Another said "the Gardiner was still the safest part of the whole ride – nobody was in danger and there wasn't any anger on the part of drivers or cyclists" (Angus 2008: para. 51). This contrasts starkly with the *Star*'s conjectural description of cyclists bringing "scores of cars and trucks behind them to a screeching halt" (Freeman 2008: para. 2).

Similarly, in consideration of the Tamil protest, where exactly can we locate the danger? A mental picture of foot traffic spilling onto a highway of speeding vehicles includes imagery of bodies bouncing off windshields, though this was not reported to be the case. A young Tamil protester explains in a *Toronto Star* interview (Theva 2009: GT1) that the highway was initially overtaken by a group of mostly young men – that at the point where it was most dangerous, it was young men that put themselves at risk. This contrasts with the mainstream characterization of elderly Tamils pushing baby strollers into the middle of speeding traffic.

Once the demonstration was established on the highway, the protest took the form of a peaceful sit-in – protesters simply refused to budge until their demands were heard. The only potential danger at this point would be a result of forcible removal by police action. Chief Blair confirms this observation in his comment:

> I was very concerned if we started to move that crowd down the ramp and if there was any kind of a stampede and people began to fall we could have had a tragedy there and even loss of life.
>
> (Marlow *et al.* 2009: para. 12)

One might ask whether there would be any reason to expect a stampede other than as a result of police use of force.

Although this discussion downplays the dangers that may or may not have existed during these demonstrations, it cannot be denied that automobility is dangerous: cars are lethal weapons, roads are dangerous places, and highways can be the deadliest. This is the reason we have voluntarily accepted such intensive systems of regulation, surveillance, and control. This is the price paid in exchange for the "freedom of movement". These cases, then, demonstrate how the dangerous landscapes of automobility make it easy to mobilize discourses of public safety in the service of repressive or disciplinary governmental technologies.

A final note on this point involves the presence of children at both the Critical Mass ride and the Tamil Gardiner event. That children were in attendance was emphasized by the media, condemned by city officials, and widely commented on by the general population as unacceptable. Then-City Councillor (and future

Mayor) Rob Ford even suggested that the Children's Aid Society should intervene in such a situation (Weese and Artuso 2009). We can easily connect these responses to the discussion of endangering safety above, but the comments also imply that the children were dragged along, that young people were not themselves concerned, and even that parents used children as "shields" from the riot police – an accusation echoing allegations against Tamil Tigers of taking refuge among civilian populations in Sri Lanka (Warmington 2009: 4). This reaction implies that young people have no business participating in political action, that they have no place in the public sphere, and reinforces media representations of public demonstrations as unruly events attended by irresponsible citizens. That child attendance outraged many and drew calls for disciplinary intervention indicates the failure of another site of governmentalization. Not only did protesters breach codes of self-governance, but also as parents they were neglecting their responsibility to properly transmit civilizing norms to their children. In this case, the family as an important site for the creation of responsible citizen-subjects (see Althusser 1971; Rose 1996) has broken down. As with other instances when governmental technologies have been evaded, there is a popular call and support for authoritative intervention.

Anti-collectivity/individualism

Neoliberal individualism as a dominant subjectivity has a number of political consequences. Two issues are relevant to this discussion. First, there is a general rejection of collective politics. As Brown notes, "the model neoliberal citizen is one who strategizes for her/himself among various social, political and economic options, not one who strives with others to alter or organize these options" (2005: 7). Second, neoliberalism sees the transfer of responsibility to each individual, who must be prepared to face the consequences of their decisions. As individuals express themselves through their autonomy in the marketplace, any failure to achieve success in life is constructed as the responsibility of the individual. The discourse of personal responsibility functions to ensure the rationalization of subjects, and also provides the opportunity for government downloading and state retrenchment (Lemke 2001).

Evidence of this anti-collective neoliberal subjectivity is seen in the language used in reaction to the two events. In Sergeant Redden's speculative account of the Critical Mass taking of the Gardiner, he notes that "they came, and then like cattle they all went up the ramp" (Freeman 2008: para. 4). At the Tamil protest, Police Chief Blair was concerned about the possibility of a "stampede". Other accounts talk of "swarms" and "swarming". This zoomorphic language attributes a herd-like mentality to the protesters, and operates as a subtle accusatory discursive device implying that protest participants are uncritical followers of the pack. In an ironic reversal, those engaged in challenging the status quo are painted as "sheep" by those content with business as usual.

Similarly we can locate anti-collective sentiment in remarks that ask "what am I supposed to do? It's not my problem", as well as in an overall consensus that it is not the business of the Canadian government to intervene in Sri Lankan affairs. *Globe and Mail* columnist Margaret Wente asks: "What are we supposed to do about it? Ban tea imports?" (2009: A14). In a combination of individualization discourse and the anti-immigrant sentiment that still pervades many parts of Canadian society, those commenting on news and blog sites advise Tamil protesters to "go back home" to fight – the proposed solution is for protesters to take individual responsibility for political outcomes through the personal decision of "returning" to the conflict zone. It is difficult to know for certain how much of the backlash against the Tamil protest was motivated by racism, but even an editorial published in the right-leaning *Toronto Sun* tabloid acknowledged the blatant anti-immigrant overtones and racial intolerance present in much of the public outcry that followed (Goldstein 2009).

Public space in the automobilized city

The actual space of the highway has particular relevance to how physical landscapes of automobility have impacted conceptions of the urban public realm. It is clear that both protest events were significant "spatial transgressions" (Cresswell 1996); people on foot or on bikes appear bizarrely "out of place" on a fast-moving elevated freeway. City Councillor and future Mayor Rob Ford had harsh words for the Tamils: "If you want to protest, fine. Get a permit like everyone else does, go to Queen's park, go to Downsview park, go to Nathan Philip's square, we are not going to tolerate any more of this hoodlumism" (Weese and Artuso 2009: 5). Likewise, then-Premiere McGuinty reportedly invited protesters to use the lawn in front of Queen's Park – exactly where Tamil protests had been occurring unnoticed for months previously. Letter writer Peter Rozzenac sums up the sentiment: "Organizers of recent Tamil protest in the GTA must understand that the QEW, Don Valley or University Ave., are not the federal seats of power" (2009: A18). Despite this popular sentiment, the visceral reaction garnered by the highway protests suggests that these actions were successful in interrupting established power channels, and thus have the potential to affect power relations.

The purpose of public demonstration is to gain visibility and representation for a cause that otherwise goes unnoticed, to attract public attention to an underrepresented issue. Traditionally this occurred in the town square – the agora – where a large group could effectively gain an audience. Tamil demonstrations at Queen's park went unnoticed for months because there was virtually no one there to notice them, and they were easy to ignore. Government and institutional buildings surround the park, a grassy area in front of the Ontario provincial legislature, which experiences little foot traffic and is circled by a ring road – a favoured route for speeding in and out of Toronto's city core. As a place of public demonstration, Queen's Park has a very minimal immediate audience. The media has a role to play in the spatial dissemination of locally contained information, but in the case of the Tamil protest, the media clearly failed in terms of airing the complaints. However, it could be argued that the shift onto the highway accomplished the protesters' goal. As one Tamil demonstrator put it: "Until the Gardiner, nobody was listening, even the media weren't listening to us. After that, everybody is trying to understand us and get the message we've been trying to get across for six months" (Mandel 2009: 4).

Public space literature calls our attention to the decline of the traditional agora. As the outdoor spaces of our city become increasingly specialized and commodified, legitimate use of public space for political purposes decreases (Mitchell 2003). Public areas are privatized, city centres are geared towards consumption, leisure use of public space is criminalized, and it becomes increasingly difficult to leave home without a wallet – that is, to use public space for non-commercial purposes. While the commodification of the urban outdoors is seen as a primary mechanism responsible for the decline of political or other alternative uses of public space, the motorization of the outdoors may be equally implicated in this decline.

If demonstrations are only permitted to occur in non-automobilized spaces, then there is an important consequence of limitation. In the absence of sufficient media coverage (whether due to editorial bias or in the interests of ad revenue), there is a reduction in the communicative potential of public demonstrations. The purpose of gaining visibility is defeated when protests are restricted to designated areas. In the modern specialized factory that is the auto-city, segregated and compartmentalized zoning performs particular functions of the productive economy. But the auto-city is poorly designed for a socially cohesive community, an enjoyable and human street life, or a space for visible representation and public discussion.

If automobility has lacerated the traditional space of the agora and rendered it defunct, the marginalized and under-represented have but one option remaining in order to communicate their message: to make themselves publicly visible in whatever outdoor spaces are available. In the decentralized auto-city the agora is everywhere – it is any space outside our doors that can be temporarily appropriated and effectively hold an audience.

The Occupy movement of 2011 adopted the strategy of demonstrating in parks and plazas, and in the streets with moderate success. The semi-permanent encampments in Zucotti Park in New York, St James Park in Toronto, and many other parks and plazas across North America and worldwide, represented the reimagining of the traditional agora. The occupation of these public spaces effectively allowed protesters to establish a visible representation for their cause, and created a space for political dialogue.

The encampments augmented their message with roaming street protests. Where the semi-permanent encampment amplifies its message with duration (occupiers in Toronto spent 40 days and nights in the park), the nomadic street demonstration spatially extends the message to a wider audience as they weave through the city.

Interestingly, much of the legal reasoning justifying the eventual eviction of occupiers hinged on the argument that protesters were privatizing public space – that they were not using the space the way it was intended to be used (see Khon

2013). One could make a similar case against the car – that over the last century the gradual encroachment and eventual dominance of the personal automobile on our city streets represents the privatization of a space that previously had many other varied uses. In this light, the act of taking to the streets could be considered an entirely legitimate act of reclamation. And the most powerful example of this will be the taking of the most sacred auto-space, *the highway* – the high citadel of the automobile.

The challenge of denaturalizing a dominant discourse is already a difficult task. In the battlefield of competing discourses, counter-narratives take root slowly and with great effort. But the task of challenging the dominance of auto-mobility is far greater. Not only does automobility dominate in language and ideas, but also the discourse of automobility is literally inscribed across the landscape. The roadways are a text we read on a daily basis. The roaring drone of automobile traffic is the incessant soundtrack of the city. The physical infrastructure of automobility serves as a massive, concrete reinforcement for the discursive structure of automobility, which in turn lends support to neoliberal auto-mentalities and political subjectivities. The important task then, for those contesting automobility and neoliberalism is not only to challenge ideology and hegemonic discourse, but to re-vision the uses of the outdoor spaces of our city through the everyday practice of reclamation – to de-motorize *roadspace*. The human occupation of sacred *autospaces* is a useful tactic in the relegitimation of alternative uses of public space, and a key site in the contestation of neoliberalism.

Conclusion

15 Post-automobility?

Dealing with the auto-city

Alan Walks

While automobility may be entrenched, the form it has taken has been neither inevitable nor pre-ordained, while the solutions to its contradictions are neither impermeable nor simple. The contemporary auto-city is not only a product of automobility, but also the field upon which the contradictions and struggles originating under automobility must be resolved.

This last chapter is concerned with how to deal with the auto-city as a social, political, ecological, and physical construct, as a marker and producer of contemporary urban inequalities, and as an object of policy reform from which a more socially just city might be built. The chapter begins by discussing the contemporary city as a key articulation of the complex non-linear system that is automobility, and possible futures of automobility under various scenarios. The literature dealing with post-car mobilities is discussed, and the implications of this for an understanding of the future of the auto-city are explored. The chapter explores how the auto-city might be reformed and transformed, and highlights the philosophical importance of the right to the city in any just and sustainable urban transformation.

The auto-city as a path-dependent non-linear complex system

As noted in the introductory chapter, scholars of automobility have characterized it as a self-organizing, non-linear, and complex socio-machinic system, composed of a hybrid assemblage of cars and other vehicles, machines, drivers and non-drivers, buildings, legal apparatuses, roads, petroleum production and distribution networks and their produces, legal rights, signs, policies, social practices, and, of course, particular cultural desires for mobility, containing and generating the economic, social, cultural, and political preconditions for its own self-expansion (Beckmann 2001, 2004; Sheller and Urry 2000; Urry 2007; Dennis and Urry 2009a). Automobility is also a path-dependent system with roots in specific places and times, stemming from the largely improvised and contingent history of automobile and petroleum development from the late 1800s through the 1910s in Western Europe and the United States (Dennis and Urry 2009a,b; Sheller and Urry 2000; Seiler 2008; Volti 1996). Under path dependence, minor events and chance occurrences become augmented through positive feedback mechanisms that produce increasing returns from certain actions, while increasingly limiting the immediate benefits of changing course (see Arthur 1994). The temporal order of processes, events, and chance occurrences matters in producing the outcomes of path dependence, and no result is inevitable. Yet, once "locked in" through such positive feedback processes, historical trajectories become somewhat irreversible. Such is the system of automobility, whose form has developed around the "steel and petroleum" car and the vast system of supports, production complexes, and social patterns of consumption that have since become naturalized and internalized in modern Western (auto-mobile) culture (Sheller 2004; Dennis and Urry 2009a).

However, as Urry (2004) notes (referencing Slater 2001), it is not the car itself that is key to understanding the dominance of automobility and its path-dependent development, but instead the way that systems of provision, social categories, political agendas, and cultural discourses come together to fix the material articulation of automobility in a stable form. Left unsaid by Urry is the fact that it is *the city* that largely articulates this complex system, and provides the material basis for automobility's self-generation. The stable form of the system rests not only, or even primarily, with the significant power given to automobile manufacturers and others with vested interests in its maintenance, nor with the common desire among modern individuals for autonomous mobility, but with the fact that automobility is concretely and spatially fixed within metropolitan geographies. While it is certainly true that transportation and mobility are two of the "breakers and makers" of cities (Banister 2005), so it is also true that cities are the makers of mobility.

It is primarily the materiality of the city that compels and directs the behaviour of urban residents, firms, and politicians, and "locks in" the predominant form of contemporary automobility. From the locations of transportation routes and corridors, building lot lines, land-use zoning, street sizes and lengths, energy distribution networks and utility corridors, to the design of public parks, the complex and interrelated patterns of urban development are difficult to alter once constructed. Firms often invest in physical plants for profitable lifespans of 50 years, while the significant cost of building roads, offices, schools, and hospitals means they will rarely if ever be moved. Social geographies of residential segregation (by class, race, immigration status, etc.) become embedded in not only socio-cultural, but economic and political geographies of the city, and in reflexive fashion come also to feed back on urban development decisions regarding transportation systems and routes, residential forms and planning, and redevelopment alternatives. Thus, not only does urban politics influence the shape of the city, but the form of the city reflexively shapes the trajectories of urban politics – they co-produce each dialectically.

It is the geography of the auto-city, far more than the car itself, that is responsible for the "Janus-faced" character of automobility – both expanding the realms of freedom and flexibility among auto-mobile publics while simultaneously compelling individuals and households to live in time-constricted and spatially distended ways (Sheller and Urry 2000). Indeed, as Mumford noted

(1953, cited in Dennis and Urry 2009b: 240), the implementation of linear transportation systems in the modern city actually "crippled the motor car, by placing on this single means of transportation the burden for every kind of travel". It is the uneven overlay between the fixed built environment, local transportation networks, urban geographies of economic production, and social resources and social interaction rooted in place that determines the distribution of mobility constraints and choices among households. This would be true with any mix of transport modes (bicycle, public transit, etc.), but the distances involved – as well as levels of flexibility and freedom – have been significantly augmented by the need to make the city function in the context of widespread automobile use, and in turn have widened the contradictions and inequalities they have begat.

The city itself is an example of path-dependent non-liner complex system. The idea of the city as a system of organized complexity was first articulated in the work of Jacobs (1961), who also noted the unintended and system-changing effects of reproducing the city for automobiles. As a particular form of city, the auto-city has advanced under its own distinct logic, and is the source of a number of the contradictions and irrationalities of automobility. This includes significant environmental contradictions related to resource depletion, and the economic limits to the expansion of automobility that this implies, not only at the local scale (sprawl) but at the global scale as well (as discussed by Walks in Chapter 4). However, as the chapters in this volume attest, the social, cultural, and political contradictions loom equally large. The auto-city is, in combination with other shifts occurring in society, responsible for producing new forms of social inequality and redistributions of wealth, due to differential access to various forms of mobility, unequal rates of indebtedness, and the rise of auto-mobile forms of citizenship that redefine and re-ascribe the rights and responsibilities of migrants and immigrants.

The auto-city produces new political incentives for maintaining and expanding the system, simultaneously promoting a neoliberal policy agenda that heightens such inequalities while impinging on the ability of urban residents to democratically deal with their contradictions. Yet, there are limits to the expansion of automobility that work to exacerbate its internal contradictions and vulnerabilities at the same time that "peak car use" leads to the emergence of new urban logics and political challenges (Cohen 2012). For instance, since the early 1990s, there has been a clear and, in some countries, quite marked decline in automobile licensing rates across the developed world (Delbosc and Currie 2013). As Beckmann (2001: 605) argues, "automobilisation has become reflexive: it has produced dangers and risks that threaten its own foundation". Because the system is non-linear and complex, so are the potential affects of any breakdown in its core logics. Due to the rising contradictions and limits of the auto-city, there are a number of potential openings that may bring with them significant reforms and modifications to the system.

The future under post-car mobilities?

Complex, path-dependent systems are subject to turning points, tipping points, and paradigm switches. These originate in non-linear fashion through the cumulative interaction of small chance occurrences, such that even seemingly minor or incremental events can have significant repercussions (such as when slight additions of heat melt a solid into liquid). Dennis and Urry (2009b) suggest certain technological, policy, and socio-cultural changes that could be producing the conditions for reaching new tipping points and affecting a switch in the current system. If these were to occur "in the 'right order'", they argue, the system could be tipped "towards forming a new mobility, the post-car system" (ibid.: 241). At present, our image of such a potential post-car world is wide open. Some have suggested that in the future high-speed underground magnetically levitated trains running in evacuated tunnels will characterize the future (Ausubel and Marchetti 2001). Others see the decline of high-speed and fossil-fuelled forms of transport, ushering in a shift towards a "low-mobility" future (Moriarty and Honnery 2008), a perspective bolstered by the identification of "peak car use" in a number of countries (Newman and Kenworthy 2011b; Metz 2010, 2013; Goodwin and Van Dender 2013; see Chapter 3 in this volume). It is not yet clear that the globe will ever reach two billion cars (Sperling and Gordon 2009), even if it could sustainably accommodate them.

Dennis and Urry (2009a,b) list five specific "small changes" that they argue have the potential to, in assemblage, trigger tipping points and system change. The first involves new fuel systems, including potentially hybrid fuels or electric vehicles, that could reduce dependence on oil and lead to new refining technologies and energy distribution systems, new industrial processes and forms of agricultural organization. The second involves new materials, possibly relying on nano-fibres and other synthetics, which could reduce the primacy of steel and reshape the functions of the car and its affects on safety and mobility. The third change involves the rise of "smart car" technologies and communications systems, including telematics and intelligent transportation systems (ITC) that turn individualized automobiles into units collectively coordinated within a networked "organic" flow of vehicles, and which manage crime through biometric user recognition and digital monitoring. This has the potential to transform not only the issues of safety and congestion, but also the individual experience and meaning of the automobile, incorporating it into a more collectivized and state- (and software-) controlled form of transportation. This fourth shift may then be linked to a de-privatization of mobility, through car-sharing, smart car-rental schemes, and cooperative automobile clubs, with access to automobiles managed online, perhaps via special quota systems to ensure equity. Finally, new "realist" policy reforms away from "predict and provide" models that privilege the automobile and towards an emphasis on energy-efficient forms of transportation such as electric buses and bicycles can be expected to "address the social implications of the car system upon communities, land use and urban architecture" (Dennis and Urry 2009b: 249). One aspect of this, they suggest, will actively involve remote control of individual vehicles (on behalf of the state) and immobilization of stationary vehicles so as to limit both criminal activities and deviant mobilities.

They suggest these five changes could produce the potential turning points for eventual refashioning of the current system, such that in the future "it will seem inconceivable that individualized mobility will be based upon the nineteenth century technologies of steel bodied cars and petroleum engines" (Dennis and Urry 2009b: 249). However, and despite their criticism of any "fix" concerned with "building a better car" (ibid.: 248), it might be noted that their list of five key changes is mostly concerned with exactly this. Only the last of the five, dealing with a new "realist" transport policy, necessarily implies changes in the way cities are built, planned, managed, or function. Even here the discussion in Dennis and Urry (2009a,b) is remarkably tame in proposing alternatives. For instance, they advocate for exclusive and separated bus lanes, running "bio-articulated buses". The main models for bus-rapid-transit (BRT) are found in Ottawa (Canada), Brisbane (Australia), Curitiba (Brazil), and Bogota (Columbia). BRT systems modelled on these examples would presumably operate alongside the state-coordinated ITC smart cars.

Yet, a future characterized by collectively rented hybrid biofuel-electric carbon-fibre smart cars and bio-articulated buses is not likely to represent any radical break with the current system and culture of automobility. Indeed, such a future more resembles a hyper-modernist sci-fi dystopian extension of the current system. Their scenarios would not significantly alter individuals' and households' relationships to the city characterized by time-compression and spatial distension, and there is not even any suggestion that they would lead to a decline in auto shares or modal splits. The BRT systems in Curitiba, Brisbane, and Ottawa largely rely on existing auto-mobilities to make urban circulation more efficient; they do not significantly alter the logic of the auto-mobile city, and the degree to which they would enhance sustainability or resilience in the context of the overall trajectory of continued urbanization and resource depletion remains unclear. Mees (2009) has shown that in the absence of transfer-free networks, BRT systems do not significantly increase connectivity or affect overall automobile use. These critiques also apply to the promotion of high-speed underground rail systems (Ausubel and Marchetti 2001), which in addition would be highly resource intensive and, in many cities, unaffordable. More likely, only a few such lines could be built in any given place, allowing households with means to appropriate the most accessible locations near the lines, and compelling those with less means to remain wed to auto-mobile patterns and forms of transport, and the auto-mental politics that comes with it. If such high-speed "public" transit lines were privately owned and operated, they would be even more likely to produce unequal affects, as private companies are typically loath to allow for free transfers with competing systems or operators: accessible cross-connecting public transit networks require public ownership and operation if they are to be equitable and successful (Mees 2009).

Nor would the scenarios put forward by Dennis and Urry deal with the unavoidable social, economic, and political contradictions, inequalities, and irrationalities of automobility and its counterpart in the auto-industrial complex, as discussed in this volume. These can only be addressed through reform of the physical, social, and political form, structure, and function of the city itself, including that regarding the ownership of urban private property and the productive process. Without fundamental reforms to the system of private property, the de-privatization of automobile consumption while potentially more equitable at one scale – the level of individual households – would not necessarily alter the logic of urban class segregation by which the wealthy appropriate the most accessible locations, nor the dominance of the firms producing the automobiles. If anything, collective auto-share programmes could easily be used by city administrations or upper levels of the state as justification for the elimination of affordable housing in the inner city and the rapid gentrification of previously working-class neighbourhoods, shunting the poor to the edges and compelling them to drive. In combination, such a scenario would not only eliminate any remaining freedom of choice of mode those with fewer means at the fringe might have, but also eliminate their potential for vehicle ownership and any residual control they might have over their own mobility within the system. It is primarily the changes that must be made to the contemporary auto-city and the socio-political economy, rather than the largely technological modernization of the automobile mentioned by Dennis and Urry, that will be the ultimate basis of any lasting sustainable transformation of the system of auto-mobile capitalism, and the erecting of a socially just post-car society.

Refashioning the auto-city

Just as there are a number of potential transformations that can be expected to influence the trajectory of future automobile-based mobilities, cities themselves will have to undergo transformations in their forms, functions, transportation networks, and mobility cultures. The Business Council for Sustainable Development (1993) suggests that developed nations need to reduce energy use and consumption by over 90 per cent by 2040 in order to live equitably within the planet's carrying capacity. Cities are key to being able to make this shift, and Dennis and Urry's "small changes" will be insufficient to meet this challenge. As noted by Low (2003), nothing short of a "paradigm shift", along the lines advocated under what Low and Gleeson (2001) call "ecosocialization" – the melding of red and green politics – will be necessary to bring about true environmental and social sustainability and resilience. Yet, given the path-dependence of the existing system and regime of automobility, and the lock-in towards the auto-city that has accompanied it, it is not difficult to imagine a future of continued high-tech hyper-automobility in which urban residents are increasingly robbed of their motility, rather than a post-car world in which urban residents are provided with sufficient alternative mobilities (including perhaps, but not dominated by, the car). Most scholarly viewpoints regarding the modelling of future transportation policy argue against traditional predict-and-provide approaches and in favour of "back-casting": starting with a vision of where one would like to end up, and then working out the details of what needs to be put in place to get there (Banister 2005). Instead of speculating on the kinds of potential incremental technological developments that might bring about tipping points and usher in unknown system changes, this section outlines three related transport objectives for building a sustainable and resilient post-car urban world – active connectedness, re-(g)localization, and biophilic urbanism – and examines their benefits and what might need to be put in place to bring them about.

Active connectedness

As numerous scholars have argued now for a number of years, urban transportation networks will need to be revisited and transformed to allow for access to a full choice of modes, seamless connectivity between routes and modes, and the promotion of active forms of travel from all sections of the city. In order for a post-automobility world to be possible, it will need to be *possible* to drive less (Handy 2006), simultaneously undercutting the primary basis through which the automobile has become a "maximum commodity" (Dawson 2011). This will necessarily involve the spatial extension of public transit systems and their modification into fully integrated transfer-free networks (Mees 2009). While high-speed rail systems running on fixed schedules may provide the backbone for such a system, those residing in areas of lower density with non-linear street forms characteristic of fringe suburban development since the 1970s will need to be served by cross-connecting networked bus lines (Mees 2009). Yet, while public transit will be required for long-distance travel, only through the extension of emission-free active travel will the city be made significantly more sustainable and resilient (Aftabuzzaman and Mazloumi 2011). An extensive system of biking and walking trails, including separated lanes on busy roadways, as well as off-road cross-connecting pathways, will need to be implemented across metropolitan regions with sub-systems of feeder pathways to each of the major transit stations and modal intersection points. Such systems must be fully continuous and networked in order to allow for the linking of all destinations and origins.

The future is likely to be one in which an increasing proportion of workers will not be able to afford to own, and in some cases even to rent or lease, an automobile, while firms will need alternatives to truck-based haulage. Not only will cities in the developed North have to learn from lower-mobility cities in the developing South (Banister 2005), but new systems that challenge the whole notion of an urban modernity built on the maximizing of trip speed and time–space compression will have to be developed. As noted in Chapter 7 in this volume, a lower-mobility future does not imply one with lower *effective speeds*, as the effective speeds of automobility are actually not significantly different from those of velo-mobility. As bicycle trailer technologies evolve to carry more weight, and hand-pulled carts re-emerge as viable ways of carting large loads, cross-town public transit options will have to be modified to allow for the moving of heavy goods, possibly with the use of flat open boxcars or specialized sections of buses and trains allowing bike trailers. Some expressways built only for automobiles will need to be re-purposed and turned into multi-modal thoroughfares, with separated lanes for cycling, transit, and automobiles and trucks. To ensure that those relying on the least powerful forms of mobility do not bear the burden of such changes, city governance will need to enshrine mobility rights for individuals walking and cycling within the overall transport system. The kinds of state "remote" controls on mobility, including those directing automobile use and behaviour, but which can also be expected to be imposed on those using other modes of transportation (including punitive taxation, tolls, fares, and licensing, as well as software-enabled monitoring, digital surveillance, and ICT, as discussed by Dennis and Urry 2009a,b), need to be *actively fought against*. Congestion charging such as that pursued in London or Singapore is not the answer – in a number of places such a policy is and will be inequitable (Banister 2005; Mees 2001, 2009; Black 2010), particularly because the contemporary logic of the auto-city as outlined in this volume is now one that disperses lower-income households to the most automobile-dependent locations, often forcing them into debt (see Chapter 4 in this volume). The basic principle is that active connectedness should expand, not restrict, the range of available choices and rights in the field of mobility and enhance the abilities of all urban residents to sustainably connect with and in the city.

Re-(g)localization

The future will require that the disorganized dispersal of places of employment, retail amenities, services, and various institutions be reversed, with these functions reconcentrated in place. This is necessary in order to deal with the social inequities involved in long-distance commuting, discussed above, as well as inevitable rising energy costs over the long term, which will make personal automobile travel more costly for the average household and drive up costs among many firms shipping over long distances (North 2010). As the fossil-fuelled auto-mobile system winds down, there will be a need to manage future time–space decompression at the global and national scales, and increasing spatial intensification of activity at the urban scale. As per Newman and Kenworthy's chapter in this volume, the latter will require new nodes to be activated for concentrating previously dispersed functions. In order to deal with the socio-spatial inequities produced by the auto-city, such nodes must be made accessible via active modes of transport, and seamlessly connected to the larger metropolitan transport network, so that accessibility is not hampered by one's choice of mode or level of income. Such patterns of reconcentration need to be made plentiful and localized, allowing for multiple centres, nodes, and cores to be accessible within a short walk, bike, or, when that is unreasonable, a short transit trip. Such a pattern would be more economically efficient, even for private vehicle travel (Zolnik 2011), as well as more equitable, more socially satisfying, and better for overall health (Freund and Martin 2007; Delmelle *et al.* 2013).

While the traditional CBD is well placed to act as one of the main nodes in such a re-localized system, reconcentration should not only, or even primarily, occur at the CBD. In many large cities of a million or more built for the automobile, it would be difficult and inefficient to reorient multiple transport systems towards a single core, or to fully recentralize activities that long ago dispersed. The long and expensive commutes such a pattern would require would also be inequitable and unaffordable, and recentralization at this scale would mean highly privileging those who could afford to live near the core while further disadvantaging many of the working class who are compelled to live near the fringe – a process already under way in many global cities. Such a monocentric scenario would also likely require imposition by an autocratic elitist state: a new Radiant Ville Contemporaine à la Le Corbusier, with its attendant lack of democracy and high levels of class segregation. The replacement for the monocentric city, largely dissolved and flattened by automobility, will need to be polymorphous, polycentric, and polynucleated, with most employment, shopping, schools, and other common amenities and institutions accessible within a much smaller average radius of daily mobility.

Localization here, it needs stressing, does not mean the retreat into disconnected urban villages, or the production of locally self-sufficient communities cut off from global trade and communication. Localization is not a communitarian panacea, and it is a fantasy to expect most communities could become economically self-sufficient or withdraw from larger (national or global) processes of production, distribution, exchange, and consumption. There remains a need to enhance communication across scales, particularly that of a political nature among urban labour forces and communities located in different parts of the world. G/localization here instead refers to an active policy preference towards the organizing of activities at the local scale for those processes for which this would improve efficiency, accessibility, and stability (Hines 2004), while maintaining open and active lines of communication and exchange. Localization will mean different things, and will in turn be best pursued at different scales, in relation to different processes (North 2010). Political connections will need to be maintained with organizations and communities at various other scales, leading to a multi-scalar glocalization of urban lifestyles, democratic controls, and the intra-urban organization of employment, residences, and mobilities.

Biophilic urbanism

Biophilic urbanism refers to a movement to "green" the city through design and planning reforms, including green roofs on schools and private buildings, "edible landscaping" of parks and forgotten open spaces, green urban courtyards, and the re-establishment and naturalization of wetlands, riparian areas, and rivers, as well as the extension of urban forests, tree canopies, and the greening of brownfield sites and utility corridors (Beatley 2010). Biophilic cities would enhance human–nature interactions, and make cities both a reflection of, and a mechanism for, producing global environmental sustainability, biodiversity, and biological resilience (Beatley 2010; Beatley and Newman 2013; see Chapter 3, this volume). Some argue that the biophilic city will enhance place attachment among residents and encourage local residents to work together for the maintenance of the ecological integrity of the system (ibid.), in contrast with the fossil-fuelled auto-mobile metropolis – often a form that is tolerated but which produces abject spaces most people wish to avoid. Through various practices, such as the restoration of wetlands that help prevent flooding and protect drinking-water sources, or the planting of locally appropriate bio-diverse forms of vegetation that provide protection against insects, disease, fire, wind storms, and/or drought, biophilic urbanism can make the city, and those living within it, more resilient to changing environmental conditions.

The biophilic city should not merely be one that includes more natural habitats and green landscapes, but is one that addresses and transforms the problematic industrial metabolism of the city. In the standard linear urban metabolism, energy and other commodities consumed in and by the city typically leave a legacy of (often toxic) wastes that then have to be carefully disposed of, enlarging the ecological footprint and making the auto-city an enemy of sustainability (Wackernagel and Rees 1996). Biophilic urbanism needs to transform more of the outputs of urban processes into useable inputs that enhance ecological integrity. Such a vision requires, among other things, low-impact development (LID), the utilization of urban land for food and crop production, and a shift from dependence on fossil-fuels to more sustainable energy forms that derive from the sun or wind. This would require the greening of transportation corridors, a shift towards hydro-electric and solar-electric public transit systems, and the promotion of active transportation. Indeed, given that a significant proportion of all airborne particulates and greenhouse gas emissions derive from oil-based transportation (Gorham 2002; DOT 2010; IEA 2012a), one of the best ways of making the city more sustainable and resilient is to promote walking and bicycling, particularly if bicycles are made from recycled materials in local low-impact production facilities. Higher levels of exercise also improve the resilience of residents both to physical health impairments and disease, as well as to stress, which then can be expected to decline with shorter commutes, more time for family, and more green surroundings (Beatley and Newman 2013).

While the combination of the above three changes to the city may involve a reduction in total levels of mobility, this need not have a negative impact on quality of life if less time spent commuting results in more time for other activities, and an increased capacity for managing both the urban environment and participating in urban politics. Of course, there is quite a difference between contemplating the changes that will need to occur, and the approaches required for actually achieving such a transformation. The next section thus considers the political task of refashioning the auto-city.

Struggling with the auto-city

Chapters 1 and 11 in this volume have demonstrated the political attributes of what might be termed the governmentality of automobility, or auto-mentality. The auto-mental position is rooted in mainstream constructions of modernity, and posits automobility as the primary solution to the problems it itself produces. Yet, after a considerable period in which many of the assumptions promoting a modernist vision of the automobile city have been taken for granted, transportation policy and automobility itself are now being increasingly politicized. How might political movements make progress in reforming and transforming the system of automobility and the form of the auto-city upon which its path-dependent trajectory has been secured?

Henri Lefebvre (1996/1967) advocated for the extension of a particular social and political right, which he called the right to the city, and argued that the future would be, among other things, characterized by struggles over urban planning, ownership, the urban environment, and ways of life. By the right to the city, Lefebvre did not mean that the political demands of current local residents should take precedence over those of future residents or residents living outside the city, but instead that the full materiality and sociality of the city – of urban everyday life – should not be commodified or impinged upon, but should be decided upon democratically, openly, and according to everyday needs and desires. Lefebvre referred to the right to the city as both a cry – against the alienation being produced by privatization, demolition, militarization, and subversion of control over the city by both private capital and the state – and a demand – for the people to reappropriate the city, and with it the promise of being able to construct a meaningful, engaged, democratic, and spontaneous future life full of possibility.

Lefebvre, it might be added, was particularly critical of the effects of the automobile and its system on the city, which he saw as imposing an authoritarian logic and rhythm, segregating the city and destroying both its diversity and its spontaneity (Lefebvre 2003/1970; Lefebvre 2004/1992; see also Walks 2013a). For Lefebvre, establishing a right to the city, and dealing with the insidious and pernicious effects of automobility, are all wrapped up together in the larger urban problematic. The kinds of transformations that are particularly important for laying the foundation of the post-auto-city are similarly part and parcel of the democratic struggle for the city, including collective local interaction and governance, greater inter-connectivity and choice of mobility, and life-supporting systems and functions. The extension of democratic rights to the city does not involve the restriction or removal of any existing rights, including those related to the automobile. One benefit of an approach based on reforms to urban socio-spatial structure is that this encourages different behaviours, mobilities, and urban ways of life without requiring the removal of any formal legal rights. The post-automobility city (biophilic, dense, clustered, and inter-connected) encourages new behaviours organically through the spatial logics of distance, time, and concentration, rather than relying on coercion, policing, or the heavy hand of the state.

Yet the question remains of how urban populations might be politically organized to demand and promote the expansion of their rights, and what political strategies might be successful in producing a city less alienating and more life-fulfilling. As Harvey (2012) notes, while the larger aims of such a political struggle will always revolve around relations of production, realization, and consumption, the smaller struggles related to the seemingly mundane issues of everyday urban life are just as important, and often more important for establishing the kinds of connections and solidarity needed for wider and more complex urban politics built around the right to the city. While sometimes issues are of such dire importance that they call for radical responses, at other times there is merit in pursuing a reformist instrumental approach that achieves improvements in incremental fashion. As in the logic of compound interest, significant achievements can be built from waves of incremental additions and modifications.

Falling into the latter category are the kinds of incremental reforms to the city that improve and reinforce overall connectivity, de-segregate land uses, and promote engagement in local governance. Projects such as the Burrard Street Bridge cycle lane in Vancouver (discussed in Chapter 13 in this volume), and bicycle path networks like the one being extended by Conservative Mayor Boris Johnson in London (UK, discussed in Chapter 11) or the one in Adelaide in South Australia, are the kinds of incremental reforms that are essential to the future promotion of urban quality of life under any post-car scenario. As in the Burrard Street Bridge process, at times this will mean moving cautiously and strategically to put in place programmes with minimal impact that can be expanded as the demand requires. Even these seemingly small reforms often entail long political battles, and in many cases they acquire important political significance that can then be rolled into more generalized support.

Changes to transportation infrastructure and urban activity patterns, however, will need to be combined with changes to urban fabrics, transportation networks, and property structures in ways that promote mixed use and multiple and active forms of mobility. Local residents and planners must insist that active transport networks and public transit accessibility be built into any and all future redevelopment plans, and in turn that such redevelopment include mixed-use tight-knit street and building patterns. The best way to do this is to code these principles into local planning ordinances and subdivision/site planning approval criteria, while allowing for flexibility in how they might be carried out in practice based on local circumstances. Similar requirements will be necessary in the field of housing through the requirement of inclusionary zoning, to ensure mixed-use and multi-nodal systems are accompanied by mixed-tenure and mixed-income communities in which those with wealth cannot displace those with fewer means through regular private property mechanisms (Schuetz *et al.* 2009; Mukhija *et al.* 2010; Mah and Hackworth 2011). The story of the removal of the Central Freeway extension in San Francisco (Chapter 12 in this volume) provides important lessons – demonstrating how political compromise can attain the incremental standing-down of auto-mobile modernism, yet at the same time how such processes have the potential to produce negative effects on social equity, as mixed-use residential and commercial redevelopment of the Hayes Valley in the wake of freeway removal led to gentrification of the area.

Of course there is the issue of how to raise awareness and acceptance of the need for equitable reforms to the forms, housing policies, and mobility structures of cities. One successful way to mobilize support and stimulate policy debate is through cycling and bus/transit riders' unions and other lobby organizations that advocate on behalf of their members. The leaders of such unions in places like Los Angeles have had success in raising awareness of problems and advocating for solutions, even without gaining any formal decision-making powers (Wexler 2000; Grengs 2002). As noted in Chapter 13, Critical Mass, the cyclist quasi-organization/movement that began in San Francisco but which has since spread to many cities across the globe, is another model. Instead of primarily seeking to influence policy or stimulate debate, by meeting locally as a group Critical Mass temporarily occupies space otherwise given over to automobile traffic. This act transgresses dominant perceptions of the proper use of urban street space and simultaneously models in real time the potential for alternative and more democratic mobilities. Critical Mass, and bicycle lobby organizations like Time's Up! and Transportation Alternatives in New York City, have had success in promoting policies that were eventually adopted by municipal administrations and built into ongoing city policy. One lesson is to be tenacious, and to follow a series of different strategies, both those that track close to mainstream opinion and those that offer novel alternatives. Over time, ideas in the latter category are often accepted into the former, and continued advocacy for what are inherently equitable and democratic objectives will have more success than intermittent, uncoordinated, and haphazard activities. Over time, such policies or strategies can then be endorsed or advocated by established political parties.

However, when an impasse is reached and democratic processes break down, more radical measures may be required. At this point, some of the lessons of workers' rights movements, as well as the Occupy movement, become pertinent, harkening back to Ghandi's peaceful occupation of key public squares, factories, state institutions, and natural resources in India. It may become necessary to occupy strategic sites, to literally reclaim the space that once was used to repress, and transform it even just temporarily into a new political space where the right to the city might be articulated, enacted, or leveraged. As Chapter 14 by Talsma in this volume attests, this could mean occupying the true "corridors of power" in the auto-city – often no longer the commercial "main streets" but the highway networks through which flow the commodities, raw resources, and workers that provide the auto-industrial complex with its economic power. Of course, one should be careful not to fetishize space (Miller and Nicholls 2013). The object is always to highlight injustice and promote political debates on issues with pertinence across the city, rather than merely to claim a space for a particular group.

In the auto-city, the most disruptive action is that which prevents the continuous movement of cars and trucks. It is thus the highway (figuratively as well as literally) that warrants consideration by contemporary urban social movements, at least as part of an ensemble of places from which to launch claims on the right to the city. The highway is both the key articulation and the key pro-genitor of auto-space. It is one of the few places in the city where, as a result of both the rules undergirding automobility and the way these spaces are con-structed, most of those who are present are compelled to be there. Those driving on the highway often view the road as a transitionary vector linking origins and destinations, even if they find ways of making the drive more pleasurable (through music, etc.), and thus experience any limitation on their ability to move as highly frustrating. Yet at the same time the potential for exit is highly constrained. Automobile drivers cannot easily climb over embankments or drive through concrete barriers (which have been put in place explicitly to fix traffic flow within prescribed lanes), and they cannot do so (nor plough through pro-testers) without breaking important legal and moral laws. Many of those who partake of the highway are compelled to do so, and hence experience access to it as a necessity of life and see it as a "natural" right. This is just as true for the firms whose business models assume unconstructed flows between production nodes, suppliers, and customers, as for individual commuters. The tight linking of the material conditions compelling particular uses of the highway and the sub-jective experience of seeing the highway as necessity and right help internalize "common-sense" understandings of what the highway is for, and thus what is out of place in such a space (Cresswell 1996).

Protests that occupy the highway and prevent the free flow of traffic are about as "out-of-place" as one can get in the auto-city. Such a transgressive event brings into focus the key contradictions between use and exchange value of con-temporary forms of mobility. While the occupation of residual degraded tradi-tional urban public spaces, now often found well away from the corridors of mobility, hardly raises an eyebrow, occupation of the highway receives imme-diate attention, propelling political grievances into the public consciousness even if such grievances are rejected out of hand or result in intense repression. This is highlighted in the visceral reaction to the spontaneous highway protests in Toronto as discussed by Talsma in Chapter 14, where occupation of the highway steered political attention onto different claims regarding the use of auto-space and the right to the city. The highway is turned into a political symbol, an object of politicization, while establishing the official state response as a new political orthodoxy.

While it is unlikely that those occupying the auto-city in this way will be suc-cessful immediately, they will not be forgotten and will compel some form of response. As during Ghandi's acts of peaceful civil disobedience, in some cases retaliation is likely to be brutal and overbearing. These responses become part of the history of mobilization. Popular opinion is likely to remain on the side of orthodoxy for a significant amount of time, as auto-mobile regimes strive to defend themselves. Many of the technological innovations mentioned by Dennis and Urry (2009a,b) may be offered as examples of how the system is being improved, refined, and reformed, yet they favour system maintenance and bolster the present regime as those invested in its expansion attempt to maintain a lock on power. It is only when the taking of the highway no longer raises the disgust and ire of the majority – when it is either so commonplace, or when the majority is no longer dependent on the highway, or when the political claims being made about rights to the city come to resonate sufficiently among the general public that they no longer can be baited into supporting a repressive backlash – that one can say with any confidence that a situation of post-automobility, or a post-car urban world, has truly arrived.

Conclusion

In producing new variegated forms of inequality, politics, and vulnerability, the global organization of auto-mobile fossil-fuelled capitalism and the systems and regimes they have begat have transformed and shifted the fields of struggle into the cities. It is thus in the cities that any lasting solutions will have to be forged, and it will be impossible to alter the structures producing inequality without transforming the spatial, social, and political structures undergirding contemporary urban automobilities. The contours of what this might look like have been outlined herein. Political, technical, and social strategies for transforming the auto-city in these ways must be diverse and multi-faceted, and must deal not only with systems of mobility, but also with the problem of private property (Blomley 2004; Harvey 2012). Mobility will need to be democratized, not through the de-privatization of the automobile, although that could also occur, but through the de-privatization and democratization of the city itself. Not only radical but moderate, incremental, as well as revolutionary, system-improving, and system-challenging strategies, movements, and agendas will be necessary for the kind of political experimentation required to produce any lasting system change, respecting Lefebvre's cry and demand for the right to the city, and to bring about the biophilic, connected, and polycentric cities of the future. Such represents the great challenge of the twenty-first century beset by rapid urbanization, globalization, the rise of mega-cities, and persistent economic and environmental vulnerability.

References

Aalbers, M. (2008) "The financialization of home and the mortgage market crisis", *Competition and Change*, 12(2): 148–168.

Aarts, L. and van Schagen, I. (2006) "Driving speed and the risk of road crashes: A review", *Accident Analysis and Prevention*, 38(2): 215–224.

ACS. (2011) *American Community Survey*, Washington: US Census Bureau.

Adam, B. (2004) *Time*, Cambridge, UK: Polity.

Adams, J. (1999) *The Social Implications of Hypermobility*, London: OECD Project on Environmentally Sustainable Transport, UCL.

ADB (Asian Development Bank). (2013) *Dhaka Clean Fuel Project Data Sheet (PDS) Overview*, Online. Available at: www.adb.org/projects/35466-013/main (accessed 20 November 2013).

Adey, P. (2010) *Mobility*, Abingdon, UK: Routledge.

Aftabuzzaman, M. and Mazloumi, E. (2011) "Achieving sustainable urban transport mobility in the post peak oil era", *Transport Policy*, 18: 695–702.

Aglietta, M. (1979) *A Theory of Capitalist Regulation*, London: New Left Books.

Ahlport, K.N., Linnan, L., Vaughn, A., Evenson, K.R., and Ward, D.S. (2008) "Barriers to facilitators of walking and bicycling to school: Formative results from the non-motorized travel study", *Health Education and Behavior*, 35: 221–244.

Albo, G., Gindin, S., and Panitch, L. (2010) *In and Out of Crisis: The Global Financial Meltdown and Left Alternatives*, New York: PM Press/Spectre.

Alcoba, N. (2010) "Calls to legalize road hockey grow", Online. Available at: http://news.nationalpost.com/2010/06/11/calls-to-legalize-road-hockey-grow (accessed 1 August 2010).

Aldred, R. (2012) "The role of advocacy and activism", in J. Parkin (ed.) *Cycling and sustainability*: 83–110, Bingley, UK: Emerald.

Alessandria, G., Kaboski, J.P., and Midrigan, V. (2010) "The great trade collapse of 2008–09: An inventory adjustment?", *IMF Economic Review*, 58(2): 254–294.

Alliance of Ontario Food Producers. (2012) *Economic Impact Analysis: Ontario Food and Beverage Processing Sector*, Online. Available at: www.aofp.ca/pub/docs/mnp-economic-report-2.pdf (accessed 8 June 2013).

Alonso, W. (1964) *Location and Land Use: Toward a General Theory of Land Rent*, Cambridge, MA: Harvard University Press.

Althusser, L. (1971) "Ideology and ideological state apparatuses", in L. Althusser (ed.) *Lenin and Philosophy and Other Essays*, New York: Monthly Review Press.

American Public Transportation Association. (2013) *Quarterly Ridership Report*, Online. Available at: www.apta.com/resources/statistics/Documents/APTA-Ridership-by-Mode-and-Quarter-1990-Present.xls (accessed 15 June 2013).

American Society of Civil Engineers. (2009) *Report Card for America's Infrastructure*, Online. Available at: www.infrastructurereportcard.org/report-cards (accessed 30 July 2012).

Amin, A. (1994) "Post-fordism: Models, fantasies and phantoms of transition", in A. Amin (ed.) *Post-Fordism: A Reader*: 1–40, Cambridge, MA: Blackwell.

Andrey, J. (2000) "The automobile imperative: Risks of mobility and mobility related risks", *The Canadian Geographer*, 44(4): 387–400.

Anggraini, R., Arentze, T.A., and Timmermans, H.J.P. (2008) "Car allocation between transit heads in car deficient households: A decision model", *European Journal of Transport and Infrastructure Research*, 8(4): 301–319.

Angus. (2008) "Cyclists shut down the Gardiner", Online posting. Available at: www.blogto.com/city/2008/05/cyclists_shut_down_the_gardiner (accessed 1 June 2010).

Ansley, F. (2010) "Constructing citizenship without a licence: The struggle of undocumented immigrants in the USA for livelihoods and recognition", *Studies in Social Justice*, 4(2): 165–178.

Arrighi, G. (1994) *The Long Twentieth Century: Money, Power and the Origin of Our Times*, London: Verso.

Arthur, W.B. (1994) *Increasing Returns and Path Dependence in the Economy*, Ann Arbor: University of Michigan Press.

Ashton, P. (2009) "An appetite for yield: The anatomy of the subprime mortgage crisis", *Environment and Planning A*, 41(11): 1420–1441.

Ausubel, J.H. and Marchetti, C. (2001) "The evolution of transport", *The Industrial Physicist*, April/May: 20–24.

Avila, E. (2004) *Popular Culture in the Era of White Flight: Fear and Fantasy in Suburban Los Angeles*, Berkeley: University of California Press.

Axisa, J., Scott, D., and Newbold, B. (2012) "Factors influencing commute distance: A case study of Toronto's commuter shed", *Journal of Transport Geography*, 24: 123–129.

Bae, C. (2004) "Transportation and the environment", in S. Hanson and G. Giuliano (eds) *The Geography of Urban Transportation*, 3rd edn, New York: Guilford Press.

Bagguley, P. (1995) "Protest, poverty and power: A case study of the anti-poll tax movement", *Sociological Review*, 43(4): 693–719.

Baing, A. (2010) "Containing urban sprawl? Comparing brownfield reuse policies in England and Germany", *International Planning Studies*, 15: 25–35.

Banister, D. (2005) *Unsustainable Transport: City Transport in the New Century*, London: Routledge.

Banister, D. (2011) "The trilogy of distance, speed and time", *Journal of Transport Geography*, 19: 950–959.

Banister, D. (2012) "Assessing the reality: Transport and land use planning to achieve sustainability", *Journal of Transport and Land Use*, 5: 1–14.

Bari, M. (2007) *A Critical Review of Dhaka Strategic Transport Plan (STP)*, Dhaka, Bangladesh.

Bari, M. and Efroymson, D. (2005a) *Rickshaw Bans in Dhaka City: An Overview of the Arguments For and Against*, Dhaka, Bangladesh: Work for Better Bangladesh (WBB) Trust.

Bari, M. and Efroymson, D. (2005b) *Vehicle Mix and Road Space in Dhaka: The Current Situation and Future Scenarios*, Dhaka, Bangladesh: Work for Better Bangladesh (WBB) Trust.

Bari, M. and Efroymson, D. (2005c) *Improving Dhaka's Traffic Situation: Lesson from Mirpur Road*, Dhaka, Bangladesh: Work for Better Bangladesh (WBB) Trust.

Barker, J. (2009) "Driven to distraction: Children's experiences of car travel", *Mobilities*, 4: 59–76.

Barnett, J. (1995) *Fractured Metropolis: Improving the New City, Restoring the Old City, Reshaping the Region*, New York: HarperCollins.

Bashevkin, S. (2006) *Tale of Two Cities: Women and Municipal Restructuring in London and Toronto*, Vancouver: University of British Columbia Press.

Basok, T. (2002) *Tortillas and Tomatoes: Transmigrant Mexican Harvesters in Canada*, Montreal and Kingston: McGill-Queen's University Press.

Bastow, S. and Martin, J. (2003) *Third Way Discourse: European Ideologies in the Twentieth Century*, Edinburgh: Edinburgh University Press.

Bauder, H. (2000) "Reflections on the spatial mismatch debate", *Journal of Planning Education and Research*, 19: 316–321.

Bauder, H. (2005) "Landscape and scale in media representations: The construction of offshore farm labour in Ontario, Canada", *Cultural Geographies*, 12(1): 41–58.

Baumert, K., Herzog, T., and Pershing, J. (2005) *Navigating the Numbers: Greenhouse Gas Data and International Climate Policy*, Washington, DC: World Resources Institute.

Baute, N. and Wallace, K. (12 May 2009) "Mayor warns defiant Tamils", *The Toronto Star*: A1.

BBC News Asia. (2013) *Bangladesh Factory Collapse Probe Uncovers Abuses*, Online. Available at: www.bbc.co.uk/news/world-asia-22635409 (accessed 23 May 2013).

Beatley, T. (2010) *Biophilic Cities: Integrating Nature into Urban Design and Planning*, Washington, DC: Island Press.

Beatley, T. and Newman, P. (2013) "Biophilic cities are sustainable, resilient cities", *Sustainability*, 5: 3328–3345.

Beauregard, R. (1993) *Voices of Decline: The Postwar Fate of U.S. Cities*, Cambridge, MA: Blackwell.

Beauregard, R. (2006) *When America Became Suburban*, Minneapolis, MN: University of Minnesota Press.

Beckmann, J. (2001) "Automobility: A social problem and theoretical concept", *Environment and Planning D: Society and Space*, 19(5): 593–607.

Beckmann, J. (2004) "Mobility and safety", *Theory, Culture and Society*, 21(4/5): 81–100.

Begum, S. and Sen, B. (2005) "Pulling rickshaws in the city of Dhaka: A way out of poverty?", *Environment and Urbanization*, 17(2): 11–25.

Beito, D.T., Gordon, P., and Tabarrok, A. (2002) *The Voluntary City: Choice, Community, and Civil Society*, Ann Arbor: University of Michigan Press.

Berger, B.M. (1960) *Working Class Suburb: A Study of Auto Workers in Suburbia*, Berkeley: University of California Press.

Bergmann, S. and Sager, T. (2008) *The Ethics of Mobilities: Rethinking Place, Exclusion, Freedom and Environment*, Aldershot: Ashgate.

Berman, M. (1988) *All That Is Solid Melts into Air: The Experience of Modernity*, New York: Penguin.

Beroud, B. and Anaya, E. (2012) "Private interventions in a public service: An analysis of public bicycle schemes", in J. Parkin (ed.) *Cycling and Sustainability*: 269–302, Bingley, UK: Emerald.

Best, A.L. (2006) *Fast Cars, Cool Rides: The Accelerating World of Youth and Their Cars*, New York: New York University Press.

Best, H. and Lanzendorf, M. (2005) "Division of labour and gender differences in metropolitan car use: An empirical study in Cologne, Germany", *Journal of Transport Geography*, 13(2): 109–121.

Binford, L. (2013) *Tomorrow We're All Going to the Harvest: Temporary Foreign Worker Programs and Neoliberal Political Economy*, Austin: University of Texas Press.

Black, W.R. (2010) *Sustainable Transportation: Problems and Solutions*, New York: Guilford Press.

Blair, R. and Wellman, G. (2011) "Smart growth principles and the management of urban sprawl", *Community Development*, 42: 494–510.

Blickstein, S.G. (2010) "Automobility and the politics of bicycling in New York City", *International Journal of Urban and Regional Research*, 34(4): 886–905.

Blickstein, S.G. and Hanson, S. (2001) "Critical Mass: Forging a politics of sustainable mobility in the information age", *Transportation*, 28(4): 347–362.

Blomley, N. (2004) *Unsettling the City: Urban Land and the Politics of Property*, London: Routledge.

Blumenberg, E. (2004a) "En-gendering effective planning: Spatial mismatch, low-income women, and transportation policy", *Journal of the American Planning Association*, 70: 269–281.

Blumenberg, E. (2004b) "Beyond the spatial mismatch: Welfare recipients and transportation policy", *Journal of Planning Literature*, 19(2): 182–205.

Blumenberg, E. (2009) "Moving in and moving around: Immigrants, travel behavior, and implications for transport policy", *Transportation Letters: The International Journal of Transportation Research*, 1(2): 169–180.

Blumenberg, E. and Evans, A.E. (2006) *Growing the Immigrant Transit Marker: Public Transit Use and California Immigrants*, Working Paper, Institute of Transportation Studies, University of California, Los Angeles.

Blumenberg, E. and Evans, A.E. (2010) "Planning for demographic diversity: The case of immigrants and public transit", *Journal of Public Transportation*, 13(2): 23–45.

Blumenberg, E. and Shiki, K. (2003) "How welfare recipients travel on public transit, and their accessibility to employment outside large urban centers", *Transportation Quarterly*, 57(2): 25–37.

Blumenberg, E. and Smart, M. (2010) "Getting by with a little help from my friends ... and family: Immigrants and carpooling", *Transportation*, 37(3): 429–446.

Blumenberg, E. and Smart, M. (2011) "Migrating to driving: Exploring the multiple dimensions of immigrants' automobile use", in K. Lucas, E. Blumenberg, and R. Weinberger (eds) *Auto Motives: Understanding Car Use Behaviours*, Bingley, UK: Emerald.

Blumenberg, E. and Song, L. (2008) "Travel behavior of immigrants in California: Trends and policy implications", paper presented at the Transportation Research Board 87th Annual Meeting.

Boarnet, M.G., Day, K., Anderson, C., McMillan, T., and Alfonzo, M. (2005) "California's safe routes to school program: Impacts on walking, bicycling, and pedestrian safety", *Journal of the American Planning Association*, 71: 301–317.

Bohm, S., Jones, C., Land, C., and Paterson, M. (2006) "Introduction: Impossibilities of automobility", *Sociological Review*, 54(s1): 3–16.

Bonham, J. (2006) "Transport: Disciplining the body that travels", in S. Bohm, C. Land, C. Jones, and M. Paterson (eds) *Against Automobility*: 54–73, London: Sociological Review Monographs, Blackwell.

Bonogoure, T. (13 May 2009) "Tamils plan human chain at Queen's park", *Globe and Mail*: A10.

Borjesson, M. and Eliasson, J. (2012) "The benefits of cycling: Viewing cyclists as travellers rather than non-motorists", in J. Parkin (ed.) *Cycling and Sustainability*: 247–268, Bingley, UK: Emerald.

Boschmann, E. (2011) "Job access, location decision, and the working poor: A qualitative study in the Columbus, Ohio metropolitan area", *Geoforum*, 42(6): 671–682.

Boudreau, J.-A. (2000) *The Megacity Saga*, Montreal: Black Rose Books.

Boudreau, J.-A., Keil, R., and Young, D. (2009) *Changing Toronto: Governing Urban Neoliberalism*, Toronto: Toronto University Press.

Bourne, L., Hutton, T., Shearmur, R., and Simmons, G. (2011) *Canadian Urban Regions: Trajectories of Growth and Change*, Don Mills, ON: Oxford University Press.

Bradsher, K. (22 April 2013) "Chinese auto buyers grow hungry for larger cars", *New York Times*: B3.

Braza, M., Shoemaker, W., and Seeley, A. (2004) "Neighborhood design and rates of walking and biking to elementary school in 34 California communities", *American Journal of Health Promotion*, 19: 128–136.

Brenner, N. and Theodore, N. (2002) "Cities and the geographies of 'actually existing neoliberalism'", *Antipode*, 34(3): 349–379.

Briand, M. and Hartz-Karp, J. (2013) *From Surviving to Thriving: The Way of Participatory Sustainability*, Washington, DC: Island Press.

Bridge, G. (2010) "Geographies of peak oil: The other carbon problem", *Geoforum*, 41(4): 523–530.

Brindle, R.E. (1994) "Lies, damned lies and 'automobile dependence': Some hyperbolic reflections", *Australian Transport Research Forum*, 94: 117–131.

Brown, L.R., Flavin, C., and Norman, C. (1979) *Running on Empty: The Future of the Automobile in an Oil-Short World*, New York: Norton.

Brown, W. (2005) "Neoliberalism and the end of liberal democracy", *Theory and Event*, 7(1): 1–25.

BRTA (Bangladesh Road Transport Authority). (2013) *Number of Registered Motor Vehicles in Dhaka (Year Wise)*, Online. Available at: www.brta.gov.bd/images/files/dhaka_statistics_20_06_13.pdf (accessed 20 November 2013).

Bruce-Briggs, B. (1977) *The War against the Automobile*, New York: Dutton.

Bruno, D. (2012) *The De-bikification of Bejing*, Online. Available at: www.theatlanticcities.com/commute/2012/04/de-bikification-beijing/1681 (accessed 9 October 2013).

BTRC (Beijing Transportation Research Center). (2011) *Bejing Transportation Annual Report*, Online. Available at: www.bjtrc.org.cn/InfoCenter/NewsAttach//aeb7c878-d31e-4f08-982f-3c17c717c87b.pdf (accessed 13 October 2013).

Buehler, R. and Pucher, J. (2010) "Cycling to sustainability in Amsterdam", *Sustain: A Journal of Environmental and Sustainability Issues*, 21: 36–40.

Buehler, R. and Pucher, J. (2011) "Sustainable transport in Freiberg: Lessons from Germany's environmental capital", *International Journal of Sustainable Transportation*, 5(1): 43–70.

Buehler, R. and Pucher, J. (2012) "Cycling to work in 90 large American cities: New evidence on the role of bike paths and lanes", *Transportation*, 39(2): 409–432.

Buliung, R.N., Mitra R., and Faulkner, G. (2009) "Active school transportation in the Greater Toronto Area, Canada: An exploration of trends in space and time (1986–2006)", *Preventive Medicine*, 48: 507–512.

Buliung, R.N., Soltys, K., Habel, C., and Lanyon, R. (2009) "Driving factors behind successful carpool formation and use", *Transportation Research Record*, 2118: 31–38.

Bunning, J., Beattie, C. Rauland, V., and Newman, P. (2013) "Low-carbon sustainable precincts: An Australian perspective", *Sustainability*, 5: 2305–2326.

Bunting, T.E. and Filion, P. (1999) "Dispersed city form in Canada: A Kitchener CMA case example", *Canadian Geographer*, 43(3): 268–287.

Burbank, M. (1995) "The psychological basis of contextual effects", *Political Geography*, 14(6/7): 621–635.

Burchell, R., Lowenstein, G., Dolphin, W., Galley, C., Downs, A., Seskin, S., Still, K., and Moore, T. (2002) *Costs of Sprawl: 2000*, Washington, DC: Transportation Research Board, National Research Council.

Business Council for Sustainable Development (BCSD) (1993) *Getting Eco-Efficient*, Report of the BCSD First Antwerp Eco-Efficiency Workshop, November, Geneva: BCSD.

Cahill, M. (2010) *Transport, Environment and Society*, Maidenhead, UK: Open University Press.

California Highway Patrol. (2002) *SAFE : A Safety and Farm Vehicle Education Program*, Online. Available at: www.popcenter.org/library/awards/goldstein/2002/02-07(W).pdf (accessed 12 May 2013).

Calthorpe, P. (1991) "The post-suburban metropolis", *Whole Earth Review*, 78: 44–51.

Calthorpe, P. and Fulton, W. (2001) *Regional City: Planning for the End of Sprawl*, Washington, DC: Island Press.

Campbell, C.J. and Laherrère, J.H. (1998) "The end of cheap oil", *Scientific American*, 278: 80–85.

Carlin, J.B., Stevenson, M.R., Roberts, I., Bennett, C.M., Gelman, A., and Nolan, T. (1997) "Walking to school and traffic exposure in Australian children", *Australian and New Zealand Journal of Public Health*, 21: 286–292.

Carlson, C. (2002) *Critical Mass: Bicycling's Defiant Celebration*, Oakland, CA: AK Press.

Caro, R.A. (1974) *The Power Broker: Robert Moses and the Fall of New York*, New York: Knopf.

Carson, R. (1962) *Silent Spring*, Boston, MA: Houghton Mifflin.

Carstensen, T.A. and Ebert, A.-K. (2012) "Cycling cultures in Northern Europe: From 'golden age' to 'rennaissance'", in J. Parkin (ed.) *Cycling and Sustainability*: 23–58, Bingley, UK: Emerald.

Cass, N., Shove, E. and Urry, J. (2005) "Social exclusion, mobility and access", *The Sociological Review*, 53(3): 539–555.

Castells, M. (1978). *City, Class, and Power*, Basingstoke, UK: Macmillan.

CBC News. (2011) *Quebec Van Rollover Kills 1, Injures 7*, Online. Available at: www.cbc.ca/news/canada/montreal/story/2011/04/18/migrant-worker-minivan-accident.html (accessed 27 October 2012).

Cervero, R. (1998) *The Transit Metropolis: A Global Inquiry*, Washington, DC: Island Press.

Cervero, R., Murphy, S., Ferrell, C., Goguts, N., Tsai, Y.H., Arrington, G.B., Boroski, J., Smith-Heimer, J., Golem, R., Peninger, P., Nakajima, E., Chui, E., Dunphy, R., Myers, M., McKay, S., and Witenstein, N. (2004) *Transit Oriented Development in America: Experiences, Challenges and Prospects*, Washington, DC: Transportation Research Board, National Research Council.

Cervero, R., Sandoval, O.S., and Landis, J. (2002) "Transportation as a stimulus of welfare-to-work: Private versus public mobility", *Journal of Planning Education and Research*, 22(1): 50–63.

Chatman, D.G. (2013) "Explaining the 'immigrant effect' on auto use: The influences of neighborhoods and preferences", *Transportation*, 40(3): 1–21.

Chatman, D.G. and Klein, N. (2009) "Immigrants and travel demand in the United States: Implications for transportation policy and future research", *Public Works Management & Policy*, 13: 312–327.

Chatterton, P. (2006) "Making autonomous geographies: Argentina's popular uprising and the 'Movimiento de Trabajadores Desocupados'", *Geoforum*, 36: 545–561.

Christie, H., Smith, S.J., and Munro, M. (2008) "The emotional economy of housing", *Environment and Planning A*, 40(10): 2296–2312.

Citizenship and Immigration Canada (CIC). (2013) *Facts and Figures 2012: Immigration Overview – Permanent and Temporary Residents, Government of Canada*, Online. Available at: www.cic.gc.ca/english/pdf/research-stats/facts2012.pdf (accessed 24 September 2013).

City of Vancouver. (2012) *Transportation 2040 Plan: A Transportation Vision for the City of Vancouver*, Online. Available at: http://vancouver.ca/streets-transportation/transportation-2040.aspx (accessed 15 November 2013).

Clapson, M. (2003) *Suburban Century: Social Change and Urban Growth in England and the USA*, New York: Berg.

Clarkson, B. (11 May 2009) "Tamils take over Gardiner", *Toronto Sun*: 5.

Clifton, K.J. and Kreamer-Fults, K. (2007) "An examination of the environmental attributes associated with pedestrian–vehicular crashes near public schools", *Accident Analysis and Prevention*, 39: 708–715.

Clifton, K.J., Burnier, C.V., and Akar, G. (2009) "Severity of injury resulting from pedestrian–vehicle crashes: What can we learn from evaluating the built environment?", *Transportation Research Part D*, 14: 425–436.

Cline, M.E., Sparks, C. and Eschback, K. (2009) "Understanding carpool use among Hispanics in Texas", paper #09-1585 presented at the Transportation Research Board Annual Meeting, Washington, DC.

Club of Rome. (1972) *The Limits to Growth: A Report on the Club of Rome's Project on the Predicament of Mankind*, New York: Universe Books.

Cohen, M.J. (2012) "The future of automobile society: A socio-technical transitions perspective", *Technology Analysis and Strategic Management*, 24(4): 377–390.

Coleman, M. (2012) "The 'local' migration state: The site-specific devolution of immigration enforcement in the U.S. South", *Law and Policy*, 34(2): 159–190.

Collins, D.C.A. and Kearns, R.A. (2002) "The safe journeys of an enterprising school: Negotiating landscapes of opportunity and risk", *Health and Place*, 7: 293–306.

Congress for the New Urbanism. (2012) *Freeways without Futures*, Chicago: Congress for the New Urbanism.

Conley, J. and McLaren, A.T. (eds) (2009) *Car Troubles: Critical Studies of Automobility and Automobility*, Aldershot: Ashgate.

Conlon, D. (2011) "Waiting: Feminist perspectives on the spacings/timings of migrant (im)mobility", *Gender, Place & Culture*, 18(3): 353–360.

Cooke, T.J. (1996) "City–suburb differences in African American male labor market achievement", *Professional Geographer*, 48(4): 458–467.

Cooper, A.R., Page, A.S., Foster, L.J., and Qahwaji, D. (2003) "Commuting to school: Are children who walk more physically active?", *American Journal of Preventive Medicine*, 25: 273–276.

Cortright, J. (2008) *Driven to the Brink: How the Gas Price Spike Popped the Housing Bubble and Devalued the Suburbs*, Chicago: CEOs for Cities.

Covenay, J. and O'Dwyer, L.A. (2009) "Effects of mobility and location on food access", *Health and Place*, 15(1): 45–55.

Cox, K. (1968) "Suburbia and voting behaviour in London Metropolitan Area", *Annals of the Association of American Geographers*, 58(1): 111–127.

Cox, K. (1969) "The voting decision in a spatial context", *Progress in Geography*, 1(1): 81–117.

Cox, K. and Jonas, A.G. (1993) "Urban development, collective consumption and the politics of metropolitan fragmentation", *Political Geography*, 12(1): 8–37.

Cravey, A.J. (2003) "Toque una ranchera, por favor", *Antipode*, 35(3): 603–621.

Cresswell, T. (1996) *In Place/Out of Place: Geography, Ideology, and Transgression*, Minneapolis: University of Minnesota Press.

Cresswell, T. (2006) *On the Move: Mobility in the Modern Western World*, London: Routledge.

Cresswell, T. (2010) "Towards a politics of mobility", *Environment and Planning D: Society and Space*, 28(1): 17–31.

Crewe, I. (4 June 2001) "How the suburbs turned red", *New Statesman*: 19–20.

Crewe, I., Sarlvik, B., and Alt, J. (1977) "Partisan dealignment in Britain, 1964–1974", *British Journal of Political Science*, 7(1): 129–190.

Crouch, C. (2009) "Privatised Keynesianism: An unacknowledged policy regime", *British Journal of Politics and International Relations*, 11(3): 382–399.

Cubukgil, A. and Miller, E.J. (1982) "Occupational status and the journey-to-work", *Transportation*, 11: 251–276.

Currie, G. and Delbose, A. (2011) "Mobility vs affordability as motivations for car-ownership choice in urban fringe, low-income Australia", in K. Lucas, E. Blumerberg, and W. Weinberger (eds) *Auto Motives: Understanding Car Use Behaviours*: 193–208, Bingley, UK: Emerald.

Currie, G., Richardson, T., Smyth, P., Vella-Brodrick, D., Hine, J., Lucas, K., Stanley, J., Morris, J., Kinnear, R., and Stanley, J. (2009) "Investigating links between transport disadvantage, social exclusion and well-being in Melbourne: Preliminary results", *Transport Policy*, 16(3): 97–105.

Curtice, J. (1995) "Is talking over the garden fence of political import?", in M. Eagles (ed.) *Spatial and Contextual Models in Political Research*: 195–210, London: Taylor & Francis.

Curtice, J. and Steed, M. (1982) "Electoral choice and the production of government: The changing operation of the electoral system in the United Kingdom since 1955", *British Journal of Political Science*, 12(2): 249–298.

CUSTReC (China Urban Transport Research Centre). (2007) *Bicycling Mode Shares in Chinese Cities*, Online. Available at: http://slocat.net/member/161 (accessed 31 August 2013).

Cutler, F. (2007) "Context and attitude formation: Social interaction, default information, or local interests?", *Political Geography*, 26(5): 575–600.

Dalley, M.L. and Ruscoe, J. (2003) *The Abduction of Children by Strangers in Canada: Nature and Scope*, Online. Available at: www.rcmp-grc.gc.ca/pubs/omc-ned/abd-rapt-eng.htm (accessed 20 June 2013).

Dalton, M.A., Longacre, M.R., Drake, K.M., Gibson, L., Adachi-Mejia, A.M., Swain, K., Xie, H., and Owens, P.E. (2011) "Built environment predictors of active travel to school among rural adolescents", *American Journal of Preventive Medicine*, 40: 312–319.

Danielson, M.D. (1976) *The Politics of Exclusion*, New York: Columbia University Press.

Danyluk, M. and Ley, D. (2007) "Modalities of the new middle class: Ideology and behaviour in the journey-to-work from gentrified neighbourhoods", *Urban Studies*, 44(10): 2195–2210.

Dargay, J., Gately, D., and Sommer, M. (2007) "Vehicle ownership and income growth, worldwide: 1960–2030", *Energy Journal*, 28: 163–190.

Davide. (12 May 2009) "Let me start off…", Weblog comment, S. Micaleff, "Tamil protest: The taking of the Gardiner expressway", *Spacing Toronto*, Online. Available at: http://spacing.ca/toronto/2009/05/10/tamil-protest-the-taking-of-the-gardiner-expressway (accessed 10 July 2010).

Davidson, G. (2013) "The suburban idea and its enemies", *Journal of Urban History*, 39(5): 829–847.

Davies, W. and McGoey, L. (2012) "Rationalities of ignorance: On financial crisis and the ambivalence of neo-liberal epistemology", *Economy and Society*, 41(1): 64–83.

Dawson, M. (2011) "Driving to carmageddon: Capitalism, transportation, and the logic of planetary crisis", in S. Best, R. Kahn, A.J. Nocella, and P. McLaren (eds) *The Global Industrial Complex: Systems of Domination*: 263–286, Toronto: Lexington Books.

Dear, M. and Flusty, S. (1998) "Postmodern urbanism", *Annals of the Association of American Geographers*, 88(1): 50–72.

Deffeyes, K.S. (2001) *Hubbert's Peak: The Impending World Oil Shortage*, Princeton, NJ: Princeton University Press.

De Genova, N.P. (2002) "Migrant 'illegality' and deportability in everyday life", *Annual Review of Anthropology*, 31(1): 419–447.

Delbosc, A. and Currie, G. (2013) "Causes of youth licensing decline: A synthesis of evidence", *Transport Reviews*, 33(3): 271–290.

Delmelle, D., Haslauer, E., and Prinz, T. (2013) "Social satisfaction, commuting and neighbourhoods", *Journal of Transport Geography*, 30: 110–116.

De Maio, P. (2003) "Smart bikes: Public transportation for the 21st Century?", *Transportation Quarterly*, 57(1): 9–11.

Denmark, D. (1998) "The outsiders: Planning and transport disadvantage", *Journal of Planning Education and Research*, 17(3): 231–245.

Dennis, K. and Urry, J. (2009a) *After the Car*, Cambridge: Polity.

Dennis, K. and Urry, J. (2009b) "Post-car mobilities", in J. Conley and A.T. McLaren (eds) *Car Troubles: Critical Studies of Automobility and Auto-mobility*: 235–252, Aldershot: Ashgate.

De Place, E. (2011) "The 'war on cars': A brief history of a rhetorical device", *Grist*, 6 January.

DHUTS (Dhaka Urban Transport Network Development Study). (2009) *Dhaka Urban Transport Network Development Study: Interim Report*, Dhaka, Bangladesh: Dhaka Transport Co-ordination Board and Japan International Cooperation Agency (JICA).

DiMaggio, C. and Li, G. (2012) "Roadway characteristics and pediatric pedestrian injury", *Epidemiologic Reviews*, 34: 46–56.

DITS. (1994) *Greater Dhaka Metropolitan Area Integrated Transport Study*, Dhaka, Bangladesh: Ministry of Planning, Government of Bangladesh and United Nations Development Program.

Dobbs, L. (2005) "Wedded to the car: Women, employment and the importance of private transport", *Transport Policy*, 12(3): 266–278.

Dobriner, W.M. (1963) *Class in Suburbia*, Englewood Cliffs, NJ: Prentice Hall.

Dodson, J. and Sipe, N. (2008) "Shocking the suburbs: Urban location, homeownership and oil vulnerability in the Australian city", *Housing Studies*, 23(3): 377–401.

Dodson, J. and Sipe, N. (2009) "A suburban crisis? Housing, credit, energy and transport", *Journal of Australian Political Economy*, 64: 199–210.

DOE (Department of Environment). (2009) *Clean Air and Sustainable Environment (CASE) Preparation Project: Final Report*, Dhaka, Bangladesh: Department of Environment (DOE), Government of Bangladesh.

Doling, J. and Ronald, R. (2010) "Home ownership and asset-based welfare", *Journal of Housing and the Built Environment*, 25(2): 165–173.

Donato, A. (12 May 2009) "Three easy steps to prevent traffic disruption due to demonstrators", *Toronto Sun*: editorial cartoon.

Donovan, K., Poisson, J., and Wallace, K. (3 October 2013) "'I don't throw my friends under the bus': Mayor Ford defends his pal after high-profile arrest on drug trafficking charges following Etobicoke raid", *Toronto Star*: A1.

Doolittle, R. and Donovan, K. (17 May 2013) "Ford in 'crack video' scandal", *Toronto Star*: A1.

DOT. (2010) *Transportation's Role in Reducing U.S. Greenhouse Gas Emissions, Vol. 1*, Washington, DC: Report to Congress, Center for Climate Change and Environmental Forecasting, US Department of Transportation.

DOT. (2012) *Traffic Safety Facts*, Washington, DC: National Highway Traffic Safety Administration, US Department of Transport.

Duany, A. and Plater-Zyberk, E. (2000) *Suburban Nation: The Rise of Sprawl and the Decline of the American Dream*, New York: North Point Press.

Dunleavy, P. (1979) "The urban basis of political dealignment: Social class, domestic property ownership, and state intervention in consumption processes", *British Journal of Political Science*, 9(4): 409–443.

Dunleavy, P. and Husbands, C.T. (1985) *British Democracy at the Crossroads: Voting and Party Competition in the 1980s*, London: Allen & Unwin.

Dunn, J. (1998) *Driving Forces: The Automobile, Its Enemies, and the Politics of Mobility*, Washington, DC: Brookings Institution.

DUTP (Dhaka Urban Transport Project). (1996) *Dhaka Urban Transport Project – Phase I: Final Report*, Dhaka, Bangladesh: Bangladesh Road Transport Authority, Government of Bangladesh.

DUTP (Dhaka Urban Transportation Plan). (2005) *The Urban Transport Policy: Final Report, Strategic Transport Plan for Dhaka*, prepared by Louis Berger Group and Bangladesh Consultant Ltd for Dhaka Transport Co-ordination Board (DTCB), Ministry of Communication, GoB, September.

Eagles, M. (1992) "Sources of variation in working class formation: Ecological, sectoral, and socialization influences", *European Journal of Political Research*, 21(3): 225–243.

Edwards, G. (2008) *Boris vs Ken: How Boris Johnson Won London*, London: Politicos Publishing.

EEA. (2006) *Urban Sprawl in Europe: The Ignored Challenge*, Copenhagen: European Environment Agency.

Efroymson, D. and Bari, M. (2005) *Improving Dhaka's Traffic Situation: Lessons from Mirpur Road*, Dhaka, Bangladesh: Roads for People.

Efroymson, D. and Rahman, M. (2005) *Transportation Policy for Poverty Reduction and Social Equity*, Dhaka, Bangladesh: Work for Better Bangladesh (WBB) Trust.

Efroymson, D., Biswas, B., and Ruma, S. (2007) *The Economic Contribution of Women in Bangladesh through Their Unpaid Labour*, Online. Available at: www.healthbridge.ca/economic contribution report.pdf (accessed 15 July 2013).

EIA (Energy Information Administration). (2013) *Annual Energy Outlook*, Washington, DC: Department of Energy.

Ellin, N. (1996) *Postmodern Urbanism*, Cambridge, MA: Blackwell.

Elliott, A. and Urry, J. (2010) *Mobile Lives*, Abingdon: Routledge.

Engel, K.C. and McCoy, P.A. (2011) *The Subprime Virus: Reckless Credit, Regulatory Failure, and Next Steps*, New York: Oxford University Press.

Engelen, E., Erturk, I., Froud, J., Johal, S., Leaver, A., Moran, M., Nilsson, A., and Williams, K. (2011) *After the Great Complacence: Financial Crisis and the Politics of Reform*, Oxford: Oxford University Press.

England, K. (1991) "Gender relations and the spatial structure of the city", *Geoforum*, 22(1): 135–147.

Environics Analytics. (2012a) *Wealthscapes Database*, Toronto: Environics Analytics.

Environics Analytics. (2012b) *Estimates of Household Expenditures from the Survey of Household Spending*, Specialized Database made available to the researchers at the University of Toronto.

EPA (Environmental Protection Agency). (2013) *Executive Order #12898*, Online. Available at: www.epa.gov/environmentaljustice (accessed 1 June 2013).

Epstein, G.A. (ed.) (2005) *Financialization and the World Economy*, Northampton, MA: Edward Elgar.

Equifax Canada. (2012/2013) *Canadian Consumer Credit Trends Reports*, Online. Available at: www.consumer.equifax.ca/about_equifax/newsroom/en_ca (accessed 31 July 2013).

Escamilla Nunez, M., Barraza-Villarreal, A., Hernandez-Cadena, L., Moreno-Macias, H., Ramirez-Aguilar, M., Sienra-Monge, J., Cortez-Lugo, M., Texcalac, J., del Rio-Navarro, B., and Romieu, I. (2008) "Traffic-related air pollution and respiratory symptoms among asthmatic children, resident in Mexico City: The EVA cohort study", *Respiratory Research*, 9: 74.

ESDC (Employment and Social Development Canada). (2013a) *Labour Market Opinion (LMO) Statistics: Annual Statistics 2012*, Online. Available at: www.hrsdc.gc.ca/eng/jobs/foreign_workers/lmo_statistics/annual2012.shtml#h2 (accessed 26 September 2013).

ESDC. (2013b) *Agreement for the Employment in Canada of Seasonal Agricultural Workers from Mexico: 2013*, Online. Available at: www.hrsdc.gc.ca/eng/jobs/foreign_workers/agriculture/seasonal/sawpmc2013.pdf (accessed 17 October 2013).

Evans, L. (2004) *Traffic Safety*, Bloomfield Hills, MI: Science Serving Society.

Ewing, R. (1996) "Transit oriented development in the sun belt", *Transportation Research Record*, 1552.

Ewing, R. and Cervero, R. (2010) "Travel and the built environment: A meta analysis", *JAPA*, 76(3): 265–294.

Ewing, R.H., Bartholomew, K., and Winkelman, S. (2008) *Growing Cooler: The Evidence of Urban Development on Climate Change*, Washington, DC: Urban Land Institute.

Ewing, R., Pendall, R., and Chen, D. (2002) *Measuring Sprawl and Its Impact*, Washington, DC: Smart Growth America.

Ewing, R., Schroeer, W., and Greene, W. (2004) "School location and student travel: Analysis of factors affecting mode choice", *Transportation Research Record*, 1895: 55–63.

Eyler, A., Brownson, R., Doescher, M., Evenson, K., Fesperman, C., Litt, J., Pluto, D., Steinman, L.E., Terpstra, J.L., Troped, P.J., and Schmid, T.L. (2007) "Policies related to active transport to and from school: A multisite case study", *Health Education Research*, 23: 963–975.

Falconer, R. and Newman, P. (2010) *Growing up: Reforming Land Use and Transport in "Conventional" Car Dependent Cities*, Saarbruecken, Germany: VDM.

Falconer, R., Newman, P., and Giles-Corti, B. (2010) "Is practice aligned with the principles? Implementing New Urbanism in Perth, Western Australia", *Transport Policy*, 17(5): 287–294.

Faraday, F. (2012) *Made in Canada: How the Law Constructs Migrant Workers' Insecurity*, Online. Available at: http://metcalffoundation.com/wp-content/uploads/2012/09/Made-in-Canada-Full-Report.pdf (accessed 1 November 2012).

Farber, S. and Paez, A. (2009) "My car, my friends, and me: A preliminary analysis of automobility and social activity participation", *Journal of Transport Geography*, 17: 216–225.

Farber, S. and Paez, A. (2011) "Running to stay in place: The time-use implications of automobility oriented land-use and travel", *Journal of Transport Geography*, 19: 782–793.

Farber, S., Neutens, T., Miller, H.J. and Li, X. (2013) "The social interaction potential of metropolitan regions: A time-geographic measurement approach using joint accessibility", *Annals of the Association of American Geographers*, 103(3): 483–504.

Faulkner, G., Buliung, R., Flora, P., and Fusco, C. (2009) "Active school transport, physical activity levels and body weight of children and youth: A systematic review", *Preventive Medicine*, 48: 3–8.

Faulkner, G.E.J., Richichi, V., Buliung, R.N., Fusco, C., and Moola, F. (2010) "What's 'quickest and easiest'?: Parental decision making about school trip mode", *International Journal of Behavioral Nutrition and Physical Activity*, 7: 62.

Fava, S.F. (1980) "Women's place in the new suburbia", in G.R. Wekerl, R. Peterson, and D. Morley (eds) *New Space for Women*: 129–150, Boulder, CO: Westview Press.

Featherstone, M., Thrift, N., and Urry, J. (eds) (2005) *Automobilities*, Thousand Oaks, CA: Sage.

Federal Reserve Bank of New York. (various years) *Quarterly Reports on Household Debt and Credit: Historic Data Tables*, Online. Available at: www.newyorkfed.org/householdcredit/historical-reports.html (accessed 30 July 2013).

Fels, M.F. and Munson, M.J. (1974) *Energy Thrift in Urban Transportation: Options for the Future*, Cambridge, MA: Ford Foundation Energy Policy Project Report.

Ferenc, L. (17 March 2009) "Human chain of Tamils circles city core", *Toronto Star*, Online. Available at: www.thestar.com/News/GTA/article/603446 (accessed 20 March 2010).

Fernandes, L. (2004) "The politics of forgetting: Class, politics, state power and the restructuring of urban space in India", *Urban Studies*, 41(12): 2415–2430.

Fesperman, C., Evenson, K.R., Rodriguez, D.A., and Salvesen, D. (2008) "A comparative case study on active transport to and from school", *Preventing Chronic Disease*, 5: 1–11.

FHA. (2008) *Highway Finance Data Collection*, Washington, DC: Office of Highway Policy Information, US Federal Highway Administration.

FHA. (2010) *Traffic Volume Trends*, Washington, DC: Federal Highway Administration, US Department of Transportation.

FHA. (2012) *Annual Vehicle-Miles Travel, 1980–2011*, Washington, DC: Federal Highway Administration, US Department of Transportation.

Filion, P. (2003) "Towards smart growth? The difficult implementation of alternatives to urban dispersion", *Canadian Journal of Urban Research*, 12(1): 48–70.

Filion, P. (2010) "Reorienting urban development? Structural obstruction to new urban forms", *International Journal of Urban and Regional Research*, 34(1): 1–19.

Filion, P. (2011) "Toronto's Tea Party: Right-wing population and planning agendas", *Planning Theory and Practice*, 12(3): 464–469.

Filion, P. and Kramer, A. (2011) "Metropolitan-scale planning in neo-liberal times: Financial and political obstacles to urban form transition", *Space and Polity*, 15(3): 197–212.

Filion, P., Bunting, T.E., and Curtis, K. (1996) *The Dynamics of the Dispersed City: Geographic and Planning Perspectives on Waterloo Region*, Waterloo, ON: Department of Geography, University of Waterloo.

Financial Express. (9 June 2013) "Reconditioned car traders' reactions to budget", Online. Available at: www.thefinancialexpress-bd.com/index.php?ref=MjBfMDZfMDlfMTNfMV8zXzE3MjI5NA (accessed 1 September 2013).

Fincham, B. (2006) "Bicycle messengers and the road to freedom", in S. Bohm, C. Land, C. Jones, and M. Paterson (eds) *Against Automobility*: 208–222, London: Sociological Review Monographs, Blackwell.

Finlayson, A. (2009) "Financialisation, financial literacy, and asset-based welfare", *British Journal of Politics and International Relations*, 11(3): 400–421.

Fischel, W.A. (2001) *The Homevoter Hypothesis: How Home Values Influence Local Government Taxation and Land-Use Policies*, Cambridge, MA: Harvard University Press.

Fishman, R. (1987) *Bourgeois Utopias: The Rise and Fall of Suburbia*, New York: Basic Books.

Fjellstrom, K. (2004) *Public Transport and Mass Rapid Transit in Dhaka*, Working paper 6, Strategic Transport Plan for Dhaka.

Fletcher, M.A. (3 September 2013) "Auto sales sizzle amid tepid economic recovery", *Washington Post*, Online. Available at: www.washingtonpost.com/business/economy/auto-sales-sizzle-amid-tepid-economic-recovery/2013/09/03/c9193990-14b9-11e3-a100-66fa8fd9a50c_story.html (accessed 4 September 2013).

Flink, J. (1975) *The Car Culture*, Cambridge, MA: MIT Press.

Flink, J. (1988) *The Automobile Age*, Cambridge, MA: MIT Press.

Florida, R. (2005) *Cities and the Creative Class*, New York: Routledge.

Florida, R. (2010) *The Great Reset: How New Ways of Living and Working Drive Post-Crash Prosperity*, New York: HarperCollins.

Florida, R. (2012) "Cities with denser cores do better", Online. Available at: www.theatlanticcities.com/jobs-and-economy/2012/11/cities-denser-cores-do-better/3911 (accessed 15 June 2013).

Foster, J.B. and Magdoff, F. (2009) *The Great Financial Crisis: Causes and Consequences*, New York: Monthly Review Press.

Fotel, T. and Thomsen, T.U. (2004) "The surveillance of children's mobility", *Surveillance and Society*, 1: 535–554.

Foucault, M. (1977) *Discipline and Punish: The Birth of the Prison* (trans. by Alan Sheridan), New York: Pantheon.

Foucault, M. (1980) *Power/Knowledge: Selected Interviews and Other Writings 1972–1977* (ed. by Colin Gordon), Brighton: Harvester.

Frank, L. and Pivo, G. (1994) "Impacts of mixed use and density on utilization of three modes of travel: Single-occupant vehicle, transit and walking", *Transportation Research Record*, 1466: 44–52.

Frank, L., Kerr, J., Chapman, J., and Sallis, J. (2007) "Urban form relationships with walk trip frequency and distance among youth", *American Journal of Health Promotion*, 21: 305–311.

Freeman, S. (2008) "Bicycle protest on Gardiner a 'crazy idea', police contend", Online. Available at: www.thestar.com/article/434942 (accessed 30 July 2010).

Freund, P. and Martin, G. (1993) *The Ecology of the Automobile*, Montreal: Black Rose.

Freund, P. and Martin, G. (1996) "The commodity that is eating the world: The automobile, the environment, and capitalism", *Capitalism Nature Socialism*, 7(4): 3–29.

Freund, P. and Martin, G. (2007) "Hyperautomobility, the social organization of space, and health", *Mobilities*, 2(1): 37–49.

Freund, P. and Martin, G. (2008) "Fast cars/fast foods: Hyperconsumption and its health and environmental consequences", *Social Theory & Health*, 6: 309–22.

Freund, P. and Martin, G. (2009) "The social and material culture of hyperautomobility", *Bulletin of Science, Technology & Society*, 29: 476–482.

Frey, W. (2012) *Demographic Reversal: Cities Thrive, Suburbs Sputter*, Washington, DC: Brookings Institution, State of Metropolitan America Series, No. 56.

Frey, W. and Berube, A. (2002) *City Families and Suburban Singles: An Emerging Household Story from Census 2000*, Washington, DC: Brookings Institution.

Frisken, F. and Wallace, M. (2008) *The Response of the Municipal Public Service Sector to the Challenge of Immigration Settlement*, Toronto: Centre for Urban and Community Studies, University of Toronto.

Frumkin, H. (2002) "Urban sprawl and public health", *Public Health Reports*, 117: 201–217.

Furness, Z. (2007) "Critical mass, urban space and velomobility", *Mobilities*, 2(2): 299–319.

Furness, Z. (2010) *One Less Car: Bicycling and the Politics of Automobility*, Philadelphia, PA: Temple University Press.

Furth, P.G. (2012) "Bicycling infrastructure for mass cycling: A transatlantic comparison", in J. Pucher and R. Buehler (eds) *City Cycling*: 105–139, Cambridge, MA: MIT Press.

Furuseth, O.J. and Smith, H.A. (2010) "Localized immigration policy: The view of Charlotte, North Carolina, a new immigrant gateway", in M.W. Varsanyi (ed.) *Taking Local Control: Immigration Policy Activism in U.S. Cities and States*: 173–192, Stanford, CA: Stanford University Press.

Fusco, C., Moola, F., Faulkner, G., Buliung, R., and Richichi, V. (2012) "Toward an understanding of children's perceptions of their transport geographies: (Non)Active school travel and visual representations of the built environment", *Journal of Transport Geography*, 20(1): 62–70.

Fyhri, A., Hjorthol, R., Mackett, R.L., Fotel, T.N., and Kyatta, M. (2011) "Children's active travel and independent mobility in four countries: Development, social contributing trends, and measures", *Transport Policy*, 18(5): 703–710.

Gainsborough, J. (2001) *Fenced Off: The Suburbanization of American Politics*, Washington, DC: Georgetown University Press.

Gallagher, R. (1992) *The Rickshaws of Bangladesh*, Dahka, Bangladesh: University Press Ltd.

Gans, H. (1967) *The Levittowners: Ways of Life and Politics in a New Suburban Community*, New York: Pantheon.

Garcia-López, M.S. and Muñiz, I. (2010) "Employment decentralisation: Polycentricity or scatteration? The case of Barcelona", *Urban Studies*, 47: 3035–3056.

Garder, P.E. (2004) "The impact of speed and other variables on pedestrian safety in Maine", *Accident Analysis and Prevention*, 36: 533–542.

Gargett, D. (2012) "Traffic growth: Modeling a global phenomenon", *World Transport Policy and Practice*, 18(4): 27–45.

Garnett, N.S. (2001) "The road from welfare to work: Informal transportation and the urban poor", *Harvard Journal on Legislation*, 38(1): 173–229.

Garreau, J. (1992) *Edge City: Life on the New Frontier*, New York: Anchor Books.

Garrett, M. and Taylor, B. (1999) "Reconsidering social equity in public transit", *Berkeley Planning Journal*, 13: 6–27.

Gartman, D. (2004) "Three ages of the automobile: The cultural logics of the car", *Theory, Culture and Society*, 21(4/5): 169–195.

Gazze, M. (11 May 2009) "Gardiner reopens after six hour shutdown due to protest", Online. Available at: www.cp24.com/servlet/an/local/CTVNews/20090510/090510_tamil_protest/20090510/?hub=CP24Home (accessed 30 July 2010).

Gee, M. (24 September 2013) "Ford's subway is crowded with caveats", *Globe and Mail*: A4.

Gehl, J. (2011) *Cities for People*, Washington, DC: Island Press.

Giddens, A. (1998) *The Third Way: The Renewal of Social Democracy*, Cambridge: Polity Press.

Giuliano, G. (2005) "Low income, public transit, and mobility", *Transportation Research Record*, 1927: 63–70.

Glackin, S., Trubka, R., Newman, P., Newton, P., and Mouritz, M. (2013) "Greening the greyfields: Trials, tools and tribulations of redevelopment in the middle suburbs", paper presented at Planning Institute of Australia National Conference, Canberra, April 2013.

Glaeser, E. (2010) *The Triumph of the City: How Our Greatest Invention Makes Us Richer, Smarter, Greener, Healthier and Happier*, London: Macmillan.

Glaeser, E. and Kahn, M. (2001) "Decentralized employment and the transformation of the American city", *Brookings-Wharton Papers on Urban Affairs*, 2: 1–63.

Gleeson, B. and Randolph, B. (2002) "Social disadvantage and planning in the Sydney context", *Urban Policy and Research*, 20(1): 101–107.

Goldring, L. and Landolt, P. (2013) *Producing and Negotiating Non-citizenship: Precarious Legal Status in Canada*, Toronto: University of Toronto Press.

Goldstein, L. (14 May 2009) "Protest backlash unearths racism", *Toronto Sun*: 22.

Goldthorpe, J.H., Lockwood, D., Bechhofer, F., and Platt, J. (1968) *The Affluent Worker: Political Attitudes and Behaviour*, Cambridge: Cambridge University Press.

Goodwin, P. (2012) "Peak travel, peak car and the future of mobility: Evidence, unresolved issues, policy implications, and a research agenda", Paris: International Transport Forum, Discussion Paper No. 2012–13.

Goodwin, P. and van Dender, K. (2013) "Peak car: Themes and issues", *Transport Reviews*, 33: 243–254.

Goonewardena, K. (2005) "The urban sensorium: Space, ideology, and the aestheticization of politics", *Antipode*, 35(1): 46–71.

Gordon, D. (14 September 2013) "GTA sprawl out of control", Online. Available at: www.thestar.com/opinion/commentary/2013/09/15/gta_sprawl_out_of_control.html (accessed 15 September 2013).

Gordon, P., Richardson, H., and Jun, M.-J. (1991) "The commuting paradox: Evidence from the top twenty", *Journal of the American Planning Association*, 57(4): 416–421.

Gorham, R. (2002) *Air Pollution from Ground Transportation*, New York: United Nations.

Gospodini, A. (2006) "Portraying, classifying and understanding the emerging landscapes in the post-industrial city", *Cities*, 23(5): 311–330.

Gotham, K.F. (2009) "Creating liquidity out of spatial fixity: The secondary circuit of capital and the subprime mortgage crisis", *International Journal of Urban and Regional Research*, 33(2): 355–371.

Gottlieb, P. and Lentnek, B. (2001) "Spatial mismatch is not always a central-city problem: An analysis of commuting behavior in Cleveland, Ohio, and its suburbs", *Urban Studies*, 38: 1161–1186.

Gowen, A. (2013) "India city's bike ban brings outcry", Online. Available at: www.thestar.com/news/world/2013/10/16/indian_citys_bike_ban_brings_outcry.html# (accessed 22 October 2013).

Graham, D. and Glaister, S. (2003) "Spatial variation in road pedestrian casualties: The role of urban scale, density and land use mix", *Urban Studies*, 8: 1591–1607.

Grant, K. (7 December 2010) "Don Cherry slams 'pinkos' in the 'left-wing media' during Ford inauguration", *Globe and Mail*: A1.

Green Leigh, N. and Hoelzel, N. (2012) "Smart growth's blind side: Sustainable cities need productive urban industrial land", *Journal of the American Planning Association*, 78(1): 87–103.

Grengs, J. (2002) "Community-based planning as a source of political change: The transit equity movement of Los Angeles' Bus Riders Union", *Journal of the American Planning Association*, 68(2): 165–178.

Grengs, J. (2010) "Job accessibility and the modal mismatch in Detroit", *Journal of Transport Geography*, 18(1): 42–54.

Greves, H.M., Lozano, P., Lenna, L., Busby, K., Cole, J., and Johnston, B. (2007) "Immigrant families' perceptions on walking to school and school breakfast: A focus group study", *International Journal of Behavioral Nutrition and Physical Activity*, 4: 64.

Grieco, M. and Urry, J. (eds) (2012) *Mobilities: New Perspectives on Transport and Mobility*, Farnham, UK: Ashgate.

Grove, A. and Burgelman, R. (2008) "An electric plan for energy resilience", *McKinsey Quarterly*, December: 1–6.

Gurran, N. (2008) "Affordable housing: A dilemma for metropolitan planning?", *Urban Policy and Research*, 26(1): 101–110.

Gusdorf, F. and Hallegatte, S. (2007) "Compact or spread-out cities: Urban planning, taxation, and the vulnerability to transportation shocks", *Energy Policy*, 35(10): 4826–4834.

Habermas, J. (1989) *The Structural Transformation of the Public Sphere: An Inquiry into a Category of Bourgeois Society* (trans. by Thomas Burger), Cambridge, MA: MIT Press.

Hahamovitch, C. (2003) "Creating perfect immigrants: Guestworkers of the world in historical perspective", *Labor History*, 44(1): 69–94.

Hall, P. (1996) *Cities of Tomorrow: An Intellectual History of Urban Planning and Design in the Twentieth Century* (updated edn), Cambridge, MA: Blackwell.

Hamer, M. (1987) *Wheels within Wheels: A Study of the Road Lobby*, London: Routledge.

Hamilton, J.D. (2009) "Causes and consequences of the oil shock of 2007–08", *Brookings Papers on Economic Activity*, 2009(1): 215–261.

Hamilton, K. and Jenkins, L. (2000) "A gender audit for public transport: A new policy tool in the tackling of social exclusion", *Urban Studies*, 37(10): 1793–1800.

Handy, S. (2005) "Smart growth and the transportation–land use connection: What does the research tell us", *International Regional Science Review*, 28: 146–147.

Handy, S. (2006) "The road less driven", *Journal of the American Planning Association*, 72(3): 274–278.

Hanlon, B. (2008) "The decline of older, inner suburbs in metropolitan America", *Housing Policy Debate*, 19(3): 423.

Hanlon, B. (2009) "A typology of inner-ring suburbs: Class, race, and ethnicity in U.S. suburbia", *City and Community*, 8(3): 221–246.

Hanson, S. (2010) "Gender and mobility: New approaches for informing sustainability", *Gender, Place & Culture*, 17(1): 5–23.

Hanson, S. and Giuliano, G. (eds) (2004) *The Geography of Urban Transportation* (3rd edn), New York: Guilford Press.

Hanson, S. and Pratt, G. (1988) "Reconceptualizing the links between home and work in urban geography", *Economic Geography*, 64(4): 299–321.

Hardt, M. and Negri, A. (2004) *Multitude: War and Democracy in the Age of Empire*, New York: Penguin.

Harvey, D. (1973) *Social Justice and the City*, Cambridge, MA: Blackwell.

Harvey, D. (1982) *Limits to Capital*, Chicago: University of Chicago Press.

Harvey, D. (1989) *The Condition of Post-Modernity: An Enquiry into the Origins of Cultural Change*, Oxford: Blackwell.

Harvey, D. (2005) *A Brief History of Neoliberalism*, Oxford: Oxford University Press.

Harvey, D. (2010) *The Enigma of Capital and the Crises of Capitalism*, London: Profile Books.

Harvey, D. (2012) *Rebel Cities: From the Right to the City to the Urban Revolution*, New York: Verso.

Harwood, D.M., Torbic, D.J., Gilmore, D.K., Bokenkoger, C.D., Dunn, J.M., Zegeer, C.V., Srinivansan, S., Carter, D., Raborn, C., Lyon, C., and Persaud, B. (2008) *Pedestrian Safety Prediction Methodology*, Washington, DC: Transportation Research Board.

Hasan, M.M.U. (2013) "Unjust mobilities: The case of rickshaw bans and restrictions in Dhaka", PhD thesis, Development Planning Unit, Bartlett Faculty of the Built Environment, University College London.

Hasan, M.M.U. and Dávila, J.D. (2012) "The politics of (im)mobility: Rickshaw bans in Dhaka, Bangladesh", paper presented at the 32nd International Geographic Conference, Cologne, Germany, August 2012.

Hayek, F. (1944) *The Road to Serfdom*, Abingdon: Routledge.

Hayek, F. (1945) "The use of knowledge in society", *American Economic Review*, 35(4): 519–530.

HDRC (Human Development Research Centre). (2004) *After Study on the Impact of Mirpur Demonstration Corridor Project (Gabtoli–Russel Square)*, Dhaka, Bangladesh: Human Development Research Centre and Dhaka Transport Coordination Board (DTCB), Government of Bangladesh.

He, D., Meng, F., Wang, M., and He, K. (2011) "Impacts of urban transportation mode split on CO_2 emissions in Jinan, China", *Energies*, 4: 685–699.

Heisz, A. and LaRochelle-Côté, S. (2005) *Work and Commuting in Census Metropolitan Areas, 1996–2001*, Statistics Canada – Catalogue No. 89-613-MIE, No. 007.

Heisz, A. and Schellenberg, G. (2004) "Public transit use among immigrants", *Canadian Journal of Urban Research*, 13(1): 170–191.

Hemily, B. (2004) *Trends Affecting Transit Effectiveness: A Review and Proposed Actions*, Washington, DC: American Public Transportation Association.

Henderson, J. (2006) "Secessionist automobility: Racism, anti-urbanism, and the politics of automobility in Atlanta, Georgia", *International Journal of Urban and Regional Research*, 30(2): 293–307.

Henderson, J. (2013) *Street Fight: The Politics of Mobility in San Francisco*, Amherst: University of Massachusetts Press.

Heron, C. and Penfold, S. (2006) "The craftmen's spectacle: Labour Day parades in Canada", in I. Radforth (ed.) *Canadian Working-Class History: Selected Readings*, Toronto: Canadian Scholars Press.

Hess, D. (2005) "Access to employment for adults in poverty in the Buffalo–Niagara Region", *Urban Studies*, 42(7): 1177–1200.

Hess, P. and Farrow, J. (2012) *Walkability in Toronto's High-rise Neighbourhoods*, Online. Available at: www.citiescentre.utoronto.ca/Assets/Cities+Centre+Centre+2013+Digital+Assets/Cities+

Centre/Cities+Centre+Digital+Assets/pdfs/publications/Walkability+Full+Report-2011Nov14-LowRes.pdf (accessed 12 May 2013).

Hickey, R., Lubell, J., Haas, P., and Morse, S. (2012) *Losing Ground: The Struggle of Moderate-Income Households to Afford the Rising Costs of Housing and Transportation*, Washington, DC: Center for Housing Policy.

Hiebert, D. (2000) "Immigration and the changing Canadian city", *Canadian Geographer*, 44(1): 25–43.

Hiebert, D. (2010) "Newcomers in the Canadian housing market", *Canadian Issues*, 8–15.

Hiebert, D., Germain, A., Murdie, M., Preston, V., Renaud, J., Rose, D., Wyly, E., Ferreira, V., Mendez, P., and Murnaghan, A.M. (2006) *The Housing Situation and Needs of Recent Immigrants in the Montréal, Toronto, and Vancouver CMAs: An Overview*, Ottawa, ON: Canada Mortgage and Housing Corporation.

Hine, J. (2011) "Mobility and transport disadvantage", in M. Grieco and J. Urry (eds) *Mobilities: New Perspectives on Transport and Society*: 21–40, Aldershot: Ashgate.

Hines, C. (2004) *A Global Look to the Local: Replacing Globalization with Democratic Localization*, London: International Institute for Environment and Development.

Hitchens, P. (2011) "This is a city built for a million people – but no one lives here: The Mongolian metropolis thrust into the 21st century in a storm of steel and concrete", Online. Available at: www.dailymail.co.uk/news/article-1391868/This-city-built-million-people-lives-here.html (accessed 13 October 2013).

Hollingworth, B., Mori, A., Cham, L., Passmore, D., Irwin, N., and Noxon, G. (2010) *Urban Transportation Indicators: Fourth Survey*, Online. Available at: www.tac-atc.ca/english/resourcecentre/readingroom/pdf/uti-survey4.pdf (accessed 1 May 2013).

Holzer, H.J. (1991) "The spatial mismatch hypothesis: What has the evidence shown?", *Urban Studies*, 28(1): 105–122.

Holzer, H.J. (1996) *What Employers Want: Job Prospects for Less-Educated Workers*, New York: Russell Sage Foundation.

Hoque, M.M., Khondokar, B., and Alam, M.J.B. (2005) "Urban transport issues and improvement options in Bangladesh", Proceedings of 2005 Canadian Transport Research Forum Conference, Hamilton, May 2005.

Horner, M. (2004) "Spatial dimensions of urban commuting: A review of major issues and their implications for future geographic research", *The Professional Geographer*, 56(2): 160–173.

Horton, D. (2006) "Environmentalism and the bicycle", *Environmental Politics*, 15(1): 41–58.

Hosken, A. (2008) *Ken: The Ups and Downs of Ken Livingstone*, London: Arcadia Books.

Hossain, S. (2008) "Rapid urban growth and poverty in Dhaka City", *Bangladesh e-Journal of Sociology*, 5(1): 1–24.

Hou, F. (2010) *Homeownership over the Life Course of Canadians: Evidence from Canadian Censuses of Population*, Ottawa: Statistics Canada, Cat. 11F0019M – No. 325, Online. Available at: www.statcan.gc.ca/pub/11f0019m/11f0019m2010325-eng.pdf (accessed 1 September 2013).

Houston, D. (2005) "Methods to test the spatial mismatch hypothesis", *Economic Geography*, 81(4): 407–434.

Huber, M.T. (2008) "Energizing historical materialism: Fossil fuels, space and the capitalist mode of production", *Geoforum*, 40(1): 105–115.

Huber, M.T. (2009) "The use of gasoline: Value, oil, and the 'American way of life'", *Antipode*, 41(3): 465–486.

Huber, M.T. (2013) *Lifeblood: Oil, Freedom and the Forces of Capital*, Minneapolis: University of Minnesota Press.

Huckfeldt, R. and Sprague, J. (1995) *Citizens, Politics, and Social Communication: Information and Influence in an Election Campaign*, Cambridge: Cambridge University Press.

Huckfeldt, R., Plutzer, E., and Sprague, J. (1993) "Alternative contexts of political behaviour: Churches, neighbourhoods, and individuals", *Journal of Politics*, 55(2): 365–381.

IEA. (2012a) *IEA Statistics 2012 Edition: CO2 Emissions from Fuel Consumption*, Paris: International Energy Agency.

IEA. (2012b) *World Energy Outlook*, Paris: International Energy Agency.

Ihlanfeldt, K.R. (1999) "The geography of economic and social opportunity in Metropolitan Areas", in A. Altshuler (ed.) *Governance and Opportunity in Metropolitan America*: 213–252, Washington, DC: National Academic Press.

Ihlanfeldt, K.R. and Sjoquist, D.L. (1998) "The spatial mismatch hypothesis: A review of recent studies and their implications for welfare reform", *Housing Policy Debate*, 9(4): 849–892.

Illich, I. (1974) *Energy and Equity*, London: Harper & Row.

Immergluck, D. (2009) *Foreclosed: High-risk Lending, Deregulation, and the Undermining of America's Mortgage Market*, Ithaca, NY: Cornell University Press.

Immergluck, D. (2010) "Neighborhoods in the wake of the debacle: Intrametropolitan patterns of foreclosed properties", *Urban Affairs Review*, 46: 3–36.

Issel, W. (1999) "Land values, human values, and the preservation of the city's treasured appearance: Environmentalism, politics, and the San Francisco freeway revolt", *Pacific Historical Review*, 68(4): 611–646.

Jacobs, J. (1961) *The Death and Life of Great American Cities*, New York: Modern Library.

Jamison, A., Eyerman, R., Cramer, J., and Laessoe, J. (1990) *The Making of the New Environmental Consciousness*, Edinburgh: Edinburgh University Press.

Jarvis, A. (2013) *Alejandro Rivera Marquez, the Latest Casualty*, Online. Available at: http://blogs.windsorstar.com/2013/05/14/alejandro-rivera-marquez-the-latest-casualty (accessed 15 May 2013).

Jarvis, H. (2003) "Dispelling the myth that preference makes practice in residential location and transport behaviour", *Housing Studies*, 18(4): 587–606.

Jarvis, H. (2005) *Work/Life City Limits*, New York: Palgrave Macmillan.

Jenkins, N.E. (2006) "'You can't wrap them up in cotton wool': Constructing risk in young people's access to outdoor play", *Health, Risk & Society*, 8(4): 379–393.

Johnson, K.M. (2009) "Captain Blake versus the Highwaymen: Or, how San Francisco won the freeway revolt", *Journal of Planning History*, 8(1): 56–83.

Johnson, M. (31 May 2008) "Wow, I'm shocked…", Weblog comment, "Cyclists shut down the Gardiner", Online. Available at: www.blogto.com/city/2008/05/cyclists_shut_down_the_gardiner (accessed 10 June 2010).

Johnston, J.D. (1997) *Driving America: Your Car, Your Government, Your Choice*, Washington, DC: AEI Press.

Johnston, R. and Pattie, C. (2000) "New Labour, new electoral system, new electoral geographies? A review of proposed constitutional changes in the United Kingdom", *Political Geography*, 19(4): 495–515.

Johnston, R. and Pattie, C. (2011) "Where did Labour's votes go? Valence politics and campaign effects at the 2010 British general election", *British Journal of Politics and International Relations*, 13(3): 283–303.

Johnston, R.J., Pattie, C.J., and Allsopp, G. (1988) *A Nation Dividing? The Electoral Map of Great Britain 1979–87*, Harlow, UK: Longman.

Johnston, R.J., Pattie, C.J., and Russell, A.T. (1993) "Dealignment, spatial polarisation and economic voting: An exploration of recent trends in British voting behaviour", *European Journal of Political Science*, 23(1): 67–90.

Jones, D.W. (1990) *California's Freeway Era in Historical Perspective*, Berkeley: Institute of Transportation Studies, University of California, Berkeley.

Jones, S.G. (1988) *Sport, Politics, and the Working Class*, Manchester: Manchester University Press.

Jones, T. and de Azevedo, L.N. (2013) "Economic, social and cultural transformation and the role of the bicycle in Brazil", *Journal of Transport Geography*, 30: 208–219.

Joshi, M.S. and MacLean, M. (1995) "Parental attitude to children's journeys to school", *World Transport Policy and Practice*, 1: 29–36.

Kain, J. (1968) "Housing segregation, Negro employment, and metropolitan decentralization", *Quarterly Journal of Economics*, 82(2): 175–197.

Katz, C. (1 November 2012) "Unequal exposure: People in poor, non-white neighborhoods breathe more hazardous particles", *Environmental Health News*, Online. Available at: www.environmentalhealthnews.org/ehs/news/2012/unequal-exposures.

Kaufmann, V. (2002) *Re-thinking Mobility: Contemporary Sociology*, Aldershot: Ashgate.

Kaufmann, V., Bergman, M., and Joye, D. (2004) "Motility: Mobility as capital", *International Journal of Urban and Regional Research*, 28(4): 745–756.

Kawabata, M. (2003) "Job access and employment among low-skilled autoless workers in US metropolitan areas", *Environment and Planning A*, 35(9): 1651–1668.

Kawabata, M. and Shen, Q. (2007) "Commuting inequality between cars and public transit: The case of the San Francisco Bay Area, 1990–2000", *Urban Studies*, 44(9): 1759–1780.

Kawahara, A. (1997) *The Origin of Competitive Strength: 50 Years of the Auto Industry in Japan and the US*, Kyoto: Kyoto University Press.

Keates, N. and Fowler, G. (16 March 2012) "The hot spot for the rising tech generation", *Wall Street Journal*.

Keil, R. (1994) "Global sprawl: Urban form after Fordism", *Environment and Planning D: Society and Space*, 12(1): 31–36.

Keil, R. (2002) "'Common-sense' neoliberalism: Progressive conservative urbanism in Toronto, Canada", *Antipode*, 34(3): 578–601.

Keller, L. (2009) *Triumph of Order: Democracy & Public Space in New York and London*, New York: Columbia University Press.

Kellerman, A. (2006) *Personal Mobilities*, Abingdon: Routledge.

Kellerman, A. (2012) *Daily Spatial Mobilities: Physical and Virtual*, Burlington, VT: Ashgate.

Kelly, J.F., Breadon, P., Mares, P., Ginnivan, L., Jackson, P., Gregson, J., and Viney, B. (2012) *Tomorrow's Suburbs*, Online. Available at: http://grattan.edu.au/static/files/assets/bb34b48d/167_tomorrows_suburbs.pdf (accessed 10 August 2013).

Kempton, R. (2007) *Provo: Amsterdam's Anarchist Revolt*, Brooklyn, NY: Autonomedia.

Kenworthy, J. and Laube, F. (2001) *The Millennium Cities Database for Sustainable Transport*, Perth and Brussels: ISTP Murdoch University and UITP Brussels.

Kenworthy, J. and Newman, P. (2001) *Melbourne in an International Comparison of Urban Transport Systems: A Report to the Department of Infrastructure, Melbourne as Part of the Melbourne Strategy*, Perth: Institute for Sustainability and Technology Policy.

Kenworthy, J., Laube, F., Newman, P., Barter, P., Raad, T., Poboon, C., and Guia, B. (1999) *An International Sourcebook of Automobile Dependence in Cities, 1960–1990*, Boulder: University Press of Colorado.

Ker, I. and Tranter, P. (2003) "A wish called wander: Reclaiming automobility from the motor car", in J. Whitelegg and G. Haq (eds) *Earthscan Reader on World Transport Policy and Practice*: 105–113, London: Earthscan.

Kern, L. (2010) *Sex and the Revitalized City: Gender, Condominium Development, and Urban Citizenship*, Vancouver: University of British Columbia Press.

Khabar South Asia. (2013) "To beat epic traffic, many Dhaka residents take to cycling", Online. Available at: http://khabarsouthasia.com/en_GB/articles/apwi/articles/features/2013/07/15/feature-06 (accessed 20 November 2013).

Khisty, C.J. (2003) "A systematic overview of non-motorized transportation for developing countries: An agenda for action", *Journal of Advanced Transportation*, 37(3): 273–293.

Khisty, J. and Ayvalik, C.K. (2003) "Automobile dominance and the tragedy of the land-use/transport system: Some critical issues", *Systematic Practice and Action Research*, 16(1): 53–73.

Khisty, J. and Zeitler, U. (2001) "Is hypermobility a challenge for transport ethics and systemicity?", *Systematic Practice and Action Research*, 14(5): 597–613.

Khon, M. (2013) "Privatization and protest: Occupy Wall Street, Occupy Toronto, and the occupation of public space in a democracy", *Perspectives on Politics*, 11(1): 99–110.

Kifer, K. (2002) *Auto Costs versus Bike Costs*, Online. Available at: www.phred.org/~alex/kenkifer/www.kenkifer.com/bikepages/advocacy/autocost.htm (accessed 10 July 2013).

Kim, S. (2009) "Immigrants and transportation: An analysis of immigrant workers' work trips", *Cityscape*, 11(3): 155–169.

Kirby, J. and Inchley, J. (2009) "Active travel to school: Views of 10–13 year old schoolchildren in Scotland", *Health Education*, 109: 169–183.

Kirsic, R. (2010) "Toronto needs to revisit cycling licenses", Online. Available at: http://thestar.

blogs.com/yourcitymycity/2010/08/toronto-needs-to-revisit-cycling-licences.html (accessed 30 August 2010).

Kneebone, E. (2009) *Job Sprawl Revisited: The Changing Geography of Metropolitan Employment*, Washington, DC: Brookings Institution.

Knox, P. (2008) *Metroburbia, USA*, New Brunswick, NJ: Rutgers University Press.

Knox, P. (1991) "The restless urban landscape: Economic and sociocultural change and the transformation of metropolitan Washington, DC", *Annals of the Association of American Geographers*, 81(2): 181–209.

Koeppel, D. (2005) *Invisible Riders*, Online. Available at: www.utne.com/2006-07-01/InvisibleRiders.aspx (accessed 18 June 2013).

Kolenda, R. and Liu, C.Y. (2012) "Are central cities more creative? The intrametropolitan geography of creative industries", *Journal of Urban Affairs*, 34: 487–512.

Kothari, R. (1993) *Growing Amnesia: An Essay on Poverty and Human Consciousness*, New Delhi: Viking.

Kramer, A. (2013) *Divergent Affordability: Transit Access and Housing in North American Cities*, unpublished PhD Thesis, University of Waterloo, Ontario.

Krippner, G. (2005) "The financialization of the American economy", *Socio-Economic Review*, 3: 173–208.

Kunstler, J.H. (1993) *The Geography of Nowhere: The Rise and Decline of America's Man-Made Landscape*, New York: Simon & Schuster.

Kunstler, J.H. (2005) *The Long Emergency*, New York: Grove Press.

Labban, M. (2010) "Oil in parallax: Scarcity, markets, and the financialization of accumulation", *Geoforum*, 41(4): 541–552.

Lang, D., Collins D., and Kearns, R. (2011) "Understanding modal choice for the trip to school", *Journal of Transport Geography*, 19: 509–514.

Lang, R., LeFurgy, J., and Nelson, A.C. (2006) "The six suburban eras of the United States", *Opolis*, 2(1): 65–72.

Larsen, J., Urry, J., and Axhausen, K. (2006) *Mobilities, Networks, Geographies*, Aldershot: Ashgate.

Larsen, K., Gilliland, J., and Hess, P.M. (2012) "Route based analysis to capture the environmental influences on a child's mode of travel between home and school", *Annals of the Association of American Geographers*, 102: 1–18.

Larsen, K., Gilliland, J., Hess, P., Tucker, P., Irwin, J., and He, M. (2009) "Identifying influences of physical environments and socio-demographic characteristics on a child's mode of travel to and from school", *American Journal of Public Health*, 99: 520–526.

LaScala, E.A., Gruenewald, P.J., and Johnson, F.W. (2004) "An ecological study of the locations of schools and child pedestrian injury collisions", *Accident Analysis and Prevention*, 36: 569–576.

Lassiter, M. (2006) *The Silent Majority: Suburban Politics in the Sunbelt South*, Princeton, NJ: Princeton University Press.

Leahy, S. (2013) "Canada's environmental activists seen as 'threat to national security'", *Guardian*, Online. Available at: www.guardian.co.uk/environment/2013/feb/14/canada-environmental-activism-threat (accessed 14 February 2013).

Lees, L., Slater, T., and Wyly, E. (2007) *Gentrification*, New York: Routledge.

Lefebvre, H. (1996) "The right to the city", in *Writings on Cities*, Oxford: Blackwell [originally published 1967].

Lefebvre, H. (2003) *The Urban Revolution* (English translation by Robert Bonono), Minneapolis: University of Minnesota Press [originally published 1970].

Lefebvre, H. (2004) *Rhythmanalysis: Space, Time and Everyday Life* (English translation by S. Elden and G. Moore), London: Continuum [originally published 1992].

Leinberger, C. (2007) *The Option of Urbanism: Investing in a New American Dream*, Washington, DC: Island Press.

Leitner, H., Sheppard, E.S., Sziarto, K., and Maringanti, A. (2007) "Contesting urban futures: Decentering neoliberalism", in H. Leitner, J. Peck, and E.S. Sheppard (eds) *Contesting Neoliberalism: Urban Frontiers*: 1–26, London: Guilford Press.

Lemke, T. (2001) "'The birth of bio-politics': Michel Foucault's lecture at the College de France on neo-liberal governmentality", *Economy and Society*, 30(2): 190–207.

Lenarcic, E. (14 May 2009) "Letter to the Editor: The rights of others", *Toronto Sun*: 21.

Le Vine, S. and Jones, P. (2012) *On the Move: Making Sense of Car and Train Travel Trends in Britain*, London: RAC Foundation.

Le Vine, S., Lee-Gosselin, M., Sivakuar, A., and Polak, J. (2013) "A new concept of accessibility to personal activities: Development of theory and application to an empirical study of mobility resource holdings", *Journal of Transport Geography*, 11: 1–10.

Levinson, D.M. and Wu, Y. (2005) "The rational locator reexamined: Are travel times still stable?", *Transportation*, 32: 187–202.

Levitt, K.P. (2013) *From the Great Transformation to the Great Financialization*, Winnipeg: Fernwood.

Lewis, T. (2013) *Divided Highways: Building the Interstate Highways, Transforming American Life*, Ithaca, NY: Cornell University Press.

Ley, D. (1996) *The New Middle Class and the Remaking of the Central City*, Oxford: Oxford University Press.

Limtanakool, N., Dijst, M., and Schwanen, T. (2006) "The influence of socioeconomic characteristics, land use and travel time considerations on mode choice for medium- and longer-distance trips", *Journal of Transport Geography*, 14(5): 327–341.

Litman, T. (2003) *Social Inclusion as a Transport Planning Issue in Canada*, Victoria, BC: Victoria Transport Policy Institute.

Litman, T. (2004) *Quantifying the Benefits of Non-motorized Transportation for Achieving Mobility Management Objectives*, Victoria Transport Policy Institute, Online. Available at: http://onestreet.org/documents/victorian_000.pdf (accessed 15 July 2013).

Litman, T. (2012) *Transportation Land Valuation: Evaluating Policies and Practices That Affect the Amount of Land Devoted to Transportation Facilities*, Victoria, BC: Victoria Transport Policy Institute.

Litman, T. (2013) *The Future Isn't What It Used to Be: Changing Trends and Their Implications for Transport Planning*, Victoria, BC: Victoria Transport Policy Institute.

Litman, T. and the Victoria Transport Policy Institute (VTPI). (2009) *Transportation Cost and Benefit Analysis: Techniques, Estimates and Implications (Transportation Cost Analysis Spreadsheet)*, Victoria, BC: Victoria Transport Policy Institute. Online. Available at: www.vtpi.org/tca (accessed 20 November 2013).

Liu, C.Y. and Painter, G. (2012a) "Immigrant settlement and employment suburbanisation in the U.S.: Is there a spatial mismatch?", *Urban Studies*, 49(5): 979–1002.

Liu, C.Y. and Painter, G. (2012b) "Travel behaviour among Latino immigrants: The role of ethnic concentration and ethnic employment", *Journal of Planning Education Research*, 32(1): 62–80.

Lo, L., Shalaby, A., and Alshalalfah, B. (2011) "Relationship between immigrant settlement patterns and transit use in the Greater Toronto Area", *Journal of Urban Planning and Development*, 137(4): 470–476.

Lomasky, L. (1997) "Autonomy and automobility", *The Independent Review*, 2(1): 5–28.

Lovejoy, K. and Handy, S. (2011) "Social networks as a source of private-vehicle transportation: The practice of getting rides and borrowing vehicles among Mexican immigrants in California", *Transportation Research Part A: Policy and Practice*, 45(4): 248–257.

Low, N. (2003) "Is urban transport sustainable?" in N. Low and B. Gleeson (eds) *Making Urban Transport Sustainable*: 1–21, New York: Palgrave Macmillan.

Low, N. and Gleeson, B. (2001) "Ecosocialization or countermodernisation? Reviewing the shifting 'storylines' of transport planning", *International Journal of Urban and Regional Research*, 25(4): 784–803.

Low, N. and Gleeson, B. (2003) *Making Urban Transport Sustainable*, New York: Palgrave Macmillan.

Lozano, R., Naghavi, M., and Foreman, K. (2013) "Global and regional mortality from 235 causes of death for 20 age groups in 1990 and 2010: A systematic analysis for the Global Burden of Disease Study 2010", *The Lancet*, 380: 2095–2128.

LTA. (2011) "Passenger transport mode share in world cities", in *Journeys: Sharing Urban Transport Solutions*, Singapore: Land Transport Authority.

Lucas, K. (ed.) (2004) *Running on Empty: Transport Social Exclusion and Environmental Justice*, Bristol: Policy Press.

Lucas, K. (2012) "Transport and social exclusion: Where are we now?", *Transport Policy*, 20: 105–113.

Lucas, K., Blumenberg, E., and Weinberger, R. (eds) (2011) *Auto Motives: Understanding Car Use Behaviours*, Bingley, UK: Emerald.

Lucy, W.H. and Phillips, D.L. (1997) "The post-suburban era comes to Richmond: City decline, suburban transition and exurban growth", *Landscape and Urban Planning*, 36(4): 259–275.

Lysandrou, P. (2011) "Global inequality, wealth concentration and the subprime crisis: A Marxian commodity theory analysis", *Development and Change*, 42(1): 183–208.

Ma, K.-R. and Kang, E.-T. (2011) "Time–space convergence and urban decentralization", *Journal of Transport Geography*, 19: 606–614.

Ma, L. and Srinivasan, S. (2010) "Impact of individuals' immigrant status on household auto ownership", *Transportation Research Record: Journal of the Transportation Research Board*, 2156(1): 36–46.

McCabe, P. (1998) "Energy resources: Cornucopia or empty barrel?", *AAPG Bulletin*, 82(11): 2110–2134.

McCann, B. and Ewing, R. (2003) *Measuring the Health Effects of Sprawl: A National Analysis of Physical Activity, Obesity and Chronic Disease*, Washington, DC: Smart Growth America.

McConnell, D. (2008) "The Tamil people's right to self-determination", *Cambridge Review of International Affairs*, 21(1): 59–76.

McDonald, N.C. and Aalborg, A.E. (2009) "Why parents drive children to school", *Journal of the American Planning Association*, 75: 331–342.

McDonald, N.C., Brown, A.L., Marchetti, L.M., and Pedroso, M.S. (2011) "U.S. school travel, 2009: An assessment of trends", *American Journal of Preventive Medicine*, 41: 146–151.

Macer, D.R.J. (2006) *A Cross-cultural Introduction to Bioethics*, Tsukuba: Eubios Ethics Institute.

McGirr, L. (2001) *Suburban Warriors: The Origins of the New American Right*, Princeton, NJ: Princeton University Press.

McGuckin, N. and Murakami, E. (1999) "Examining trip-chaining behavior: Comparison of travel by men and women", *Transportation Research Record*, 1693: 79–85.

McIntosh, J., Newman, P., and Glazebrook, G. (2013) "Why fast trains work: An assessment of a fast regional rail system in Perth, Australia", *Journal of Transportation Technologies*, 3: 37–47.

McIntosh, J., Newman, P., Trubka, R., and Kenworthy, J. (forthcoming) "Framework for land value capture from transit in car dependent cities", *Journal of Land Use and Transport Planning*, accepted.

McKenzie, E. (1994) *Privatopia: Homeowner Associations and the Rise of Residential Private Government*, New Haven, CT: Yale University Press.

McKinsey Global Institute. (2012) *Debt and Deleveraging: Uneven Progress on the Path to Growth*, Washington, DC: McKinsey Global Institute.

McLafferty, S. and Preston, V. (1996) "Spatial mismatch and employment in a decade of restructuring", *Professional Geographer*, 48(4): 417–467.

McMillan, T.E. (2007) "The relative influence of urban form on a child's travel mode to school", *Transportation Research Part A*, 41: 69–79.

McNally, D. (2006) "Democracy against capitalism: The revolt of the dispossessed", in *Another World Is Possible: Globalization and Anti-capitalism*: 267–335, Winnipeg: Arbeiter Ring Publishing.

McNally, D. (2009) "From financial crisis to world-slump: Accumulation, financialisation, and the global slowdown", *Historical Materialism*, 17(1): 35–83.

McNeill, D. and While, A. (2001) "The new urban economies", in R. Paddison (ed.) *Handbook of Urban Studies*: 295–307, Thousand Oaks, CA: Sage.

McQuaid, R.W. and Chen, T. (2012) "Commuting times: the role of gender, children and part-time work", *Research in Transportation Economics*, 34(1): 66–73.

McShane, C. (1994) *Down the Asphalt Road: The Automobile and the American City*, New York: Columbia University Press.

Madre, J.-L., Collet, R., Villareal, I., and Bussiere, Y. (2012) "Are we heading towards a reversal of the trend for ever-greater mobility?", International Transport Forum, Discussion Paper No. 2012–16.

Magalhaes, L., Carrasco, C., and Gastaldo, D. (2010) "Undocumented migrants in Canada: A scope literature review on health, access to services, and working conditions", *Journal of Immigration and Minority Health*, 12: 132–151.

Mah, J. and Hackworth, J. (2011) "Local politics and inclusionary housing in three large Canadian cities", *Canadian Journal of Urban Research*, 20(1): 57–80.

Mahmud, S.M.S. and Hoque, S.M. (2012) "Management of rickshaw in Dhaka City for ensuring desirable mobility and sustainability: The problems and options", paper presented at Conference CODATU XV, Addis Ababa, October 2012.

Mahmud, S.M.S., Hossain, M.H., Hoque, M., and Hoque, M.H. (2006) "Pedestrian safety problems, existing facilities and required strategies in the context of Dhaka Metropolitan city", Proceedings of the International Conference on Road Safety in Developing Countries, Bangladesh University of Engineering and Technology, Dhaka, Bangladesh.

Mahoney, J. (19 August 2010) "Rob Ford and a decade of controversy", *Globe and Mail*.

Major, A. (2012) "Neoliberalism and the new international financial architecture", *Review of International Political Economy*, 19(4): 536–561.

Mandel, M. (14 May 2009) "Passion can't be denied", *Toronto Sun*: 4.

Maniruzzaman, K.M. and Mitra, R. (2005) "Road accidents in Bangladesh", *IATSS Research*, 29(2): 71–73.

Maoh, H. and Tang, Z. (2012) "Determinants of normal and extreme commute distance in a sprawled midsize Canadian city: Evidence from Windsor, Canada", *Journal of Transport Geography*, 25: 50–57.

Mapes, J. (2009) *Pedaling Revolution: How Cyclists Are Changing American Cities*, Corvallis: Oregon State University Press.

Marchetti, C. (1994) "Anthropological invariants in travel behaviour", *Technical Forecasting and Social Change*, 47(1): 75–78.

Marcuse, P. and van Kempen, R. (2000) *Globalizing Cities*, Oxford: Blackwell.

Marinucci, C. (19 March 2012) "Republicans seek inroads in liberal San Francisco", *San Francisco Chronicle*: A1.

Marlow, I., Stanc, H., and Nicole, B. (10 May 2009) "Tamil protest moves off Gardiner to Queen's Park", *Toronto Star*, Online. Available at: www.thestar.com/news/gta/article/632136 (accessed 30 June 2010).

Marks, P. (1990) *Bicycles, Bangs and Bloomers: The New Women in the Popular Press*, Lexington: University of Kentucky Press.

Martin, G. (1999) *Hyperautomobility and Its Social-Material Impacts*, Guildford: Centre for Environmental Strategy, University of Surrey, Working Paper Series.

Martin, G. (2007) "Global motorization, social ecology, and China", *Area*, 39: 66–73.

Martin, G. (2009) "The global intensification of motorization and its impacts on urban social ecologies", in J. Conley and A. McLaren (eds) *Car Troubles: Critical Studies of Automobility and Automobility*, Farnham: Ashgate.

Martin, R.L. (1988) "Industrial capitalism in transition: The contemporary reorganization of the British space-economy", in D. Massey and J. Allen (eds) *Uneven Re-development*: 202–231, London: Hodder & Stoughton.

Martin, R.W. (2004) "Spatial mismatch and the structure of American metropolitan areas, 1970–2000", *Journal of Regional Science*, 44(3): 467–488.

Marx, K. (1972) *Capital: A Critique of Political Economy, Vol. 1*, London: Dent [originally published 1884].

Mathieu, E. and Yutangco, P. (27 April 2009) "Tamil protest closes section of University Ave", *Toronto Star*, Online. Available at: www.thestar.com/news/gta/article/624852 (accessed 30 August 2010).

Mees, P. (2001) *A Very Public Solution: Transport in the Dispersed City*, Melbourne: University of Melbourne Press.

Mees, P. (2009) *Transport for Suburbia: Beyond the Automobile Age*, London: Earthscan.

Menchetti, P. (2005) *Cycle Rickshaws in Dhaka Bangladesh, Social Movements and Collective Action*, Rosanne Rutten, Universiteit van Amsterdam Collegekaartnummer, the Netherlands.

Mensah, J. (1995) "Journey to work and job search characteristics of the urban poor: A gender analysis of a survey data from Edmonton, Alberta", *Transportation*, 22: 1–19.

Mercado, R.G., Paez, A., Farber, S., Roorda, M.J., and Morency, C. (2012) "Explaining transport mode use of low-income persons for journey to work in urban areas: A case study of Ontario and Quebec", *Transportmetrica*, 8: 157–179.

Merriman, P. (2012) *Mobility, Space, and Culture*, Abingdon: Routledge.

Metz, D. (2010) "Saturation of demand for daily travel", *Transport Reviews*, 30(5): 659–674.

Metz, D. (2013) "Peak car and beyond: The fourth era of travel", *Transport Reviews*, 1–16.

Midgley, P. (2011) *Bicycle-sharing Schemes: Enhancing Sustainable Mobility in Urban Areas*, New York: United Nations Department of Economics and Social Affairs, Background Paper No. 8, Online. Available at: www.un.org/esa/dsd/resources/res_pdfs/csd-19/Background-Paper8-P. Midgley-Bicycle.pdf (accessed 30 July 2013).

Millard-Ball, A. and Schipper, L. (2010) "Are we reaching peak travel? Trends in passenger transport in eight industrialized countries", *Transport Reviews*, 31(3): 1–22.

Miller, B. and Nicholls, W. (2013) "Social movements in urban society: The city as a space of politicization", *Urban Geography*, 34(4): 452–473.

Miller, D. (ed.) (2001) *Car Cultures*, Oxford: Berg.

Miller, E.J. and Shalaby, P.E. (2003) "Evolution of personal travel in Toronto area and policy implications", *Journal of Urban Planning and Development*, 129(1): 1–26.

Miller, W.L. (1977) "Social class and party choice in England: A new analysis", *British Journal of Political Science*, 8(3): 257–284.

Mindali, O., Raveh, A., and Saloman, I. (2004) "Urban density and energy consumption: A new look at old statistics", *Transportation Research Part A*, 38: 143–162.

Mitchell, D. (2003) *The Right to the City: Social Justice and the Fight for Public Space*, New York: Guilford Press.

Mitchell, D. (2005) "The S.U.V. model of citizenship: Floating bubbles, buffer zones, and the rise of the 'purely atomic' individual", *Political Geography*, 24(1): 77–100.

Mitchell, H., Kearns, R.A., and Collins, D.C.A. (2007) "Nuances of neighbourhood: Children's perceptions of the space between home and school in Auckland, New Zealand", *Geoforum*, 38: 614–627.

Montgomerie, J. (2006) "The financialization of the American credit card industry", *Competition and Change*, 10(3): 301–319.

Montgomerie, J. (2007) "The logic of neoliberalism and the political economy of consumer debt-led growth", in S. Lee and S. McBride (eds) *Neo-liberalism, State Power, and Global Governance*: 157–172, Dordrecht: Springer.

Moore, O., Morrow, A., and Rogers, K. (13 July 2013) "Rob Ford now supportive of tax hike; Calling it an 'investment,' the mayor will accept a 0.25-per-cent property tax increase to fund a subway for Scarborough", *Globe and Mail*: A6.

Moos, M. (2012) *Housing and Location of Young Adults, Then and Now: Consequences of Urban Restructuring in Montréal and Vancouver*, PhD thesis, Department of Geography, University of British Columbia. Online. Available at: https://circle.ubc.ca/handle/2429/41101.

Moos, M. and Skaburskis, A. (2010) "The globalization of urban housing markets: Immigration and changing housing demand in Vancouver", *Urban Geography*, 31(6): 724–749.

Moriarty, P. and Honnery, D. (2008) "Low-mobility: The future of transport", *Futures*, 40: 865–872.

MOUD (Ministry of Urban Development). (2008) *Study on Traffic and Transportation Policies and Strategies in Urban Areas in India*, final report, Delhi: Government of India.

Mountz, A. (2011) "The enforcement archipelago: Detention, haunting, and asylum on islands", *Political Geography*, 30(3): 118–128.

Mueller, B.A., Rivara, F.P., Lii, S.M., and Weiss, N.S. (1990) "Environmental factors and the risk for childhood pedestrian–motor vehicle collision occurrence", *American Journal of Epidemiology*, 132: 550–560.

Mukhija, V., Regus L., Slovin, S., and Das, A. (2010) "Can inclusionary zoning be an effective and efficient housing policy? Evidence from Los Angeles and Orange Counties", *Journal of Urban Affairs*, 32(2): 229–252.

Mumford, L. (1961) *The City in History*, Harmondsworth: Penguin.

Murdie, R. and Teixeira, C. (2003) "Towards a comfortable neighbourhood and appropriate housing: Immigrant experiences in Toronto", in M. Lanphier and P. Anisef (eds) *World in a City*: 132–191, Toronto: University of Toronto Press.

Murdie, R. and Teixeira, C. (2006) "Urban social space", in T. Bunting and P. Filion (eds) *Canadian Cities in Transition: Local through Global Perspectives*: 154–170, Don Mills, ON: Oxford University Press.

Murphy, L. (2011) "The global financial crisis and the Australian and New Zealand housing markets", *Journal of Housing and the Built Environment*, 26(3): 335–351.

Murphy, T.P. and Rehfuss, J. (1976) *Urban Politics in the Suburban Era*, Homewood, IL: Dorsey Press.

Murray, L. (2009) "Making the journey to school: The gendered and generational aspects of risk in constructing everyday mobility", *Health, Risk & Society*, 11: 471–486.

Muth, R. (1969) *Cities and Housing*, Chicago: University of Chicago Press.

Myers, D. (1997) *Changes over Time in Transportation Mode for Journey to Work: Effects of Aging and Immigration – Decennial Census Data for Transportation Planning: Case Studies and Strategies for 2000*, Vol. 2, Washington, DC: Case Studies, Transportation Research Board.

Nadal, L. (2007) "Bike sharing sweeps Paris off its feet", *Sustainable Transport*, 19: 8–12.

Naess, P., Naess, T., and Strand, A. (2009) "Oslo's farewell to urban sprawl", *European Planning Studies*, 19: 113–139.

Nair, H.V. (2013) "Cycle rickshaws may be back on Delhi main roads", *Hindustan Times*, Online. Available at: www.hindustantimes.com/India-news/NewDelhi/Cycle-rickshaws-may-be-back-on-Delhi-main-roads/Article1-1065516.aspx (accessed 13 October 2013).

National Center for Safe Routes to School (NCSRS). (2010) *Personal Security and Safe Routes to School*, Online. Available at: www.safekidscanada.ca/Professionals/Documents/33563-PersonalSecuritySafeRoutesToSchool.pdf (accessed 26 August 2012).

NBS. (2011) *Statistical Data*, Beijing: National Bureau of Statistics of China, Online. Available at: www.stats.gov.cn/english/statisticaldata (accessed 28 May 2013).

Neff, J.W. (1996) *Substitution Rates between Transit and Automobile Travel*, paper presented at the Association of American Geographers Annual Meeting, Charlotte, North Carolina, April.

Newman, C.E. and Newman, P.W.G. (2006) "The car and culture", in P. Beilhartz and T. Hogan (eds) *Sociology: Place, Time and Division*, Oxford: Oxford University Press.

Newman, P. (1995) "The end of the urban freeway", *World Transport Policy and Practice*, 1(1): 12–19.

Newman, P. and Kenworthy, J. (1989) *Cities and Automobile Dependence: A Sourcebook*, Aldershot: Gower.

Newman, P. and Kenworthy, J. (1999) *Sustainability and Cities: Overcoming Automobile Dependence*, Washington, DC: Island Press.

Newman, P. and Kenworthy, J. (2006) "Urban design to reduce automobile dependence: How much development will make urban centres viable?", *Opolis*, 2(1): 35–52.

Newman, P. and Kenworthy, J. (2011a) "Evaluating the transport sector's contribution to greenhouse gas emissions and energy consumption", in R. Salter, S. Dhar, and P. Newman (eds) *Technologies for Climate Change Mitigation: Transport Sector*, Roskilde: Riso Centre on Energy, Climate and Sustainable Development.

Newman, P. and Kenworthy, J. (2011b) "Peak car use: Understanding the demise of automobile dependence", *World Transport Policy and Practice*, 17: 31–42.

Newman, P. and Matan, A. (2013) *Green Urbanism in Asia*, Singapore: World Scientific Publications.

Newman, P., Beatley, T., and Boyer, H. (2009) *Resilient Cities: Responding to Peak Oil and Climate Change*, Washington, DC: Island Press.

Newman, P., Kenworthy, J., and Glazebrook, G. (2008) "How to create exponential decline in car use in Australian cities", *Australian Planner*, 45(3): 17–19.

Newman, P., Kenworthy, J., and Glazebrook, G. (2013) "Peak car and the rise of global rail", *Journal of Transportation Technologies*, 3(4): 272–287.

Newton, P., Newman, P., Glackin, S., and Trubka, R. (2012) "Greening the greyfields: Unlocking the development potential of middle suburbs in Australian cities", *World Academy of Science, Engineering and Technology*, 71: 138–157.

Ng, W.-S., Chen, Y., and Schipper, L. (2010) "China motorization trends", *Journal of Transportation and Landuse*, 3(3): 5–25.

Nicolaides, B. (2006) "How Hell moved from the city to the suburb: Urban scholars and changing perceptions of authentic community", in K.M. Kruse and T. Sugrue (eds) *The New Suburban History*, Chicago: University of Chicago Press.

Norcliffe, G. (2001) *The Ride to Modernity: The Bicycle in Canada, 1869–1900*, Toronto: University of Toronto Press.

Norcliffe, G. (2011) "Neoliberal mobility and its discontents: Working tricycles in China's cities", *City, Culture and Society*, 2: 235–242.

Nordlinger, E.A. (1967) *The Working Class Tories*, London: MacGibbon & Kee.

Norman, J., MacLean, H., and Kennedy, C. (2006) "Comparing high and low residential density: Life-cycle analysis of energy use and greenhouse gas emissions", *Journal of Urban Planning and Development*, 132: 10–21.

North, P. (2010) "Eco-localisation as a progressive response to peak oil and climate change: A sympathetic critique", *Geoforum*, 41(4): 585–594.

Norton, P.D. (2008) *Fighting Traffic: Dawn of the Motor Age in the American City*, Cambridge, MA: MIT Press.

Ohnmacht, T., Maksim, H., and Bergman, M.M. (2009) *Mobilities and Inequality*, Farnham: Ashgate.

Oldenziel, R. and de la Bruhèze, A.A. (2011) "Contested spaces: Bicycle lanes in urban Europe, 1900–1995", *Transfers*, 1(2): 29–49.

Oliver, J.E. (2001) *Democracy in Suburbia*, Princeton, NJ: Princeton University Press.

Ong, P. and Blumenberg, E. (1998) "Job access, commute and travel burden among welfare recipients", *Urban Studies*, 35: 77–93.

Owen, D. (18 October 2004) "Green Manhattan: Why New York is the greenest city in the US", *The New Yorker*.

Pabon Lopez, M. (2004) "More than a license to drive: State restrictions on the use of driver's licenses by noncitizens", *Southern Illinois University Law Journal*, 29(1): 91–128.

Packer, J. (2008) *Mobility without Mayhem: Safety, Cars and Citizenship*, Durham, NC: Duke University Press.

Paez, A., Mercado, R.G., Farber, S., Morency, C., and Roorda, M. (2009) "Relative accessibility deprivation indicators for urban settings: Definitions and application to food deserts in Montreal", *Urban Studies*, 47(7): 1415–1438.

Paine, C. (2006) *Who Killed the Electric Car?*, Culver City, CA: Sony Pictures (motion picture).

Panitch, L. and Gindin, S. (2012) *The Making of Global Capitalism: The Political Economy of American Empire*, New York: Verso.

Parks, V. (2004) "Access to work: The effects of spatial and social accessibility on unemployment for native-born Black and immigrant women in Los Angeles", *Economic Geography*, 80(2): 141–172.

Passel, J.S. (2006) *The Size and Characteristics of the Unauthorized Migrant Population in the US*, Washington, DC: Pew Hispanic Center, Online. Available at: www.pewhispanic.org (accessed 26 September 2013).

Paterson, M. (2007) *Automobile Politics: Ecology and Cultural Political Economy*, Cambridge: Cambridge University Press.

Paul, C. and Miranda, M. (2011) "Air pollution and global public health", in R. Parker and M. Sommer (eds) *Routledge Handbook in Global Public Health*, New York: Routledge.

PEAC. (2011) *Air Quality: A Follow-up Report*, London: Parliamentary Environmental Audit Committee, Online. Available at: www.publications.parliament.uk/pa/cm2010 (accessed 24 June 2013).

Peck, J. (2010) *Constructions of Neoliberal Reason*, Oxford: Oxford University Press.

Peck, J. (2011) "Neoliberal suburbanism: Frontier space", *Urban Geography*, 32(6): 884–919.

Pelligrino, G. (2011) *The Politics of Proximity: Mobility and Immobility in Practice*, Farnham, UK: Ashgate.

Perry, J.A. (2012) "Barely legal: Racism and migrant farm labour in the context of Canadian multiculturalism", *Citizenship Studies*, 16(2): 189–201.

Peters, D. (1999) "Gender issues in transportation: A short introduction", presentation notes for the UNEP "Deals on Wheels" Seminar, San Salvador, 28–30 July.

Phillips, K.P. (1969) *The Emerging Republican Majority*, New Rochelle, NY: Arlington House.

Pinker, S. (2011) *The Better of Our Angels: Why Violence Has Declined*, London: Viking.

Pisarski, A. (2006) *Commuting in America III: The Third National Report on Commuting Patterns and Trends*, Washington, DC: Transportation Research Board, Online. Available at: http://onlinepubs.trb.org/onlinepubs/nchrp/ciaiii.pdf (accessed 23 October 2013).

Plaut, P. (2004) "Non-commuters: The people who walk to work or work at home", *Transportation*, 31: 229–255.

Plaut, P. (2006) "The intra-household choices regarding commuting and housing", *Transportation Research Part A*, 40: 561–571.

Plumert, J.M., Kearney, J.K., and Cremer, J.F. (2004) "Children's perception of gap affordances: Bicycling across traffic-filled intersections in an immersive virtual environment", *Child Development*, 75: 1243–1253.

Pooley, C.A., Turnbull, J., and Adams, M. (2005) "The journey to school in Britain since the 1940s: Continuity and change", *Area*, 37: 43–53.

Posner, J.C., Liao, E., Winston, F.K., Cnaan, A., Shaw, K.N., and Durbin, D.R. (2002) "Exposure to traffic among urban children injured as pedestrians", *Injury Prevention*, 8: 231–235.

Pratt, G. (1986) "Housing tenure and social cleavages in urban Canada", *Annals of the Association of American Geographers*, 76(3): 366–380.

Preibisch, K. (2010) "Pick-your-own labor: Migrant workers and flexibility in Canadian agriculture", *International Migration Review*, 44(2): 404–441.

Presser, H.B. and Cox, A.G. (1997) "The work schedules of low-educated American women and welfare reform", *Monthly Labor Review*, 120(4): 25–34.

Preston, V. and McLafferty, S. (1993) "Gender differences in commuting at suburban and central locations", *Canadian Journal of Regional Science/Revue canadienne des sciences régionales*, 16(2): 237–259.

Preston, V. and McLafferty, S. (1999) "Spatial mismatch research in the 1990s: Progress and potential", *Papers in Regional Science*, 78(4): 387–402.

Price, A.E., Pluto, D.M., Ogoussanc, O., and Banda, J.A. (2011) "School administrators' perceptions of factors that influence children's active travel to school", *Journal of School Health*, 81: 741–748.

Price, G. (2013) "Relax Vancouver drivers, you can handle another bike lane", *Globe and Mail*, Online. Available at: www.theglobeandmail.com/commentary/relax-vancouver-drivers-you-can-handle-another-bike-lane/article13498165 (accessed 30 July 2013).

Public Health Agency of Canada. (2009) *Child and Youth Unintentional Injury: 2009 Edition – Spotlight on Consumer Safety*, Online. Available at: www.phac-aspc.gc.ca/publicat/cyi-bej/2009 (accessed 19 August 2013).

Pucher, J. (1981) "Equity in transit finance: Distribution of transit subsidy benefits and costs among income classes", *Journal of the American Planning Association*, 47(4): 387–407.

Pucher, J. and Buehler, R. (2006) "Why Canadians cycle more than Americans: A comparative analysis of bicycling trends and policies", *Transport Policy*, 13: 265–279.

Pucher, J. and Buehler, R. (2008) "Making cycling irresistible: Lessons from the Netherlands, Denmark and Germany", *Transport Reviews*, 28(4): 495–528.

Pucher, J., Buehler, R., and Seinen, M. (2011) "Bicycling renaissance in North America? An update and re-appraisal of cycling trends and policies", *Transportation Research Part A*, 45: 451–475.

Pucher, J., Korattyswaropam, N., Mittal, N., and Ittyerah, N. (2005) "Urban transport crisis in India", *Transport Policy*, 12: 185–198.

Pucher, J., Peng, Z.-R., Mittal, N., Zhu, Y., and Korattyswaroppam, N. (2007) "Urban transport trends and policies in China and India: Impacts of rapid economic growth", *Transport Reviews*, 27(4): 379–410.

Puentes, R. (2012) *Have Americans Hit Peak Travel? A Discussion of the Changes in US Driving Habits*, Washington, DC: Brookings Institution, Discussion Paper No. 2012–14.

Puentes, R. and Tomer, A. (2009) *The Road Less Traveled: An Analysis of Vehicle Miles Traveled*, Washington, DC: Brookings Institution.

Purvis, C. (2003) *Commuting Patterns of Immigrants*, CTPP 2000 Status Report, Washington, DC: Federal Highway Administration, Bureau of Transportation Statistics, Federal Transit Administration, US Department of Transportation.

Pushkarev, B. and Zupan, J. (1997) *Public Transportation and Land Use Policy*, Bloomington, IN, and London: Indiana Press.

Pye, D. (2003) *Fellowship is Life: The Story of the Clarion Cycling Club*, Bolton: Clarion Publishing.

Quastel, N., Moos, M., and Lynch, N. (2012) "Sustainability as density and the return of the social: The case of Vancouver, British Columbia", *Urban Geography*, 33(7): 1055–1084.

Racioppi, F., Eriksson, L., Tingvall, C., and Villaveces, A. (2004) *Preventing Road Traffic Injury: A Public Health Perspective for Europe*, Copenhagen: World Health Organization.

Rahman, M.M., Glen, D., and Bunker, J.M. (2009) "Is there a future for non-motorized public transport in Asia?", Proceedings of the 8th International Conference of the Eastern Asia Society for Transportation Studies (EASTS), Surabaya, Indonesia, 16–19 November 2009.

Rajan, S.C. (1996) *The Enigma of Automobility: Democratic Politics and Pollution Control*, Pittsburgh, PA: University of Pittsburgh Press.

Rajan, S.C. (2006) "Automobility and the liberal disposition", *Sociological Review*, 54(1): 113–129.

Rajkumar, D., Berkowitz, L., Vosko, L.F., Preston, V., and Latham, R. (2012) "At the temporary–permanent divide: How Canada produces temporariness and makes citizens through its security, work, and settlement policies", *Citizenship Studies*, 16(3/4): 483–510.

Randolph, B., Pinnegar, S., and Tice, A. (2013) "The first home owner boost in Australia: A case study of outcomes in the Sydney housing market", *Urban Policy and Research*, 31(1): 55–73.

Rattner, S. (2010) *Overhaul: An Insider's Account of the Obama Administration's Emergency Rescue of the Auto Industry*, New York: Houghton Mifflin Harcourt Publishing.

Rauland, V. and Newman, P. (2011) "Decarbonising Australian cities: A new model for creating low carbon, resilient cities", paper presented at the 19th International Congress on Modelling and Simulation (MODSIM) Conference, Perth, 12–16 December 2011.

Redshaw, J. (2008) *In the Company of Cars: Driving as a Social and Cultural Practice*, Aldershot: Ashgate.

Reinhart, A. (12 May 2009) "Canadian diaspora divided on protests against war in homeland", *Globe and Mail*: A12.

Replogle, M. (1991) "Non-motorized vehicles in Asia: Lessons for sustainable transport planning and policy", in *Non-motorized Vehicles in Asian Cities*, World Bank Technical Report 162, Environmental Defense Fund, Washington, DC: World Bank.

Replogle, M. (1992) *Non-Motorised Vehicles in Asian cities*, Technical Paper 162, Washington, DC: World Bank.

Rérat, P. and Lees, L. (2011) "Spatial capital, gentrification and mobility: Evidence from Swiss core cities", *Transactions of the Institute of British Geographers*, 36(1): 126–142.

Retting, R.A., Ferguson, S.A., and McCartt, A.T. (2003) "A review of evidence-based traffic engineering measures designed to reduce pedestrian–motor vehicle crashes", *American Journal of Public Health*, 93: 1456–1463.

Ridgley, J. (2008) "Cities of refuge: Immigration enforcement, police, and the insurgent genealogies of citizenship in U.S. sanctuary cities", *Urban Geography*, 29(1): 53–77.

Rietveld, P., Zwart, B., van Wee, B., and van den Hoorn, T. (1999) "On the relationship between travel time and travel distance of commuters", *Annals of Regional Science*, 33: 269–287.

Rivera, F.P. and Barber, M. (1985) "Demographic analysis of childhood pedestrian injuries", *Pediatrics*, 76: 375–381.

Roberts, D.J. and Mahtani, M. (2010) "Neoliberalizing race, racing neoliberalism: Placing 'race' in neoliberal discourses", *Antipode*, 42(2): 248–257.

Roberts, I. (1995) "Effect of environmental factors on risk of injury of child pedestrians by motor vehicles: A case control study", *British Medical Journal*, 310: 91–94.

Romanos, M.C. (1978) "Energy price effects on metropolitan spatial structure and form", *Environment and Planning A*, 10(1): 93–104.

Ronald, R. (2008) *The Ideology of Home Ownership*, New York: Palgrave Macmillan.

Roney, J.M. (2008) *Bicycle Production: Bicycles Pedaling into the Spotlight*, Washington, DC: Earth Policy Institute, Online. Available at: www.earth-policy.org/indicators/C48 (accessed 30 August 2013).

Rose, D. and Villeneuve, P. (1994) "Gender and occupational restructuring in Montréal in the 1970s", in A. Kobayashi (ed.) *Women, Work and Place*: 130–161, Montreal and Buffalo: McGill-Queen's University Press.

Rose, D. and Villeneuve, P. (1998) "Engendering class in the metropolitan city: Occupational pairings and income disparities among two earner couples", *Urban Geography*, 19(2): 123–159.

Rose, N. (1996) "Governing advanced liberal democracies", in A. Barry, T. Osborne, and N. Rose (eds) *Foucault and Political Reason: Liberalism, Neoliberalism, and Rationalities of Government*: 37–64, Chicago: University of Chicago Press.

Rosen, G. and Walks, A. (2013) "Condominium development and the private transformation of the metropolis", *Geoforum*, 49: 160–172.

Rosenbloom, S. (1978) "The need for study of women's travel issues", *Transportation*, 7: 347–350.

Rothman, L., Howard, A.W., Camden, A., and Macarthur, C. (2012) "Pedestrian crossing location influences injury severity in urban areas", *Injury Prevention*, 18: 365–370.

Rozzenac, P. (13 May 2009) "Organizers of recent Tamil protest...", Letter to the Editor, *Toronto Star*: A18.

Rubin, J. (2012) *The End of Growth*, Toronto: Random House.

Ryan, L. and Turton, H. (2007) *Sustainable Automobile Transport: Shaping Climate Change Policy*, Cheltenham: Edward Elgar.

CfGRS (Commission for Global Road Safety). (2009) *Make Roads Safe: A Decade of Action for Road Safety*, Online. Available at: www.makeroadssafe.org/publications/Documents/decade_of_action_report_lr.pdf

Said, C. (11 August 2011) "S.F. apartment rent rises as vacancy rates fall", *San Francisco Chronicle.*

Saldanha, A. (2009) "So what is race?", *Insights*, 2(12): 2–11.

Sanchez, T. (1999) "The connection between public transit and employment", *Journal of the American Planning Association*, 65(3): 284–296.

Santos, A., McGuckin, N., Nakamoto, H.Y., Gray, D., and Liss, S. (2011) *Summary of Travel Trends: 2009 National Household Travel Survey*, Washington, DC: US Department of Transportation, Federal Highway Administration, Online. Available at: http://nhts.ornl.gov/2009/pub/stt.pdf (accessed 1 September 2013).

Sassen, S. (2006) *Cities in a World Economy*, 3rd edn, Thousand Oaks, CA: Pine Forge.

Satzewich, V. (1991) *Racism and the Incorporation of Foreign Labour: Farm Labour Migration to Canada since 1945*, New York: Routledge.

Saunders, P. (1978) "Domestic property and social class", *International Journal of Urban and Regional Research*, 2(2): 233–251.

Saunders, P. (1990) *A Nation of Homeowners*, London: Unwin Hyman.

Savage, M. (1987) "Understanding political alignments in contemporary Britain: Do localities matter?", *Political Geography Quarterly*, 6(1): 53–76.

Schafran, A. (2013) "Origins of an urban crisis: The restructuring of the San Francisco Bay area and the geography of foreclosures", *International Journal of Urban and Regional Research*, 37(2): 663–688.

Schiller, P.L., Bruun, E.C., and Kenworthy, J.R. (2010) *An Introduction to Sustainable Transportation: Policy, Planning and Implementation*, London: Earthscan.

Schlich, R., Schonfelder, S., Hanson, S., and Axhausen, K.W. (2004) "Structures of leisure travel: Temporal and spatial variability", *Transport Reviews*, 24(2): 219–237.

Schlossberg, M., Greene, J., Phillips, P.P., Johnson, B., and Parker, B. (2006) "School trips: Effects of urban form and distance on travel mode", *Journal of the American Planning Association*, 72: 337–346.

Schuetz, J., Metzer, R., and Been, V. (2009) "31 flavours of inclusionary zoning: Comparing policies from San Francisco, Washington, D.C., and suburban Boston", *Journal of the American Planning Association*, 75(4): 441–456.

Schwartz, H. (2008) "Housing, global finance, and American hegemony: Building conservative politics one brick at a time", *Comparative European Politics*, 6: 262–284.

Schwartz, H. (2009) *Subprime Nation: American Power, Global Capital, and the Housing Bubble*, Ithaca, NY: Cornell University Press.

Scott, A.J. (2008) *Social Economy of the Metropolis: Cultural-Cognitive Capitalism and the Global Resurgence of Cities*, Oxford: Oxford University Press.

Scott, A.J. (2011) "Emerging cities of the third wave", *City: Analysis of Urban Trends, Culture, Theory, Policy, Action*, 15(3/4): 289–321.

Seif, H. (2010) "'Tired of illegals': Immigrant driver's licenses, constituent letters, and shifting restrictionist discourse in California", in M.W. Varsanyi (ed.) *Taking Local Control: Immigration Policy Activism in U.S. Cities and States*, Stanford, CA: Stanford University Press.

Seiler, C. (2008) *Republic of Drivers: A Cultural History of Automobility in America*, Chicago: University of Chicago Press.

Sellers, J. (1999) "Public goods and the politics of segregation: An analysis and cross-national comparison", *Journal of Urban Affairs*, 21(2): 237–262.

Sellers, J. (2005) "Re-placing the nation", *Urban Affairs Review*, 40(4): 419–445.

Sellers, J. (2013) "Place, institutions and the political ecology of US metropolitan areas", in J. Sellers, D. Kubler, M. Walter-Rogg, and A. Walks (eds) *The Political Ecology of the Metropolis*: 37–86, Colchester, UK: ECPR Press.

Sellers, J. and Walks, A. (2013) "Introduction: The metropolitanization of politics", in J. Sellers, D. Kubler, M. Walter-Rogg, and A. Walks (eds) *The Political Ecology of the Metropolis*: 3–36, Colchester, UK: ECPR Press.

Sellers, J.M., Kubler, D., Walter-Rogg, M., and Walks, R.A. (eds) (2013) *The Political Ecology of the Metropolis*, Colchester, UK: ECPR Press.

Seto, K., Fragkais, M., Guneralp, B., and Reilly, M. (2011) "A meta-analysis of global urban land expansion", *PLoS ONE*: 6.

Sewell, J. (2009) *The Shape of the Suburbs: Understanding Toronto's Sprawl*, Toronto: University of Toronto Press.

SFCTA (San Francisco County Transportation Authority). (2011) *The Role of Shuttle Services in San Francisco's Transportation System*, San Francisco: SFCTA.

SFMTA (San Francisco Municipal Transportation Agency). (2010) *Transportation Fact Sheet*, San Francisco: SFMTA.

SFMTA. (2011) *Climate Action Strategy for San Francisco's Transportation System*, San Francisco: SFMTA.

SFPD (San Francisco Planning Department). (1971) *Transportation: Conditions, Problems, and Issues*, San Francisco: San Francisco Planning Department.

SFPD. (2007) *Planning Department Assessment of Proposition H, Parking Regulations, Memo to Department of Elections*, San Francisco: San Francisco Planning Department.

SFPD. (2008) *Market and Octavia: An Area of the General Plan of San Francisco*, San Francisco: San Francisco Planning Department.

Shaheen, S.A., Guzman, S., and Zhang, H. (2012) "Bikesharing across the globe", in J. Pucher and R. Buehler (eds) *City Cycling*, Cambridge, MA: MIT Press.

Sharma, N. (2006) *Home Economics: Nationalism and the Making of "Migrant Workers" in Canada*, Toronto: University of Toronto Press.

Shearmur, R. (2006) "Travel from home: An economic geography of commuting distance in Montréal", *Urban Geography*, 27(4): 330–359.

Shearmur, R. and Coffey, W. (2002) "A tale of four cities: Intrametropolitan employment distribution in Toronto, Montréal, Vancouver, and Ottawa – Hull, 1981–1996", *Environment and Planning A*, 34(4): 575–598.

Shearmur, R., Coffey, W., Dube, C., and Barbonne, R. (2007) "Intrametropolitan employment structure: Polycentricity, scatteration, dispersal and chaos in Toronto, Montréal and Vancouver, 1996–2001", *Urban Studies*, 44(9): 1713–1738.

Shefali, M.K. (2000) *Study on Gender Dimension in Dhaka Urban Transport Project*, Dhaka, Bangladesh: Nari Uddug Kendra.

Sheller, M. (2004) "Automotive emotions: Feeling the car", *Theory, Culture & Society*, 21(4/5): 221–242.

Sheller, M. and Urry, J. (2000) "The city and the car", *International Journal of Urban and Regional Research*, 24(4): 737–757.

Shoup, D. (2005) *The High Cost of Free Parking*, Chicago: Planners Press, The American Planning Association.

Singh, S.K. (2005) "Review of urban transportation in India", *Journal of Public Transportation*, 8(1): 79–97.

Singhal. (13 May 2009) "Over the past two weeks…", Letter to the Editor, *Globe and Mail*: A14.

Sivak, M. and Schoettle, B. (2012) "Update: Percentage of young persons with a driver's license continues to drop", *Traffic Injury Prevention*, 13: 341.

Skaburskis, A. and Moos, M. (2008) "The redistribution of residential property values in Montréal, Toronto and Vancouver: Examining neoclassical and Marxist views on changing investment patterns", *Environment and Planning A*, 40(4): 905–927.

Slack, C. (2007) "Municipal targeting of undocumented immigrants' travel in the post 9/11 suburbs: Waukegan, Illinois case study", *Georgetown Immigration Law Journal*, 22: 485–505.

Smart, M. (2010) "US immigrants and bicycling: Two-wheeled in Autopia", *Transport Policy*, 17(3): 153–159.

Smith. (10 May 2009) "Don't these people realize…", Weblog comment, J. Lewington and K. Makin, "Tamil protest surprises Toronto", *Globe and Mail*, Online. Available at: http://v1.theglobeandmail.com/servlet/story/RTGAM.20090510.wtamilprotest0510/CommentStory/Front (accessed 10 July 2010).

Soja, E. (2000) *Postmetropolis: Critical Studies of Cities and Regions*, Malden, MA: Blackwell.

Soole, D.W., Watson, B.C., and Fleiter, J.J. (2013) "Effects of average speed enforcement on speed compliance and crashes: A review of the literature", *Accident Analysis and Prevention*, 54: 46–56.

Soron, D. (2009) "Driven to drive: Cars and the problem of 'compulsory consumption'", in J. Conley and A.T. McLaren (eds) *Car Troubles: Critical Studies of Automobility and Auto-mobility*: 181–198, Aldershot, UK: Ashgate.

Sperling, D. and Gordon, D. (2009) *Two Billion Cars: Driving Towards Sustainability*, Oxford: Oxford University Press.

Standing Advisory Committee for Trunk Road Assessment. (1994) *Trunk Roads and the Generation of Traffic*, London: Department of Transport.

Stanford, J. (2010) "The geography of auto globalization and the politics of auto bailouts", *Cambridge Journal of Regions, Economy, and Society*, 3(3): 383–405.

Stanley, J. and Barrett, S. (2010) *Moving People: Solutions for a Growing Australia*, Australia: Australasian Railway Association, Bus Industry Confederation and UITP.

Statistics Canada. (1996) *Census of Canada Public Use Microdata, 1996 Individuals*, Online. Available through using the Canadian Census Analyzer at CHASS at the University of Toronto http://sda.chass.utoronto.ca/sdaweb/html/canpumf.htm (accessed 1 September 2013).

Statistics Canada. (2003) *2001 Census: Analysis Series – Where Canadians Work and How They Get There*, Catalogue no. 96F0030XIE2001010, Ottawa: Ministry of Industry.

Statistics Canada. (2006) *Census of Canada Public Use Microdata, 2006 Individuals*, Online. Available through using the Canadian Census Analyzer at CHASS at the University of Toronto: http://sda.chass.utoronto.ca/sdaweb/html/canpumf.htm (accessed 1 September 2013).

Statistics Canada. (2007) *Portrait of the Canadian Population in 2006*, Catalogue no. 97-550-XIE, Ottawa: Ministry of Industry.

Statistics Canada. (2009) *Census Dictionary: Occupation (Based on the National Occupational Classification for Statistics 2006 [NOC-S 2006])*, Online. Available at: www12.statcan.gc.ca/census-recensement/2006/ref/dict/pop102-eng.cfm (accessed 20 November 2013).

Steinbach, R., Green, J., Datta, J., and Edwards, P. (2011) "Cycling and the city: A case study of how gendered, ethnic and class identities can shape healthy transport choices", *Social Science and Medicine*, 72(7): 1123–1130.

Steinherr, A. (1998) *Derivatives: The Wild Beast of Finance*, New York: Wiley.

Stevenson, M.R. (1997) "Childhood pedestrian injuries: What can changes to the road environment achieve?", *Australian and New Zealand Journal of Public Health*, 21: 33–37.

Stokes, G. and Lucas, K. (2011) *National Travel Survey Analysis*, Working Paper No. 1053, Transport Studies Unit, Oxford University.

Stoll, M. and Covington, K. (2012) "Explaining racial/ethnic gaps in spatial mismatch in the U.S.: The primacy of racial segregation", *Urban Studies*, 49(11): 2501–2521.

Strange, L. and Brown, R. (2002) "The bicycle, women's rights, and Elizabeth Cady Stanton", *Women's Studies*, 31(5): 609–626.

STP (Strategic Transport Plan). (2005) *Strategic Transport Plan for Dhaka: Final Report*, Dhaka Transport Co-ordination Board, Ministry of Communication, Government of Bangladesh.

Tal, G. and Handy, S. (2010) "Travel behavior of immigrants: An analysis of the 2001 National Household Transportation Survey", *Transport Policy*, 17: 85–93.

Taylor, B.D. (2006) "Putting a price on mobility: Cars and contradictions in planning", *Journal of the American Planning Association*, 72(3): 279–284.

Taylor, B.D. and Ong, P.M. (1995) "Spatial mismatch or auto-mobile mismatch: An examination of race, residence, and commuting in U.S. metropolitan areas", *Urban Studies*, 32(9): 1453–1473.

Taylor, B.D., Miller, D., Iseki, H., and Fink, C. (2009) "Nature and/or nurture? Analyzing the determinants of transit ridership across US urbanized areas", *Transportation Research Part A*, 43: 60–77.

TCRP. (2002) *Transit-Oriented Development and Joint Development in the United States: A Literature Review*, Washington, DC: Transit Cooperative Research Program, US Federal Transit Administration.

Teaford, J.C. (1997) *Post-Suburbia: Government and Politics in the Edge Cities*, Baltimore, MD: Johns Hopkins University Press.

Theva, G. (12 May 2009) "Why I blocked the Gardiner", *Toronto Star:* GT1.

Tim. (31 May 2008). *Cyclists Shut Down the Gardiner*, Online. Available at: www.blogto.com/city/2008/05/cyclists_shut_down_the_gardiner (accessed 1 June 2010).

Timperio, A., Crawford, D., Telford, A., and Salmon, J. (2004) "Perceptions about local neighborhood and walking and cycling among children", *Preventive Medicine*, 38: 39–47.

Tiwari, G. and Jain, D. (2013) *Promoting Low Carbon Transport in India: NMT Infrastructure in India – Investment, Policy and Design*, United Nations Environment Program (UNEP), Online. Available at: www.unep.org/transport/lowcarbon/publications.asp (accessed 7 April 2014).

Tiwari, G. and Jain, H. (2008) "Bicycles in urban India", *Urban Transport Journal*, December: 59–68.

Tomic, P. and Trumper, R. (2012) "Mobilities and immobilities: Globalization, farming, and temporary work in the Okanagan Valley", in P.T. Lenard and C. Straehle (eds) *Legislated Inequality: Temporary Labour Migration in Canada*: 73–94, Montreal and Kingston: McGill-Queen's University Press.

Topping, D. (30 May 2008) "Here we are, let's take the Gardiner", Online. Available at: http://torontoist.com/2008/05/critical_mass_takes_over_the_gardiner.php (accessed 1 June 2010).

Topping, D. (11 May 2009) "Tamils take the Gardiner", Online. Available at: http://torontoist.com/2009/05/tamils_take_to_the_gardiner (accessed 1 June 2010).

Tossell, I. (2012) *The Gift of Ford*, Toronto: Random House/Hazlit.

Transport Canada. (2010) *Canadian Motor Vehicle Traffic Collision Statistics*, Online. Available at: www.tc.gc.ca/eng/roadsafety/tp-1317.htm#3 (accessed 1 November 2013).

Tranter, P. (2004) *Effective Speeds: Car Costs Are Slowing Us Down*, Canberra: Australian Greenhouse Office, Department of the Environment and Heritage.

Tranter, P. (2010) "Speed kills: The complex links between transport, lack of time and urban health", *Journal of Urban Health*, 87: 155–166.

Tranter, P. (2011) "The urban speed paradox: Time pressure, cars and health", *Dissent*, 36(Spring): 9–12.

Tranter, P. (2012) "Effective speed: Cycling because it faster", in J. Pucher and R. Buehler (eds) *City Cycling*: 57–74, Cambridge, MA: MIT Press.

Tranter, P. and Ker, I. (2007) "A wish called $quander: (In)effective speed and effective well-being in Australian cities", Proceedings of the States of Australian Cities, 2007 National Conference, University of South Australia and Flinders University, Adelaide, Australia, 28 November.

Trivector Traffic. (2006) *Changes in Travel Habits in Stockholm County*, Stockholm: City of Stockholm, Online. Available at: www.stockholmsforsoket.se/upload/Sammanfattningar/English/Changes%20in%20travel%20habits%20in%20Stockholm%20County.pdf (accessed 1 September 2013).

Trubka, R., Newman, P., and Bilsborough, D. (2010) "Costs of urban sprawl (2): Greenhouse gases", *Environment Design Guide*, 84: 1–16.

TSA. (2012) *The Statistical Abstract*, Washington, DC: US Department of Commerce.

TTI. (2005) *Roadway Congestion in Major Urban Areas*, College Station: Texas Transportation Institute, Texas A & M University.

Tuan, Y.F. (1979) *Landscapes of Fear*, Minneapolis: University of Minnesota Press.

Tudor-Locke, C., Ainsworth, B.E., Thompson, R.W., and Matthews, C.E. (2002) "Comparison of pedometer and accelerometer measures of free-living physical activity", *Medicine and Science in Sports and Exercise*, 34: 2045–2051.

Turcotte, M. (2011) *Commuting to Work: Results of the 2010 General Social Survey*, Statistics Canada, Catalogue no. 11-008-X, Ottawa: Ministry of Industry.

Turner, T. and Niemeier, D. (1997) "Travel to work and household responsibility: New evidence", *Transportation*, 24: 397–419.

Tverberg, G.E. (2012) "Oil supply limits and the continuing financial crisis", *Energy*, 37: 27–34.

Ulfarsson, G.F. and Shankar, V.N. (2008) "Children's travel to school: Discrete choice modeling of correlated motorized and nonmotorized transportation modes using covariance heterogeneity", *Environment and Planning B: Planning and Design*, 35: 195–206.

UN (United Nations). (2009) *Planning Sustainable Cities: Global Report on Human Settlements*, Nairobi: UN, Human Settlements Programme (UN-Habitat).

UN. (2011) *World Population Prospects: The 2010 Revision*, New York: Population Division, United Nations.

UN. (2012) *World Urbanization Prospects, 2011 Revision*, New York: United Nations, Department of Economic and Social Affairs, Population Division, Working paper no. ESA/P/WP.224, Online. Available at: http://esa.un.org/unup/pdf/WUP2011_Highlights.pdf (accessed 1 July 2013).

UN. (2013) *World Population Prospects, 2012 Revision*, New York: United Nations, Department of Economic and Social Affairs, Population Division, Working paper no. ESA/P/WP.228, Online. Available at: http://esa.un.org/unpd/wpp/Documentation/pdf/WPP2012_HIGHLIGHTS.pdf (accessed 1 September 2013).

UN ESCAP. (1997) *Integration of Non-Motorized Transport in the Urban Transport System of Dhaka, Part 1: Non-motorized Transport in Dhaka*, United Nations Economic and Social Commission for Asia and the Pacific (ESCAP).

Unger, D.J. (8 April 2013) "How North Sea oil helped Margaret Thatcher", *Christian Science Monitor*, Online. Available at: www.csmonitor.com/Environment/Energy-Voices/2013/0408/How-North-Sea-oil-helped-Margaret-Thatcher (accessed 1 September 2013).

United States Census Bureau. (various years) *Intercensal Population and Housing Unit Estimates* and *Department of Housing and Urban Development (HUD) Median Family Income Estimates*, Online. Available at: www.census.gov/popest and http://2010.census.gov/2010census (accessed 1 July 2013).

Urry, J. (2000) *Sociology beyond Societies: Mobilities for the 21st Century*, London: Routledge.

Urry, J. (2004) "The 'system' of automobility", *Theory, Culture and Society*, 21(4/5): 25–39.

Urry, J. (2007) *Mobilities*, London: Polity.

Urry, J. (2009) *The Great Crash of 2008: Pouring Oil on Troubled Waters*, Lancaster: Lancaster University Working Paper, Online. Available at: www.lancaster.ac.uk/fass/doc_library/sociology/Urry_the_great_crash.pdf (accessed 1 August 2013).

Urry, J. (2012) "Social networks, mobile lives, and social inequalities", *Journal of Transport Geography*, 21: 24–30.

US Federal Highway Administration. (2012) *The Road to Civil Rights: The League of American Wheelmen*, Washington, DC: Federal Highway Administration, US Department of Transportation, Online. Available at: www.fhwa.dot.gov/highwayhistory/road/s05.cfm (accessed 30 August 2013).

Uteng, T.P. (2009) "Gender, ethnicity, and constrained mobility: Insights into the resultant social exclusion", *Environment and Planning A*, 41(5): 1055–1071.

Valentine, G. (1997) "'Oh yes I can' 'Oh no you can't': Children and parents' understanding of kid's competence to negotiate public space safely", *Antipode*, 29: 65–89.

Valenzuela, A., Schweitzer, L., and Robles, A. (2005) "Camionetas: Informal travel among immigrants", *Transportation Research Part A: Policy and Practice*, 39(10): 895–911.

Vandersmissen, M.H., Villeneuve, P., and Thériault, M. (2003) "Analyzing changes in urban form and commuting time", *The Professional Geographer*, 55(4): 446–463.

Van Rijn, N. (10 May 2009) "Gardiner chaos averted", *Toronto Star*, Online. Available at: www.thestar.com/news/gta/2009/05/11/gardiner_chaos_averted.html (accessed 1 August 2013).

Varsanyi, M.W. (2006) "Interrogating 'urban citizenship' vis-a-vis undocumented migration", *Citizenship Studies*, 10(2): 229–249.

Varsanyi, M.W. (ed.) (2010) *Taking Local Control: Immigration Policy Activism in U.S. Cities and States*, Stanford, CA: Stanford University Press.

Vidyattama, Y., Tanton, R., and Nepal, B. (2013) "The effect of transport costs on housing-related financial stress in Australia", *Urban Studies*, 50(9): 1779–1795.

Virilio, P. (1986) *Speed and Politics: An Essay on Dromology*, New York: Semiotext(e).

Vojnovic, I. (2000) "Shaping metropolitan Toronto: A study of linear infrastructure subsidies, 1954–66", *Environment and Planning B: Planning and Design*, 27: 197–230.

Volti, R. (1996) "A century of automobility", *Technology and Culture*, 37(4): 663–685.

Wachs, M. and Taylor, B. (1998) "Can transportation strategies help meet the welfare challenge?", *Journal of the American Planning Association*, 64(1): 15–17.

Wackernagel, M. and Rees, B. (1996) *Our Ecological Footprint: Reducing Human Impact on the Earth*, Gabriola Island, BC: New Society Publishers.

Walker, R. (2010) "The golden state adrift", *New Left Review*, 66(6): 5–30.

Walks, A. (2001) "The social ecology of the post-Fordist/global city? Economic restructuring and socio-spatial polarisation in the Toronto urban region", *Urban Studies*, 38(3): 407–447.

Walks, A. (2004a) "Place of residence, party preferences, and political attitudes in Canadian cities and suburbs", *Journal of Urban Affairs*, 26(3): 269–295.

Walks, A. (2004b) "Suburbanization, the vote, and changes in federal and provincial political representation and influence between inner cities and suburbs in large Canadian urban regions, 1945–1999", *Urban Affairs Review*, 39(4): 411–440.

Walks, A. (2005a) "City-suburban electoral polarization in Great Britain, 1950 to 2001", *Transactions of the Institute of British Geographers*, 30(4): 472–499.

Walks, A. (2005b) "The city–suburban cleavage in Canadian federal politics", *Canadian Journal of Political Science*, 38(2): 383–413.

Walks, A. (2006) "The causes of city–suburban political polarization? A Canadian case study", *Annals of the Association of American Geographers*, 96(2): 390–414.

Walks, A. (2007) "The boundaries of suburban discontent? Urban definitions and neighbourhood political effects", *Canadian Geographer*, 51(2): 160–185.

Walks, A. (2008) "Urban form, everyday life, and ideology: Support for privatization in three Toronto neighbourhoods", *Environment and Planning A*, 40(2): 258–282.

Walks, A. (2010a) "Bailing out the wealthy: Responses to the financial crisis, ponzi neoliberalism, and the city", *Human Geography*, 3(1): 54–84.

Walks, A. (2010b) "Electoral behaviour behind the gates: Partisanship and political participation among Canadian gated community residents", *Area*, 42(1): 7–24.

Walks, A. (2011) "Economic restructuring and trajectories of socio-spatial polarization in the 21st century Canadian city", in T. Hutton, R. Shearmur, L.S. Bourne, and R.J. Simmons (eds) *Trajectories of Change in Canada's Urban Regions*: 125–159, Oxford: Oxford University Press.

Walks, A. (2014) "Canada's housing bubble story: Mortgage securitization, the state, and the global financial crisis", *International Journal of Urban and Regional Research*, 38(1): 256–284.

Walks, A. (2013a) "Suburbanism as a way of life, slight return", *Urban Studies*, 50(8): 1471–1488.

Walks, A. (2013b) "Mapping the urban debtscape: The geography of household debt in Canadian cities", *Urban Geography*, 34(2): 153–187.

Walks, A. (2013c) *Income Inequality and Polarization in Canada's Cities: An Examination and New Form of Measurement*, Toronto: University of Toronto Cities Centre, Research Paper no. 227.

Walks, A. (2013d) "The political ecology of metropolitan Britain", in J. Sellers, D. Kubler, M. Walter-Rogg, and A. Walks (eds) *The Political Ecology of the Metropolis*: 125–160, Colchester, UK: ECPR Press.

Walks, A. (2013e) "Metropolitan political ecology and contextual effects in Canada", in J. Sellers, D. Kubler, M. Walter-Rogg, and A. Walks (eds) *The Political Ecology of the Metropolis*: 87–124, Colchester, UK: ECPR Press.

Walks, A. and Bourne, L. (2006) "Ghettos in Canada's cities? Racial segregation, ethnic enclaves and poverty concentration in Canadian urban areas", *The Canadian Geographer*, 50(3): 273–297.

Walks, A. and Maaranen, R. (2008) "Gentrification, social mix, and social polarization: Testing the linkages in large Canadian cities", *Urban Geography*, 29(4): 293–326.

Walsh, M. (2008) "Gendering mobility: Women, work and automobility in the United States", *History*, 93(311): 376–395.

Wang, Y. (2011) "Disparities in pediatric obesity in the United States", *Advances in Nutrition*, 2: 23–31.

Ward, C. (1977) *The Child in the City*, London: Architectural Press.

Warmington, J. (12 May 2009) "Only cowards use kids as shields", *Toronto Sun*: 4.

Webster, C. (2002) "Property rights and the public realm: Gates, green belts, and Gemeinshaft", *Environment and Planning B: Planning and Design*, 29(3): 397–412.

Weese, B. (1 June 2008) "Gardiner bike-in 'not planned'", *Toronto Sun*: 16.

Weese, B. and Artuso, A. (12 May 2009) "Mayor slams Tamil protest", *Toronto Sun*: 5.

Weisbaum, H. (18 April 2013) *Car Loans Stretch out to 8 Years, Costing Buyers More*, CNBC News Today, Online. Available at: www.cnbc.com/id/100654029 (accessed 1 August 2013).

Weitz, J. and Crawford, T. (2012) "Where the jobs are going: Job sprawl in U.S. metropolitan regions, 2001–2006", *Journal of the American Planning Association*, 78(1): 53–69.

Wei Wei, A. (2003) *Forever*, Toronto: Art Gallery of Ontario.

Wells, C.W. (2012) *Car Country*, Seattle: University of Washington Press.

Wells, M.J. (2004) "The grassroots reconfiguration of US immigration policy", *International Migration Review*, 38(4): 1308–1347.

Wen, L.M., Fry, D., Rissel, C., Dirkis, H., Balafas, A., and Merom, D. (2008) "Factors associated with children being driven to school: Implications for walk to school programs", *Health Education Research*, 23: 325–334.

Wente, M. (14 May 2009) "Tamils deserve straight talk", *Globe and Mail*: A14.

Wexler, H. (2000) *Bus Riders Union*, Los Angeles: Strategy Centre (motion picture).

Whitelegg, J. (1993) *Transport for a Sustainable Future: The Case for Europe*, London: Bellhaven Press.

Whitelegg, J. and Williams, N. (2000) "Non-motorised transport and sustainable development: Evidence from Calcutta", *Local Environment*, 5(1): 7–18.

WHO (World Health Organization). (2008) *Overview Fact Sheet: Children and Road Traffic Injury*, Online. Available at: www.who.int/violence_injury_prevention/child/injury/world_report/Road_traffic_injuries_english.pdf (accessed 2 July 2013).

WHO. (2009) *Global Status Report on Road Safety: Time for Action*, Geneva: World Health Organization.

WHO. (2013) *Global Status Report on Road Safety 2013: Supporting a Decade of Action*, Geneva: World Health Organization.

Whyte, W.H. (1956) *The Organization Man*, New York: Simon & Schuster.

Wolman, H. and Marckini, L. (1998) "Changes in central-city representation and influence in Congress since the 1960s", *Urban Affairs Review*, 34(2): 291–312.

Wood, R.C. (1958) *Suburbia: Its People and Their Politics*, Boston, MA: Houghton Mifflin.

Work Safe BC. (2007) *Three Farm Workers Killed in 15-Passenger Van Rollover*, Vancouver: Work-Safe BC, Online. Available at: www2.worksafebc.com/PDFs/investigations/IIR2007137101997.pdf (accessed 1 October 2012).

World Bank. (1995) *Non-Motorized Vehicles in Ten Asian Cities: Trends, Issues, and Policies*, Transportation, Water and Urban Development Department, Transport Division, World Bank.

World Bank. (1998) *Poverty Reduction and Social Assessments*, World Bank Group, Online. Available at: www.worldbank.org/transport/roads/pov&sa.htm#importance (accessed 15 July 2013).

World Bank. (2002) *The Role of Non-motorized Transport: Cities on the Move – A World Bank Urban Transport Strategy Review*, Washington, DC: World Bank, Online. Available at: http://siteresources.worldbank.org/INTURBANTRANSPORT/Resources/cities_on_the_move.pdf (accessed 15 July 2013).

World Bank. (2009) *Project Appraisal Document on a Proposed Credit to the People's Republic of Bangladesh for the Clean Air and Sustainable Environment Project*, Social, Environment and Water Resources Management Unit, Sustainable Development Department, South Asia Region.

World Bank. (2010) *Country Assistance Strategy for the People's Republic of Bangladesh*, Bangladesh Country Management Unit, South Asia Region, World Bank.

World Bank. (2012) *World Development Indicators*, Washington, DC: World Bank.

World Bank. (2013) *Turn Down the Heat: Climate Extremes, Regional Impacts and the Case for Resilience*, Washington, DC: World Bank.

Worthington, P. (15 May 2009) "Weary of Tamil disruption", *Toronto Sun*: 22.

Williams, R. (2006) "Generalized ordered logit/partial proportional odds models for ordinal dependent variables", *Stata Journal*, 6: 58–82.

Wilson, W.J. (2009) "The political and economic forces shaping concentrated poverty", *Political Science Quarterly*, 123: 555–571.

Wyly, E. (1996) "Race, gender, and spatial segmentation in the twin cities", *Professional Geographer*, 48(4): 431–444.

Wyly, E. (2009) "Strategic positivism", *Professional Geographer*, 61(3): 310–322.

Wyly, E. (2011) "Positively radical", *International Journal of Urban and Regional Research*, 35(5): 889–912.

Wyly, E., Moos, M., Kabahizi, E., and Hammel, D. (2009) "Cartographies of race and class: Mapping the class-monopoly rents of American subprime mortgage capital", *International Journal of Urban and Regional Research*, 33(2): 343–364.

Yardley, J. (8 May 2004) "China races to reverse its falling production of grain", *New York Times*.

Yiannakoulias, N. and Scott, D.M. (2013) "The effects of local and non-local traffic on child pedestrian safety: A spatial displacement of risk", *Social Science and Medicine*, 80: 96–104.

Yiannakoulias, N., Bland, W., and Scott, D.M. (2013) "Altering school attendance times to prevent child pedestrian injuries", *Traffic Injury Prevention*, 14(4): 405–412.

Yiannakoulias, N., Smoyer-Tomic, K., Hodgson, J., Spady, D.W., Rowe, B.H., and Voaklander, D.C. (2002) "The spatial and temporal dimensions of child pedestrian injury and Edmonton", *Canadian Journal of Public Health*, November–December: 447–451.

Zegeer, C.V. and Bushell, M. (2012) "Pedestrian crash trends and potential counter measures from around the world", *Accident Analysis and Prevention*, 44: 3–11.

Zhang, L., Hong, J., Nasri, A., and Shen, Q. (2012) "How built environment affects travel behavior: A comparative analysis of the connections between land use and vehicle miles traveled in US cities", *Journal of Transport and Land Use*, 5: 40–52.

Zhao, P. (2011) "Car use, commuting and urban form in a rapidly growing city: Evidence from Beijing", *Transportation Planning and Technology*, 34: 509–527.

Zhao, P., Chapman, R., Randal, E., and Howden-Chapman, P. (2013) "Understanding resilient urban futures: A systemic modelling approach", *Sustainability*, 5: 3202–3223.

Zohir, S.C. (2003) *Integrating Gender into World Bank Financed Transport Programs: Case Study – Bangladesh*, Dhaka, Bangladesh: Dhaka Urban Transport Project.

Zolnik, E.J. (2011) "The effects of sprawl on private-vehicle commuting distances and times", *Environment and Planning B: Planning and Design*, 38(6): 1071–1084.

Zuniga, K.D. (2012) "From barrier elimination to barrier negotiation: A qualitative study of parents' attitudes about active travel for elementary school trips", *Transport Policy*, 20: 75–81.

Index

Page numbers in *italics* denote tables, those in **bold** denote figures.

climate change: and effective speed 140; and the resilient city 41; and sea levels 35

Clinton, Bill 32

CO₂ emissions: China and the US's contributions 32; reducing through use of public transit 32

collective consumption 204

commuting: and effective mobility 131; fastest-growing form 109; gendered perspective 172–3; North American experience 105–8; post-freeway experience in San Francisco 234–5; US experience of time spent **108**

commuting distance: household restructuring and *122*, 126; income and 120, 132; relationship with occupation 115, 120

commuting patterns: automobiles' continued prevalence 111; in Canada's global cities 112–26; the changing metropolitan context 105–12; of immigrants 121; impact of decentralization 103; modelling under metropolitan restructuring 121–6; North American growth trends in metropolitan flows *110*; and the post-Fordist transformation of production 108–12; relationship between income, education and 126; shaping characteristics 112

compulsory consumption 6

congestion charging 213, 217, 280

Conley, J. 4

consumption of motor vehicles 26–8

Copenhagen 244–5

Cortright, J. 66

Critical Mass 251–2, 259, 264–6, 285

cultural activities, locational tendencies 109

Curitiba 277

cycle lanes 12, 213, 215, 233–4, 237–8, 248–50, 252

cycle-rickshaws: city with the largest fleet 153; impact of state-sponsored removal 164; non-recognition of contribution to mobility 155–6; socio-economic overview 166; street bans in Dhaka 162

cycling: bike-sharing schemes 153, 245–6, 252; and community tensions 242; drivers' frustrations with 263–4; Dutch history 243–4; effective speed 141–4, 148; gendered perspective 238, 246; household debt in the cycling nations

68; impact of China's state encouragement 239; patterns of migrants into North America 171; perceptions of 257; political perspective 238–43, 247, 250–1; post-car mobility and 282; post-war planning invisibility 239; rates of in San Francisco 221; social class perspective 238; *see also* the bicycle; velo-mobility

Dallas 33

Dalley, M.L. 82

deaths, transport related *48–9*

debt, automobility and among nations 59–68 (*see also* auto-indebtedness)

decentralization 17, 103–4, 108–9, 112

deindustrialization 73, 104, 115, 126, 205–6, 216

Delhi 33, 156

Denmark 68, 243, 245

Dennis, K. 276–7

Detroit 112

Dhaka 40; *see also* NMT in Dhaka, Bangladesh

Dodson, J. 66, 72

downtown revitalization: drivers 121; impact on mobility patterns 121

driver's licences 28, 182, 186–91, 257

driving: decline in among youth 28; isolating effect 258; by poorer households 43; *see also* commuting patterns

Dubai 173

Dunleavy, P. 204

Dunn, J. 12–13, 205

earthquakes: Los Angeles 224; San Francisco 222

Ebert, A.-K. 243

The Ecology of the Automobile (Freund/Martin) 4

ecosocialization 278

edge cities 103

education, and commuting distance 126 (*see also* school travel)

effective speed: calculating 133–5; climate change and 140; comparison by mode and motility 141–5; concept analysis 132–3; improving 145; income and 145–7; international comparison 135–40; and Marx's theory of value 133; policy implications 144; social and political potential 133; variables affecting 140

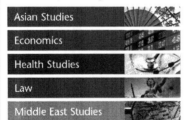